THE NATURAL HISTORY OF THE WILD CATS

THE NATURAL HISTORY OF THE WILD CATS

ANDREW KITCHENER

COMSTOCK PUBLISHING ASSOCIATES a division of
CORNELL UNIVERSITY PRESS Ithaca, New York

Line illustrations by Sally Alexander and Andrew Kitchener

Library of Congress Cataloging-in-Publication Data

Kitchener, Andrew,
The natural history of the wild cats/Andrew Kitchener.
p. cm.
Includes bibliographical references and index.
ISBN 0-8014-2596-4 (cloth : alk. paper)
ISBN 0-8014-8498-7 (pbk. : alk. paper)
1. Felidae. I. Title. II. Series.
QL737.C23K58 1991
599.74′428—dc20 90-45833

Typeset by Florencetype Ltd., Kewstoke, Avon.
Printed in the United States of America

Cornell University Press strives to utilize environmentally responsible suppliers and materials to the fullest extent possible in the publishing of its books. Such materials include vegetable-based, low-VOC inks and acid-free papers that are also either recycled, totally chlorine-free, or partly composed of nonwood fibers.

Paperback printing 10 9 8 7 6 5 4 3 2 1

CONTENTS

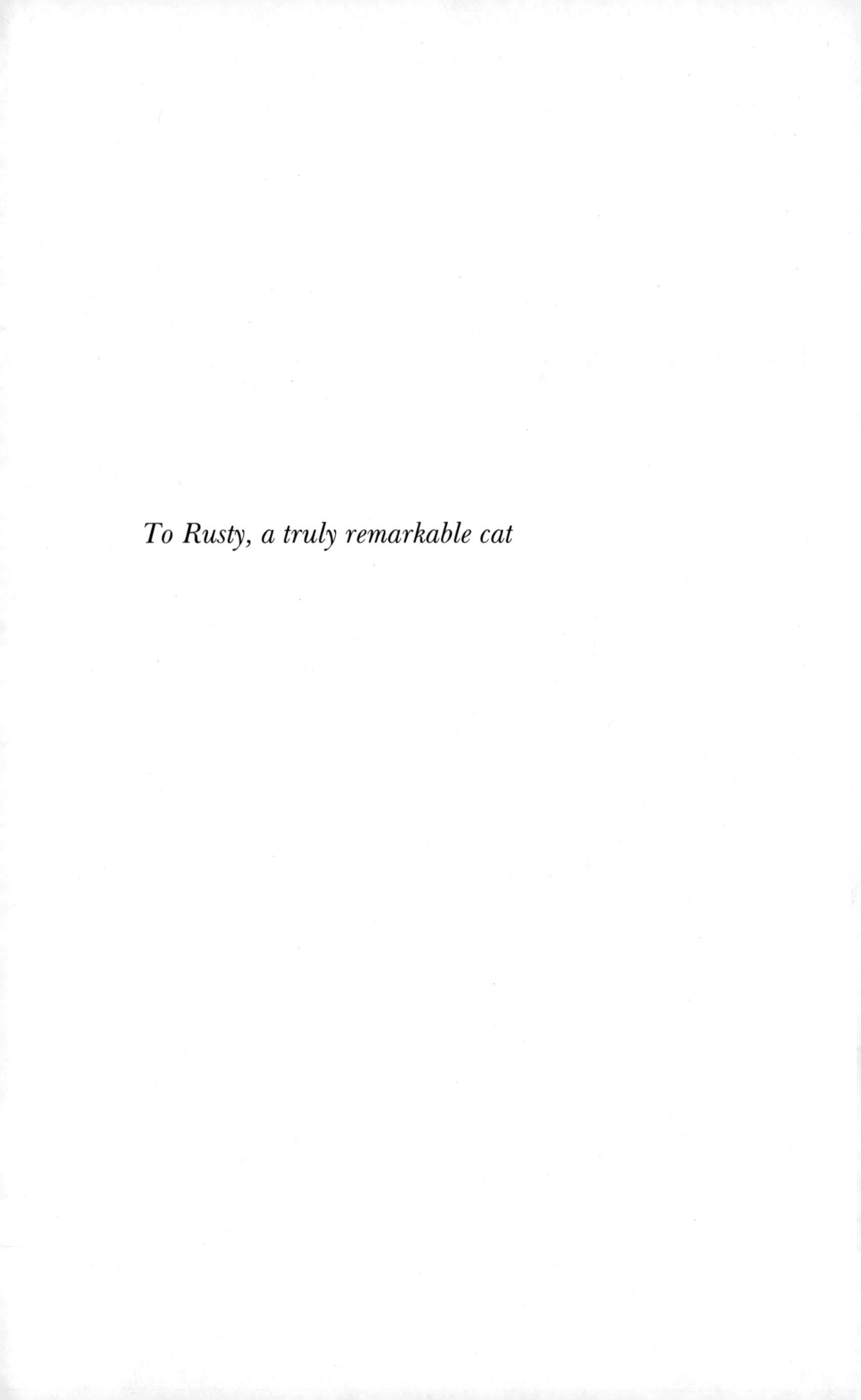

To Rusty, a truly remarkable cat

COLOUR PLATES

FIGURES

TABLES

ACKNOWLEDGEMENTS

Many people have contributed generously to this book, none more so, although indirectly, than the dedicated people who have carried out their painstaking research over many years into the lives of the world's most specialised carnivores.

In particular, I would like to thank A.A. Geertsema, Shigeki Yasuma, Simon Davis, the Howletts and Port Lympne Zoos and Kilverstone Wildlife Park, who have provided photographs. I am also indebted to Terry and Judith Moore of the Cat Survival Trust, who have loaned me literature I have needed from time to time.

Any review must rely primarily on information that has already been published. I would particularly like to acknowledge the following books and journals as important sources of information, illustrations and tabulated data in this book: *The Carnivores* by R.F. Ewer (New York: Cornell); *Carnivore Behaviour, Ecology and Evolution* by J.L. Gittelman (ed.) (London: Chapman and Hall); *Cats of the World* by S.D. Miller and D.D. Everett (eds.) (Washington DC: National Wildlife Federation); *The Domestic Cat* by D.C. Turner and P. Bateson (eds.) (Cambridge: University Press); *The Serengeti Lion* by G. Schaller (Chicago: University Press); *American Naturalist*; *Journal of Mammalogy*; *Journal of Wildlife Management*; *Journal of Zoology, London*; *Mammalia*; *Mammalian Species*; *Nature; Scientific American*; and *Zeitschrift fur Säugetierkunde*.

I am very grateful to Juliet Clutton-Brock, Bob Savage, Chris Wemmer and Martyn Gorman, who have read and commented on selected chapters. Mel and Fiona Sunquist also supplied much needed encouraging noises about an earlier draft of this book. To them all, thank you. I would like to thank Ernest Neal and the staff at Christopher Helm (particularly Robert Kirk) who have waited so patiently for the manuscript to arrive on their desks. Finally I would like to express my very sincere thanks to all my friends and family, especially Amanda, who have put up with my three-year moan of 'But I've got to do my book'.

SERIES EDITOR'S FOREWORD

In recent years there has been a great upsurge of interest in wildlife and a deepening concern for nature conservation. For many there is a compelling urge to counterbalance some of the artificiality of present-day living with a more intimate involvement with the natural world. More people are coming to realise that we are all part of nature, not apart from it. There seems to be a greater desire to understand its complexities and appreciate its beauty.

This appreciation of wildlife and wild places has been greatly stimulated by the world-wide impact of natural-history television programmes. These have brought into our homes the sights and sounds both of our countryside and of far-off places that arouse our interest and delight.

In parallel with this growth of interest there has been a great expansion of knowledge and, above all, understanding of the natural world—an understanding vital to any conservation measures that can be taken to safeguard it. More and more field workers have carried out painstaking studies of many species, analysing their intricate behaviour, relationships and the part they play in the general ecology of their habitats. To the time-honoured techniques of field observations and experimentation has been added the sophistication of radio-telemetry, whereby individual animals can be followed, even in the dark and over long periods, and their activities recorded. Infra-red cameras and light-intensifying binoculars now add a new dimension to the study of nocturnal animals. Through such devices great advances have been made.

This series of volumes aims to bring this information together in an exciting and readable form so that all who are interested in wildlife may benefit from such a synthesis. Many of the titles in the series concern groups of related species such as otters, squirrels and rabbits so that readers from many parts of the world may learn about their own more familiar animals in a much wider context. Inevitably more emphasis will be given to particular species within a group as some have been more extensively studied than others. Authors too have their own special interests and experience and a text gains much in authority and vividness when there has been personal involvement.

Many natural history books have been published in recent years which have delighted the eye and fired the imagination. This is wholly good. But it is the intention of this series to take this a step further by exploring the subject in greater depth and by making available the results of recent research. In this way it is hoped to satisfy to some extent at least the curiosity and desire to know more which is such an encouraging characteristic of the keen naturalist of today.

Ernest Neal
Bedford

PREFACE

The first felid influence on my life was when, as a six-year-old, I was happily building a sandcastle with my sister on Sandown Beach on the Isle of Wight. Suddenly, I looked over my shoulder to see a rather large puma hurtling towards me. Despite the fact that it was attached to a lead held by a man, and that I had read in my *How and Why Wonder Book* on *Wild Animals* that pumas do not usually harm people, I felt extremely vulnerable and ran away as fast as I could with my rather bewildered sister in tow. The horror of that moment has never left me, but I have come to terms with cats.

This book is a review of the current and not so current scientific literature concerning all 37 or so species of cats. Many books have been written about wild cats, but they mostly concern the big cats and are usually species-by-species accounts. Within these covers I have tried to synthesise as much as I could of our current knowledge of wild cats in a comparative approach, but emphasising the much neglected smaller cats. In recent years there has been a considerable amount of fieldwork on small species of wild cat. The miniaturisation and development of radiotelemetry has at last allowed the researcher into the tropical rain forests to study the wild cats there. In the future our knowledge of these smaller cats will expand enormously. This book represents merely a progress report of what we know today.

Much has been written elsewhere about the super senses and the athletic bodies of cats. In Chapter 1, these adaptations for hunting prey are discussed in the order in which they come into play during killing and feeding. The cat family's history stretches back nearly 50 million years and has involved various evolutionary experiments about just how to build the world's most perfect predator. None are more famous than the sabretooths, whose lifestyles are described in Chapter 2, together with recent developments in elucidating the evolutionary relationships between the living cat species of today.

Despite the fact that a considerable amount of fieldwork has now been carried out on some of the smaller cats, the majority remain virtually unknown except for skins and skulls in museum collections. Therefore all the cat species are described in Chapter 3, with their known distributions. Measurements and weights are given in the Appendix. In Chapter 4 it is back to the comparative approach, with a detailed look at hunting, killing and feeding behaviour of many different cats. On a wider scale, the effects of predators on prey populations and *vice versa* are also discussed at length. Cat diets are described in Chapter 5. This chapter is intended more as a reference than a good read. More often than not, it is the diet of a cat that is better known than the cat itself.

In Chapter 6 the social organisation of wild cats is discussed. Wild cats retain a very flexible approach to using their environment depending on changing external factors. By using a variety of smells, wild cats manage to maintain an exclusive, or where necessary a shared, living space. For most male cats this results in a polygynous mating system, but without the blood and guts involved in the annual fighting of deer and antelopes. There is also a lengthy discussion of the reasons why some cats, the cheetah, lion and domestic cat, have gone in for a social life. I cannot say that I have the ultimate answers, but there are plenty of ideas to be going on with.

In Chapter 7 the reproductive behaviour and life histories of wild cats are described. Much of what we know about mating and of rearing kittens or cubs is based on the domestic cat or captive wild cats, except for a few studies of the bigger cats in open habitats. There is much scope here for ingenious future research. Finally, in Chapter 8 various aspects of the relationship between cats and humans are discussed. The fur trade, conservation, man-eating, domestication and re-introductions to the wild are all described. It is to be hoped that the relationship between our own species and the Felidae will take a much needed turn for the better, if the wild cats are to thrive well into the next century.

One of the most confusing things about wild cats concerns their common names. I have not used scientific names throughout the text, which means that confusion could become worse still. The wild cat species in Europe and Africa, *Felis silvestris*, is commonly known as the 'wild cat'. However, I often refer to all cat species (excluding the domestic cat, *Felis catus*) as 'wild cats'. To avoid this confusion, I have called *F. silvestris* the 'wildcat' to distinguish it from the other wild cats.

HOW TO BE A CARNIVORE

Almost every feature of a cat's body is related to the way it detects and catches its prey. Cats have no option but to be efficient hunters because they are the most exclusive of meat-eaters, and compared with other food resources, meat is in limited supply within a given area (Gittelman and Harvey, 1982). Unlike other carnivores, cats cannot supplement their diet with plant matter because their gut physiology is so specialised for a diet of meat. Larger cats face an additional problem because they often prey on animals larger than themselves, and it is important that they kill their prey quickly to avoid being injured in any subsequent struggles.

Cats hunt either by patrolling their home ranges until they come across a potential victim, or by waiting in ambush for them to pass by (Kruuk, 1986). They need keen senses to detect their prey before being detected themselves, and must be well concealed while stalking or waiting to strike in order to maximise their chances of getting close enough to capture a potential prey animal. Cats must be able to move quickly and quietly as they approach to capture and kill their prey. The need to overcome all of these problems is reflected in specialisations in the form and structure of cats.

Below, the various specialisations of the feline body will be explored to show how cats have become the most ruthless and efficient of living carnivores.

CONCEALMENT—THE PELAGE

The furry coat or pelage of cats not only serves to insulate them from temperature changes in the environment, but also shows a wide variety of different lengths, patterns and colours which seem to be related to cats' needs to remain concealed from their prey while hunting in the wide variety of habitats in which the different species are found. The Pallas's cat and snow leopard both have fluffy coats for keeping them warm, but the colour and markings of their coats are vital in helping them to remain 'hidden' in their respective steppe and montane habitats.

The background colour of the coat is usually similar to the habitat in which the cat is normally found, and there is often a variety of spots, stripes and blotches which help to match the cat more closely to its habitat or break up its outline to make it difficult to see. Desert or semi-desert cats have light-coloured coats, which show more spots and stripes with increasing vegetational cover and decreasing aridity. For example, the sand cat of the deserts of Asia and North Africa is sandy coloured with very few markings, which matches its desert environment. The closely related African wildcat has a similar coloration to the dry grassy habitat in which it lives. Its stripes help it to 'disappear' in the uneven pattern of light and shade in tall grasses. These are examples of cryptic coloration.

Many forest and woodland cats have a bold pattern of stripes, blotches and spots which seems very conspicuous to us in the artificiality of zoos. However, the dappled light falling through vegetation has a similar contrasting effect. This disruptive coloration helps to break up the outline of the cat and make it hard for prey animals to detect in the cats' forest home. The ocelot and margay from South and Central America are examples of this type of coloration, but many Old World cats have similar coat patterns, including the leopard cat, fishing cat and rusty-spotted cat from southern Asia.

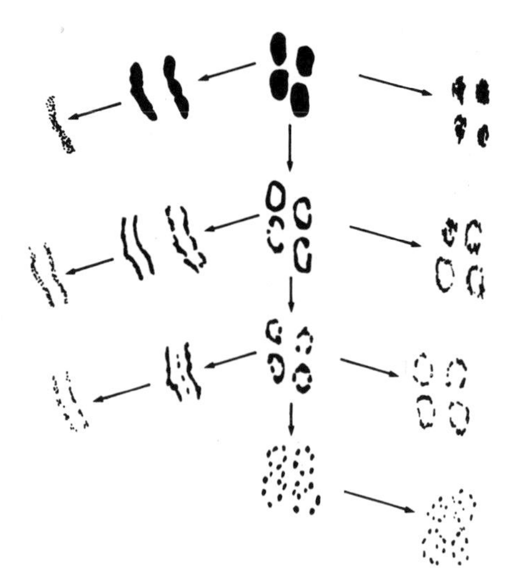

Figure 1.1 The derivation of the different coat patterns in cats from a dark-spotted ancestor as proposed by Weigel (1961) (after Ewer, 1973) (A.K.)

The evolution of cat coat patterns is still not completely understood. Weigel (1961) believes that all coat patterns evolved from a dark-spotted type. The spots subsequently broke up and fused in a variety of ways to give the modern array of coat markings (Figure 1.1). Kingdon (1977) considers that the pattern of spots in leopards has evolved in the opposite direction: small flecks coalescing to produce large, distinct spots and rosettes. Recently, Murray (1988) has discussed the embryological development of spotted and striped cats by a reaction diffusion mechanism. Further elucidation of this mechanism may help to resolve how cat coat patterns have evolved.

INTRASPECIFIC VARIATION

In widely distributed species there is often a wide variation in coat colour and patterns to match the variety of habitats they inhabit. For example, the European wildcat (*Felis silvestris silvestris*) is darker and more heavily banded than the African wildcat, reflecting its forest home; the African wildcat (*F. s. lybica*) is lighter coloured and lightly banded where it remains concealed on the African savanna, but the Indian desert cat (*F. s. ornata*) is very light in colour with small dark spots to match the semi-deserts of South West Asia. Another wide ranging species, the leopard cat (*Felis bengalensis*), shows an equally wide variation in coat colour and markings from the cold north to the hot south of its range (see Chapter 3). Leopards show a similar variation in the boldness of their spots and rosettes with changing habitat; desert leopards (e.g. *Panthera pardus nimr*) are much paler and less heavily spotted than forest leopards (e.g. *P. p. leopardus*) (Harrison, 1968; Rosevear, 1974).

Some cat species are polymorphic in coat colour. In the past this has led to considerable confusion, because these different coat colours have resulted in the description of different species of cat. One example is the jaguarundi which has both a grey and reddish form. In the nineteenth century the reddish or erythrystic form was described as a separate species, the eyra (*F. eyra*), until both forms were found in the same litter. Even more confusing is the polymorphism in the African golden cat (van Mensch and van Bree, 1969). Again there are red and grey forms with varying degrees of spotting, but one individual kept at London Zoo changed from red to grey in a period of four months (Pocock, 1907).

MELANISM AND OTHER COLOUR MUTATIONS

Melanism is a common coat colour mutation in cats. It has been recorded in leopards, lions, tigers, jaguars, caracals, tiger cats, pumas, bobcats, margays, ocelots and servals (Robinson, 1970, 1976a; Ewer, 1973; York, 1973).

Melanism in leopards and domestic cats is usually due to a

recessive gene at the agouti locus for coat colour, and this is probably the case in all cats (see Chapter 7) (Robinson, 1970; 1976a). Melanistic leopards or black panthers are very common in the rain forests of southern Asia. Presumably, so little light penetrates the forest canopy that a dark coloration confers either a selective advantage in concealment or no obvious disadvantage in this habitat.

Leucism or a lack of dark pigments has been recorded quite widely in different cat species, including the famous white tigers of Rewa and the white lions of Timbavati (Robinson, 1969a). The genetic causes of leucism do, however, vary (see Chapter 7).

A spectacularly marked cheetah with a prominent mane was described in 1927 as the king cheetah, *Acinonyx rex* (Pocock, 1927). However, it is now known that these unusual markings are also due to a recessive gene and both normal and king cheetahs can occur in the same litter (van Aarde and van Dyk, 1986).

DETECTION OF PREY—THE SENSES

Cats detect their prey usually by sight and sound. Smell is rarely used in hunting, but is very important in intraspecific communication. Whiskers and other sensory hairs are used as a tactile sense to help cats feel their way while stalking and catching prey.

SIGHT

Most cats are crepuscular and nocturnal in their activity, but they also hunt to a lesser extent during the day depending on the activity of their prey (e.g. Schaller, 1967, 1972; Corbett, 1979; Geertsema, 1985). Consequently their eyes should be as sensitive as possible to cope with low light levels, but still be able to function in bright illumination. For example, domestic cats are able to see light one sixth as intense as our own eyes, but they are still able to see perfectly well in broad daylight (Ewer, 1973). Cats' eyes show a number of adaptations which allow them to function successfully in such low levels of light.

One way to increase the sensitivity of the eye to light is to increase the size of the pupil to allow more light to enter it (Walls, 1942; Ewer, 1973). However, this creates two problems. Firstly, a larger lens is needed so that light is not distorted by the edge of the lens as it passes through the larger aperture. Also, the curvature of the lens must be increased to refract the incoming light more in order to focus it on the relatively nearer light-sensitive retina at the back of the eye. The cornea is also more curved to help refract the light so that the anterior chamber of the eye is relatively larger than in human eyes (Walls, 1942; Ewer, 1973) (see Figure 1.2). Thus, it is possible to distinguish between the eyes of cats that hunt predominantly at night and those that may hunt commonly during the day. The lynx has a large anterior chamber to its eye reflecting its mostly nocturnal habits, whereas the puma has an eye structure intermediate between the lynx's and our own, which shows that it is suited to working well both in daylight and at night (Walls, 1942; Ewer, 1973) (see Figure 1.2).

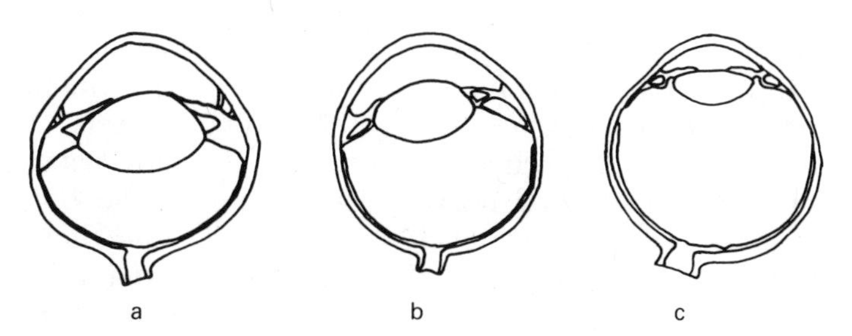

Figure 1.2 Sagittal sections through the eyes of lynx, puma and human. The nocturnal lynx (**a**) has a large anterior chamber to its eye, a larger lens and a wider pupil than the less nocturnal puma (**b**) and the diurnal human (**c**) (after Walls, 1942; Ewer, 1973) (A.K.)

The light-sensitive area at the back of vertebrate eyes is called the retina. There are two types of light receptor cells in the retinae of mammals. Cones are sensitive to high levels of light and are used in colour vision, whereas rods function in low levels of light and do not detect colour at all. Not surprisingly the eyes of cats consist primarily of rods, but there is a cone-rich patch at the centre of the retina as in our own eyes (Ewer, 1973; Hughes, 1977, 1985). It is unlikely that cats use their cones for colour vision even though there are at least two types of cones in their retina, one of which is sensitive to green light and the other to blue light (Ewer, 1973). Cats can be trained to use their colour vision, but this is a very lengthy process. It seems likely that cats have two types of cone so that they can see over a wider range of wavelengths of light in daylight when their rods will not function as well.

Cats show yet another adaptation to improve their ability to see at night. When a torch or headlamp is shone into the eyes of cats, they shine back eerily. This is because of a layered structure at the back of the eye behind the retina called the *tapetum lucidum*, which is found in many other vertebrates although its structure and origin may vary considerably (Locket, 1977; Wen, Sturman and Shek, 1985). When the mirror-like *tapetum* is caught in a beam of light, it reflects the light back through the retina to help produce a brighter image. It is therefore a structure which has evolved many times to help the eyes of mammals function in low light levels (Lockett, 1977).

The density of the receptor cells in the retina may vary quite considerably. In our own eyes, there are low concentrations of rods at the periphery of the retina which increase towards the centre of the retina and which are replaced by very high densities of cones at the very centre of the eye (Hughes, 1977). This high concentration is centred on a small depression called the fovea or yellow spot, which helps increase the density of cones even further to allow for even better colour vision. The high concentration of cones improves our ability to resolve detail in daylight, but the relatively low rod concentration makes us very poor at resolving detail in low light levels. This resolving power is called visual acuity and would, for example, make the difference between looking at the pages of a newspaper at a distance and seeing them as grey sheets, and being able to distinguish the individual letters on the pages. Cats also have a high concentration of nerve cells leading to the optic nerve emanating from the centre of the eye, but instead of there being a central spot as in our own eyes, there is instead a broadly horizontal streak (the visual streak), which increases visual acuity horizontally,

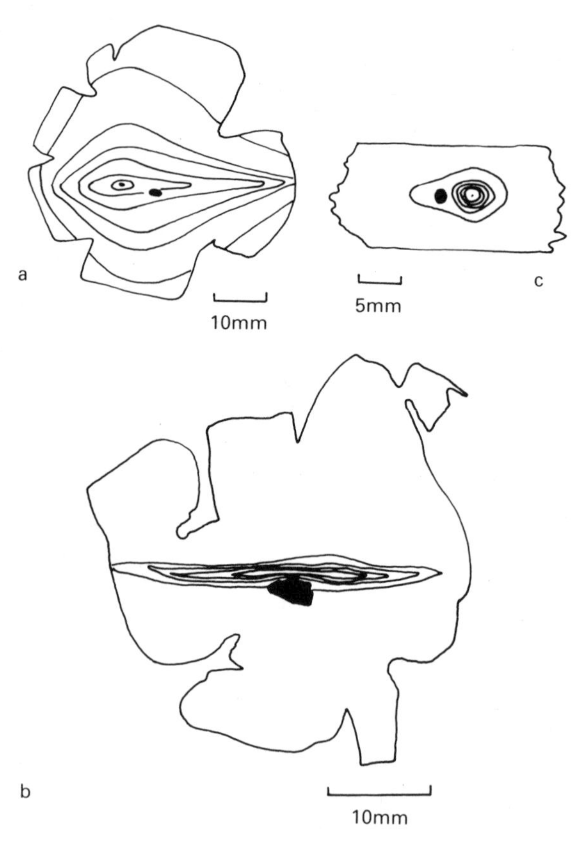

Figure 1.3 Diagram of the visual streaks on the retinae of the domestic cat (**a**), cheetah (**b**) and human (**c**). The horizontal visual streak of the cheetah reflects its need to be able to detect prey moving against the horizon in open habitats (Hughes, 1977, 1985) (A.K.)

so that cats are better able to spot their prey moving in this plane (Figure 1.3) (Hughes, 1977, 1985). This has reached an extreme case in the cheetah, which chases its prey in open habitats. The visual streak is a very thin, concentrated band of nerve cells which allows the cheetah to pick out its prey moving against the horizon. (Figure 1.3) (Hughes, 1977).

However, cats have poorer visual acuity than humans. They have sacrificed visual acuity for greater sensitivity by adding together the input from many more receptor cells to stimulate fewer optic nerve cells at very low light levels (Ewer, 1973).

When cats stalk and capture their prey they must be able to judge distances. This ability is also needed by arboreal cats when jumping from branch to branch. Consequently, cats have the most highly developed binocular vision of all carnivores (Figure 1.4) (Vakkur and Bishop, 1963; Hughes, 1976). Indeed it is almost as good as our own binocular vision.

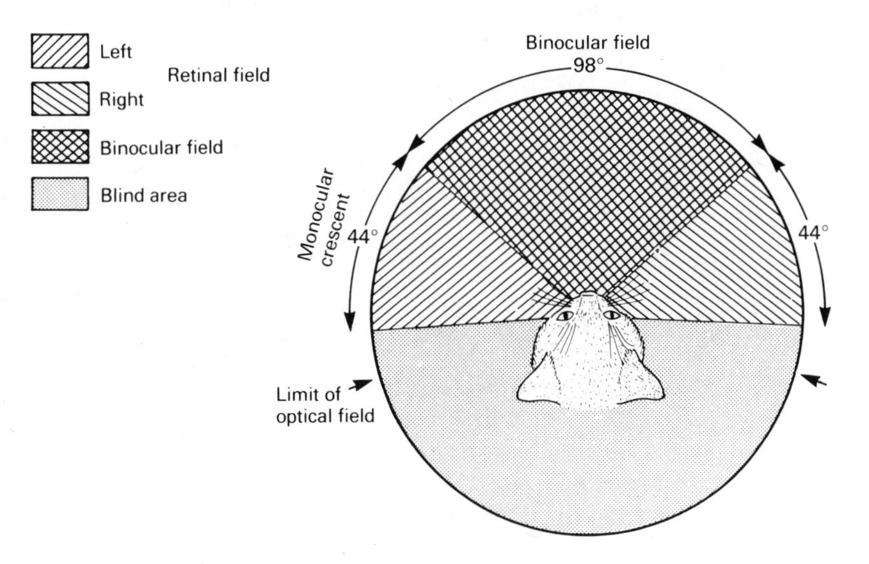

Figure 1.4 The visual field of the domestic cat (Hughes, 1976). Apart from primates, cats have the most highly developed binocular vision of mammals (S.A.)

Cats may hunt by night or day, although they are predominantly nocturnal and crepuscular. Only the cheetah hunts regularly during the day. For this reason, the pupils of cats' eyes must be able to undergo very large changes of size. They must be very large to allow as much light as possible to enter the eye at night, and as small as possible during the day so that they are not dazzled. In our own eyes the round pupil is controlled by a circular ciliary muscle (Figure 1.5) (Walls, 1942; Ewer, 1973). The trouble with this sort of arrangement is that the aperture of the pupil cannot be changed very much. The muscle fibres would have to contract to zero length to close up the pupil almost completely. Instead, the cats of the domestic cat lineage (see Chapter 2) have an elliptical pupil in which the ciliary muscle fibres interlace and pull across each other so that the pupil can almost be completely closed up in bright light, leaving only two pinholes at either end of the vertical slit (Figure 1.5) (Walls, 1942; Ewer, 1973). Other cats, and in particular the big cats, have broadly elliptical pupils which appear round when dilated and form an elliptical pupil in daylight. The Pallas's cat's pupils close up to a small round hole in bright daylight.

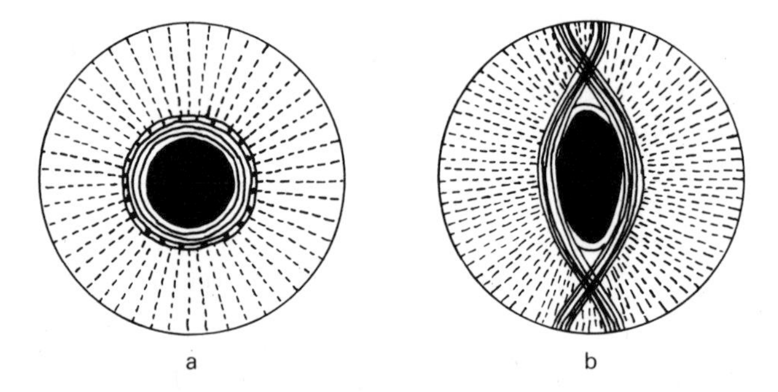

Figure 1.5 Human pupils (**a**) are controlled by a circular ciliary muscle, but this arrangement would not allow the felid pupil to undergo the extreme changes in size needed for seeing well in both daylight and at night. Instead the ciliary muscles of a cat's eyes (**b**) interlace and draw across each other to close the pupil to a vertical slit (Ewer, 1973) (A.K.)

HEARING

Cats can hear over a very wide range of different frequencies from 200 Hz up to 100 kHz, which is much greater than the upper limit of 20 kHz for human hearing (Ewer, 1973). However, at the upper end of the range the loudness or intensity of the sound would have to be so great that for practical reasons the upper limit has been recorded at 70 kHz for the jaguarundi and 65 kHz for the domestic cat (Ewer, 1973). But what use is this? Rodents are a common prey item of cats and many species use ultrasound to communicate (e.g. Sales and Pye, 1974). The advantage of ultrasound compared to the use of sound in our hearing range is that it does not carry very far and is therefore very useful for short range communication in dense habitats. Rodent ultrasonic communication occurs in the range of 20–50 kHz, which is well within the range of cat hearing, whereas we would be unable to hear them. Cats can, therefore, hear the 'quiet' squeaks of their rodent prey.

The problem for the cat in picking up these very high frequency communication sounds and also lower frequency sounds caused by movement is that they are of low intensity. To combat this, cats have very large external ear flaps or pinnae. Servals have relatively the largest external pinnae of any cat, which helps them pick up the sounds of their rodent prey scurrying through tall grass (Geertsema, 1976, 1985). Sand cats also have large external pinnae, which help them pick up the faint sounds of their rodent prey in the desert. The hot dry air of the desert absorbs sound so that large ears are needed to pick up faint sounds (Savage, 1977). But desert animals also show another specialisation which is as yet poorly understood. There is a chamber surrounding the three ear bones or auditory ossicles called the auditory bulla. In cats it is formed from two bones which fuse to form a bony septum dividing the bulla in two (Hunt, 1974; Savage, 1977). Desert animals including jerboas, kangaroo rats, fennec foxes and sand cats all have enlarged auditory bullae compared with mammals of a similar body size, which do not live in deserts. Klimova and Chernys (1980) found that forest cats have relatively much smaller bullae than desert or savanna cats. One theory to explain this is that there is less air resistance in a large bulla, which allows for the vibration of the auditory ossicles at specific resonant frequencies equivalent to the movements of prey or predator (Webster, 1962; Savage, 1977). Only in a desert is it quiet enough to be able to use hearing successfully in hunting in this way. Therefore, the sand cat can use its sensitive hearing to pick out specifically the sounds produced by jerboas and gerbils as they move along.

SMELL

Cats rarely use their sense of smell in hunting, but it is still used as a very important means of intraspecific communication. But the sense of smell of cats is not as highly developed compared with other carnivores.

The olfactory bulbs are the areas of the brain that process the information from the olfactory epithelium in the nose. They are relatively much smaller in cats compared with other carnivores (Figure 1.6) (Radinsky, 1975). Dogs have a highly developed sense of smell and have olfactory bulbs which occupy on average 5 per cent of the brain volume compared with only 2.9 per cent in cats.

The olfactory epithelium occupies one of the two main groups of turbinal or scroll bones in the nasal cavity (Savage, 1977). The group at the front is called the maxilloturbinals and they are responsible for warming and filtering the air as it is breathed in. Behind these are the ethmoturbinals and these bear the olfactory epithelium. The ethmoturbinals of dogs have a greater surface area (125 cm^2) than those of cats (13.9 cm^2) as would be expected for the longer muzzles of dogs (Dodd and Squirrell, 1980). However, the density of receptor cells on the cat olfactory epithelium is twice as great as in the dog, so that overall dogs only have about twice as many olfactory receptors (Dodd and Squirrell, 1980).

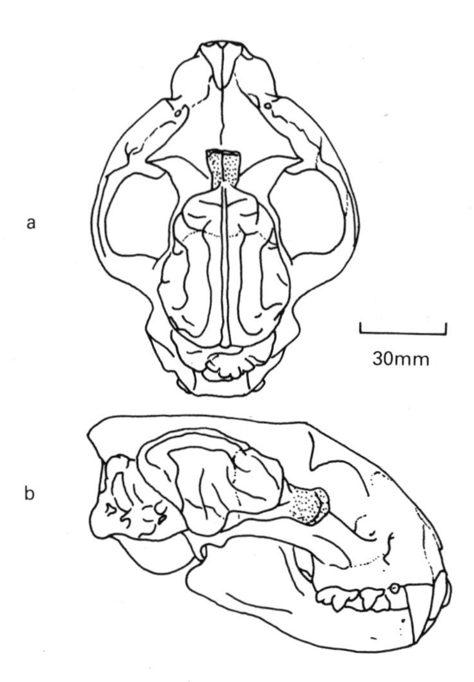

Figure 1.6 Dorsal (**a**) and lateral (**b**) views of the skull of a fishing cat showing the position of the brain. The stippled structures at the front of the brain are the olfactory bulbs (after Radinsky, 1975) (A.K.)

The cheetah has very large nasal passages which is unusual for a cat. This is not because the cheetah has a well-developed sense of smell. Instead, this specialisation seems to be related to the way in which cheetahs hunt. After running down their prey, cheetahs suffocate them by grabbing their throats and blocking their tracheae or windpipes. The problem for the cheetah is to regain its breath after its short, but exhausting dash. The wide nasal passages allow it to breathe easily while its prey dies (Ewer, 1973; Kingdon, 1977).

WHISKERS

Whiskers are specialised hairs which are adapted to a tactile sensory function. In carnivores they commonly occur in four different groups: on the cheeks (genal), above the eyes (superciliary), on the muzzle (mystacial) and below the chin (inter-ramal) (Figure 1.7). Cats lack the inter-ramal tuft of whiskers. Pocock (1917) suggested that they did not need these whiskers because they would rarely need to bend down to sniff the ground to track their prey.

The mystacial whiskers are particularly well-developed and their orientation depends on what the cat is doing. In sniffing they are retracted against the side of the face, at rest they are extended laterally and when walking they are extended forwards (Figure

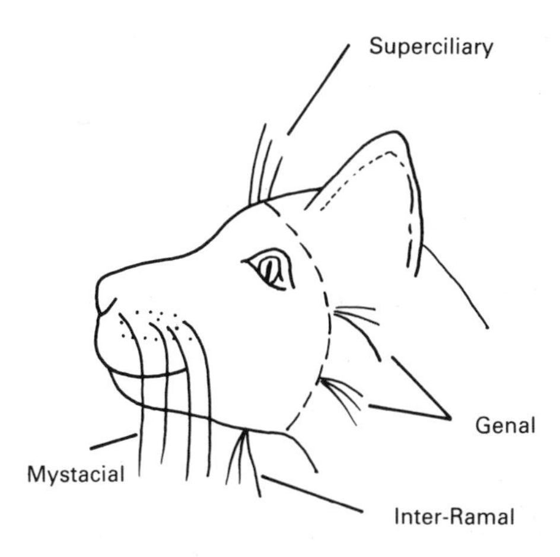

Figure 1.7 The whiskers of a generalised carnivore (after Tabor, 1983) (A.K.)

1.8) (Weigel, 1975). At the time of prey capture they are extended like a net in front of the mouth so that the cat can judge exactly where the prey is for an accurate killing bite. Recently it has been discovered that the visual and tactile senses of the domestic cat have a similar arrangement in adjacent areas of the brain, so that it would seem that cats can interpret their surroundings in terms of a complementary visuo-tactile sense (Stein, Magalhaes-Castro and Kruger, 1976). The cheetah is the most diurnal of all the cats. Its whiskers are accordingly reduced compared with those of other cats (Ewer, 1973).

Tylotrichs are another type of sensory hair which occur all over the body (Ryder, 1973).

Figure 1.8 Position of the whiskers when a cat is (**a**) at rest, (**b**) walking and (**c**) when it is sniffing, biting, bowing its head in greeting, or is defensive (Weigel, 1975) (A.K.)

LOCOMOTION AND PREY CAPTURE

In the final dash for their prey, cats need both speed to approach and also power for the capture. However, these two requirements place two different constraints on the design of the skeleton, which cats have been forced to compromise upon, or overcome.

In a limb built for power, the lower part of the limb, from the elbow or knee down, must be relatively short, so that the leverage produced by the muscles of the upper limb and limb girdles is powerful, but relatively slow. We can see this in the forelimbs of moles, which are renowned for their amazing burrowing strength, but which are rather cumbersome when attempting to run. In comparison, limbs built for speed have a longer lower limb compared to the upper limb, achieved by shortening of the humerus and femur, elongation of the radius, ulna, tibia, fibula, metacarpals and metatarsals (metapodials) of the lower leg, and by a tip-toe (digitigrade) stance (Figure 1.9). To give greater stability to the

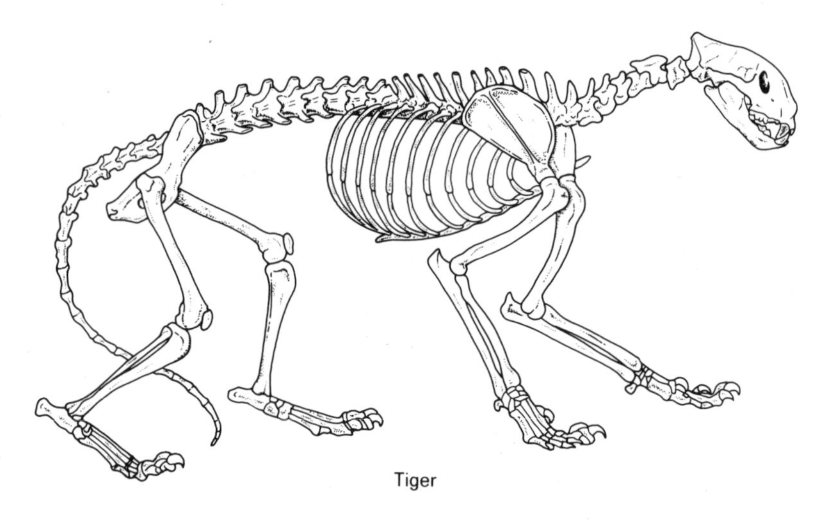

Figure 1.9 Like those of many other carnivores, the limbs of cats are adapted for a cursorial lifestyle. They have a digitigrade stance, and elongated, fused lower limb bones (after Ewer, 1973) (S.A.)

lower leg, the metapodials are usually bound tightly together and the carpals (and tarsals) are fused (Ewer, 1973), although this latter adaptation may be more useful in prey capture when the paw is slightly flexed. The cheetah's tibia and fibula are firmly bound together with fibrous tissue which allows very little rotation about the lower leg (Ewer, 1973). In complete contrast, the arboreal margay has a very flexible ankle joint which can supinate through 180°. The margay can easily climb through trees using both its fore and hind legs. Unlike the domestic cat, it can run head first down a tree trunk,

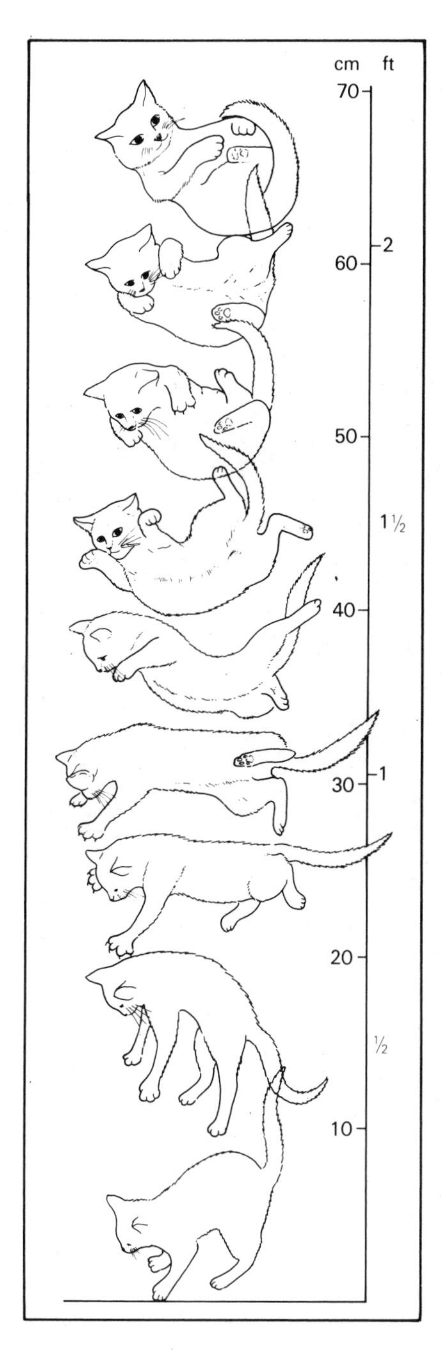

Figure 1.10 The articulations of vertebrae of cats are smooth and rounded to allow great flexibility and twisting so that falling cats are, for example, able to land on their feet (after Burton, 1978) (S.A.)

easily being able to grasp the trunk with all its feet (Figure 1.12). Like the cheetah, the serval also has long legs, but these are not designed for running fast, because the metapodials are loosely held together. Instead, the serval's long legs seem to be an adaptation for hunting in tall grass. The serval springs through the grass locating and pouncing on prey animals, often using its long legs to stun and kill them (Geertsema, 1985).

The problem for most cats is to retain a powerful enough limb to bring down their prey, but also to have a limb built for speed to catch the prey in the first place. How do they do this? Like the digitigrade ungulates such as deer and antelopes, cats do show some elongation of the metapodials and other lower limb bones, but not to the same degree. Cats increase their stride length and, hence, speed further by flexing and extending their vertebral columns during running. This skeletal adaptation reaches an extreme for cursoriality in the cheetah so that it can run fast enough to overtake its favourite food item, gazelles (Figure 1.11). Hildebrand (1959, 1961) has measured the contribution of the flexing and extending vertebral column to the cheetah's running to be 11 per cent of the stride length of 6.9 m at 56 kph.

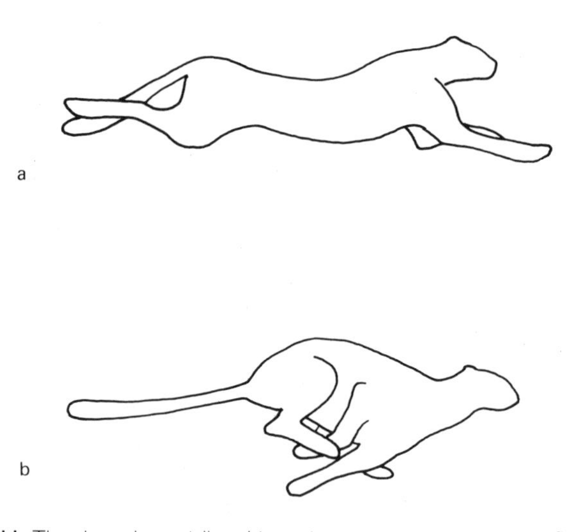

Figure 1.11 The cheetah partially achieves its great running speed by flexing (**a**) and straightening (**b**) its backbone to produce a greater stride length (Hildebrand, 1961) (A.K.)

The articulations between the vertebrae are very flexible, allowing rotation about the long axis of the vertebral column (Savage, 1977). This gives rise to, for example, a cat's extraordinary ability always to land on its feet after a fall (Figure 1.10).

Cats further increase their stride length by not having a clavicle (collar bone), which is usually reduced or absent (Figure 1.13) (Savage, 1977). This frees the shoulder joint and allows the scapula

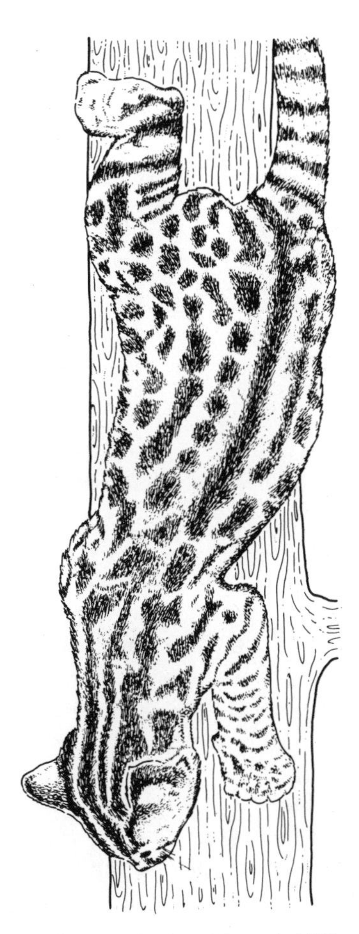

Figure 1.12 The margay's ankle can be bent through 180° to allow it to climb easily through the trees (after Ewer, 1973) (S.A.)

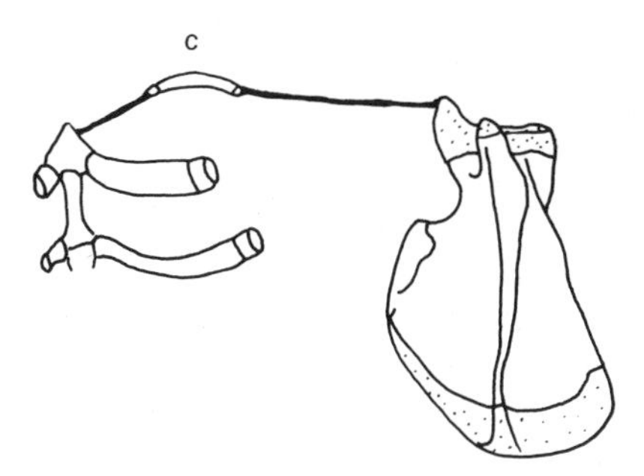

Figure 1.13 The clavicle usually acts as a brace to strengthen the thorax of mammals. In cats the clavicle (c) is much reduced and attached by a ligament to the sternum (left) and scapula (right). The reduced clavicle allows free fore and aft forelimb movements to increase stride length (after Ewer, 1973) (A.K.)

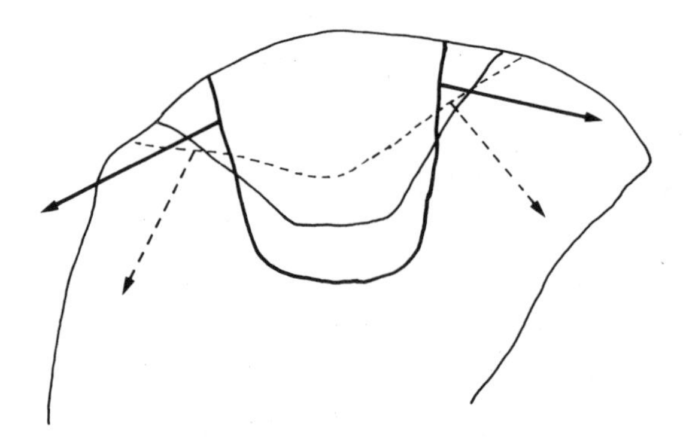

Figure 1.14 Diagram of the upper part of the medial surface of the scapula, to show the differing shape of the area of insertion of the serratus magnus muscles and the levator anguli scapulae muscles in the leopard (dotted), lion (continuous) and cheetah (bold) (Hopwood, 1947). See text for explanation (A.K.)

(shoulder blade) to increase the effective length of the limb. In cats, the scapula pivots about its mid-point and is moved back and forth by two muscle groups which insert on its inside face. But the shape and size of the scapula and the way in which these muscles insert vary between cats (Hopwood, 1947). In the highly cursorial cheetah, the scapula is rectangular and the muscles that move it insert in a narrow deep strip so that their action is to pull the scapula back and forth for running (Figure 1.14). In the arboreal leopard, the scapula is shorter and more fan-shaped and the muscles insert in a wide, shallow strip on the border of the scapula. The line of action of the muscles is ideal for abduction of the limbs for climbing (Figure 1.14). The lion's arboreal and cursorial abilities are intermediate between the running cheetah and the climbing leopard, and consequently it shows an intermediate scapula structure and pattern of muscle insertion. Hildebrand (1959, 1961) has calculated that the movement of the scapula contributes 12 cm or 2 per cent of the stride length of a cheetah at 56 kph.

PREY CAPTURE—CLAWS AND PAWS

At rest a cat's claws are held retracted by ligaments in a sheath of skin to protect them from wear, so that they are as sharp as possible for prey capture and tree climbing (Figure 1.15) (Ewer, 1973; Gonyea and Ashworth, 1975; Savage, 1977). In order to use its claws, a cat must extend the terminal phalanges and protract the claws using muscular action (Figure 1.15).

Cheetahs are often said to have non-retractile claws. In fact their claws are weakly retractile, but there is no sheath of skin to protect them (Figure 1.17) (Pocock, 1916b; pers. obs.). Cheetahs seem to use their claws as running spikes to increase their grip while in pursuit of their prey.

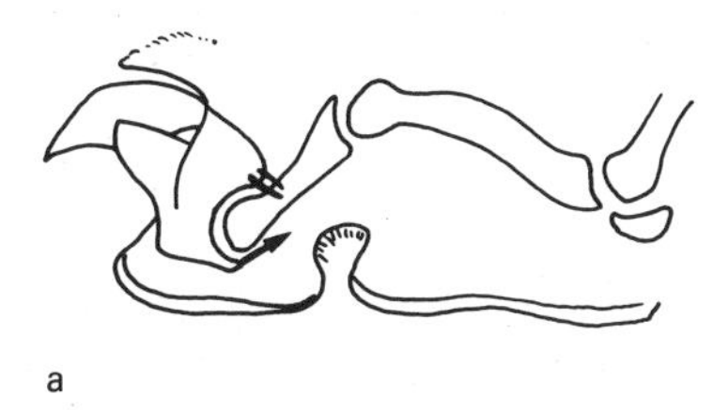

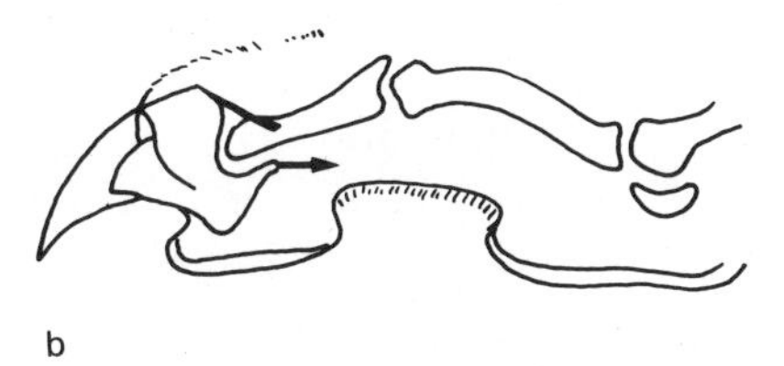

Figure 1.15 The protraction mechanism for the felid claw: (a) at rest the claw is held retracted by ligaments inside a sheath of skin; (**b**) when needed the lower ligament pulls the claw forward and the phalanges straighten so that the claw protracts from the sheath (Ewer, 1973) (A.K.)

Figure 1.16 The underside of the paw of a sand cat to show that it is protected by fur from the extremes of temperature in the desert. The fur also helps to spread the weight of the cat so that it can move easily over the shifting sands (Harrison, 1968) (S.A.)

Cats have five claws on the forefeet and only four on the hind feet. The first claw on the forefoot is reduced to a small dew claw which does not reach the ground. This dew claw is, however, highly developed in the cheetah, where it is used to help drag down its prey (Figure 1.17) (Wrogemann, 1975).

The paws of cats have soft pads which allow for a quiet approach in stalking. The mechanics of cat paws have recently been investigated by Alexander, Bennett and Ker (1986). When running, cats must put their feet down in such a way as to avoid a chattering effect due to the viscoelastic properties of their foot pads, in order to maintain a sure grip.

Cats from very hot or very cold climates have fur-covered foot pads to protect them from extremes of either hot or cold. The Canada lynx's furry feet not only keep it warm, but act as snowshoes, spreading the cat's weight over a larger area of soft snow. Furriness reaches its extreme in the sand cat, where the foot pad is totally obscured by protective fur, which also helps to spread the weight of this desert predator as it moves across shifting sands (Figure 1.16) (Harrison, 1968).

Cheetahs have very hard foot pads as would be expected for running on hard ground. Ridges, which are thought to act like the tread on car tyres, run along the foot pads (Figure 1.17) (Pocock, 1916b; Ewer, 1973).

KILLING—TEETH AND JAWS

Cats show a number of specialisations of their teeth and jaws for both killing and eating their prey. The need to shorten the length of the jaws for a powerful bite has led to a reduction in the number of teeth from the typical number of 42 found in the closely related civets and mongooses to only 30 or 28 in the Felidae, to give the dental formula below:

$$I = \frac{3}{3} \quad C = \frac{1}{1} \quad P = \frac{2\text{–}3}{2} \quad M = \frac{1}{1}.$$

Some short-faced species of cats have lost one of the upper premolars to give a total of 28 teeth. These include the lynxes, golden cats and the caracal (Ewer, 1973; Savage, 1977).

The main features of felid dentition are the canines, which are used in killing the prey, and the carnassials comprising the fourth upper premolar (P^4) and the first lower molar (M_1), which are used to slice through meat. The incisors help to grab hold of the prey and are used in plucking fur or feathers, and pulling meat off bones. There is usually a gap or diastema behind the canines, which allows them to sink in as deeply as possible during the killing bite.

This is absent in the cheetah, which mainly kills its prey by suffocation (Pocock, 1916b).

The canine teeth differ from those of other carnivores in being broader and more robust (van Valkenburgh and Ruff, 1987). For example, cats have relatively longer and stronger canines than dogs. When cats kill their prey, they sink their canines into the nape of the neck and dislocate the cervical vertebrae of their victim. Strong canines are needed in case the teeth smash against bone. Dog canines are more slender and elliptical in cross-section because their teeth are not designed for stabbing.

The carnassials of cats are the most scissor-like of all carnivores (Figure 1.18). The loss of the other molars and premolars has

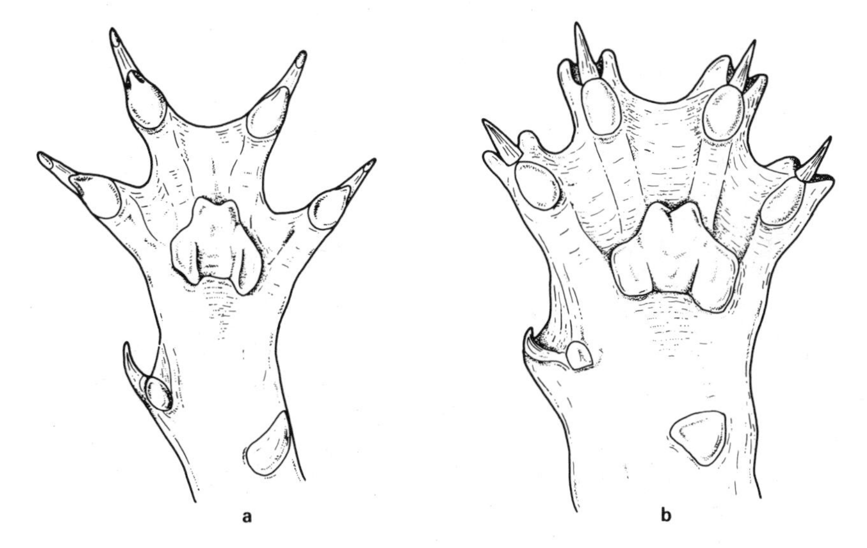

Figure 1.17 (**a**) The cheetah has no fleshy sheaths to protect its claws when they are partially retracted. Cheetah claws are usually worn blunt like dog's claws. (**b**) The leopard's claws are withdrawn into fleshy sheaths. They remain sharp for prey capture and tree climbing (Pocock, 1916b) (S.A.)

resulted in an increase in the length and height of the fourth upper premolar and the first lower molar until they are like knife blades. There is a notch in the carnassials to hold the slippery meat while it is sliced up. The conformation of the jaws of cats means that only the carnassials on one side of the jaw can be used at any one time. Consequently, the articulation of the mandible is cylindrical to allow the jaws to swing from side (Figure 1.19) (Ewer, 1973).

Greaves (1983) has modelled the jaws of carnivores to see if they have evolved an optimal design for the most powerful killing bite and chewing action. This model shows that the resultant force of the jaw muscles should act at a distance that is 60 per cent of the length between the jaw articulation and the carnassial, in order to

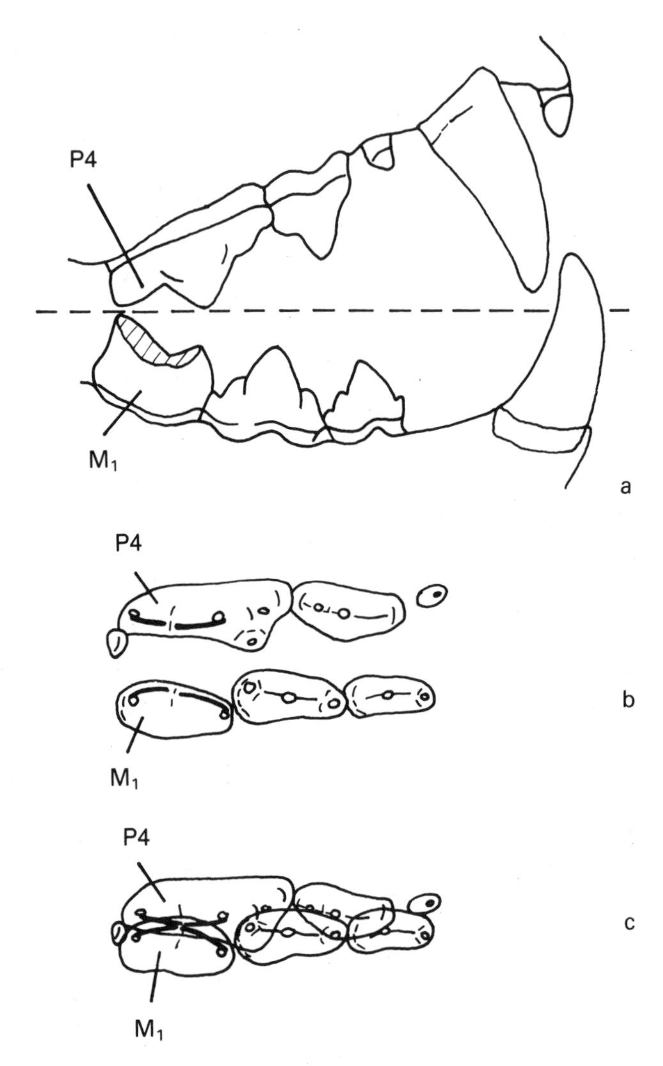

Figure 1.18 (**a**) Lateral view of the jaws of a cat to show the position of the upper (P^4) and lower (M_1) carnassials. (**b**) Plan view of the cheek teeth of a cat before chewing. Open circles represent cusps and lines show ridges. The bold lines show the two shearing ridges of the carnassial teeth. (**c**) Plan view of the cheek teeth when the jaws are closed to show how the blade-like ridges of the carnassial teeth shear past each other (after Savage, 1977) (A.K.)

ensure the most powerful bite coupled with a reasonably wide gape. Measurement of real carnivore jaws shows that they have indeed evolved this optimal design.

Cats have a very powerful bite because of the reduced length of their jaws, which gives their main jaw-closing muscle (at a wide gape), the temporalis, a greater mechanical advantage. The lower jaw has a coronoid process onto which the temporalis inserts, which accentuates this effect further (Figure 1.19). The temporalis

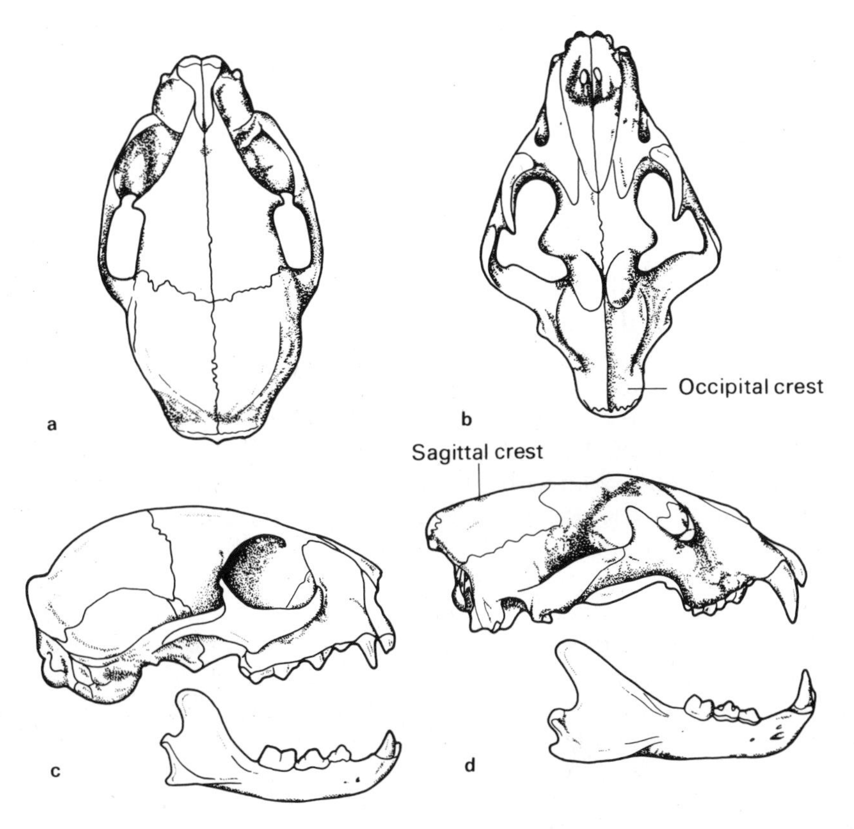

Figure 1.19 The skulls of small cats such as the jaguarundi (**a, c**) are smooth and have sufficient surface area for the attachment of the temporalis jaw muscles. The skulls of big cats (lions) (**b, d**) have well-developed sagittal crests for the attachment of the temporalis muscles and an occipital crest for the attachment of neck muscles (after Ewer, 1973) (S.A.)

originates from the side of the cranium and there is plenty of room for attachment in small cats. However, brain size does not increase as fast as body size in larger cats so that there is not enough surface area to which the temporalis can be attached (Radinsky, 1975). Therefore, large cats have developed bony ridges on their skulls to provide a greater surface area for muscle attachment (Figure 1.19).

The other main jaw muscle, the masseter, which runs from the zygomatic arch to the mandible, is of greater importance when the carnassials are used to slice up meat and in the later stages of the killing bite when the jaws are almost closed (Ewer, 1973).

THE GUT

Compared with other carnivores, cats have relatively short digestive tracts. This is because they feed on more easily-digestible meat than all other carnivores.

Recently David Houston of Glasgow University observed Egyptian vultures feeding on lion faeces in East Africa, but never on wild dog or hyaena faeces (Houston, 1988). Houston believed that this might be due to the low digestive efficiency of the relatively shorter gut of the lion, which allowed more nutritious material to end up in its faeces. It is known, for example, that domestic cats have a mean digestive efficiency for total energy of 79 per cent, compared with 89 per cent for the domestic dog. To test this idea, digestive efficiencies of lions, leopards, vultures and hyaenas were tested by adding a chemical marker (0.1 per cent chromic oxide) to the minced meat that these animals were fed. Chromic oxide is not absorbed by the gut, so that an increase in its concentration in faeces can be used to calculate the digestive efficiency of the animal.

Lions have a digestive efficiency similar to the domestic cat (79 per cent), but the leopard's was slightly higher (81 per cent). These low values compared with 90 per cent for the spotted hyaena and 91 per cent for the Rüppell's vultures. Therefore, it would be an advantage for vultures to feed on lion faeces if other food were in short supply.

Houston (1988) suggests that because the typical hunting behaviour of cats requires rapid acceleration, they minimise body weight, and hence inertia, by having a short, light gut. In contrast, hyaenas or wild dogs hunt by sustained, long chases where inertia is not important, but longer guts for high digestive efficiency are needed to supply sufficient energy for the chase.

2

CATS OF THE PAST

More than 50 million years ago there were two main groups of carnivorous mammals: the now-extinct creodonts and the ancestors of today's living carnivores (Figure 2.1) (Savage, 1977). Both groups had carnassial teeth for slicing up raw meat, but different teeth had evolved this function in the two groups. Typically, creodonts had carnassials formed from the first and second upper molars and the second and third lower molars, rather than just the fourth upper premolar and first lower molar as found in true carnivores (Savage, 1977). Creodonts were the predominant carnivores during the Palaeocene and Eocene (60–35 million years ago) and they radiated into a number of different forms which resembled some modern carnivores in their morphology, and possibly also their lifestyle. Some oxyaenid creodonts were bear-like (e.g. *Sarkastodon*, *Patriofelis*), fox-like (e.g. *Sinopa*) and even cat-like (*Paleonictis*) (Savage and Long, 1987). The hyaenodontid creodonts evolved into sabretooth, civet-like, dog-like and sea otter-like forms.

At that time, the Order Carnivora was represented by a group of small carnivores which are usually collectively called the miacids (Savage, 1977). Contemporary omnivores with carnivorous diets included the arctocyonid and mesonychid condylarths, but these did not develop carnassials. Marsupial carnivores have also developed the carnassial specialisation a number of different times, independently.

By the end of the Eocene, most of the creodonts and all of the carnivorous condylarths had become extinct (Figure 2.1) (Radinsky, 1977, 1978; Savage, 1977). It is not known why this happened. The usual story is that miacids with larger brains (i.e. cleverer animals) were instrumental in the takeover from the creodonts as the world's dominant carnivores. However, Radinsky (1977, 1982) found that brain size increased similarly in the two groups. Hyaenodontid

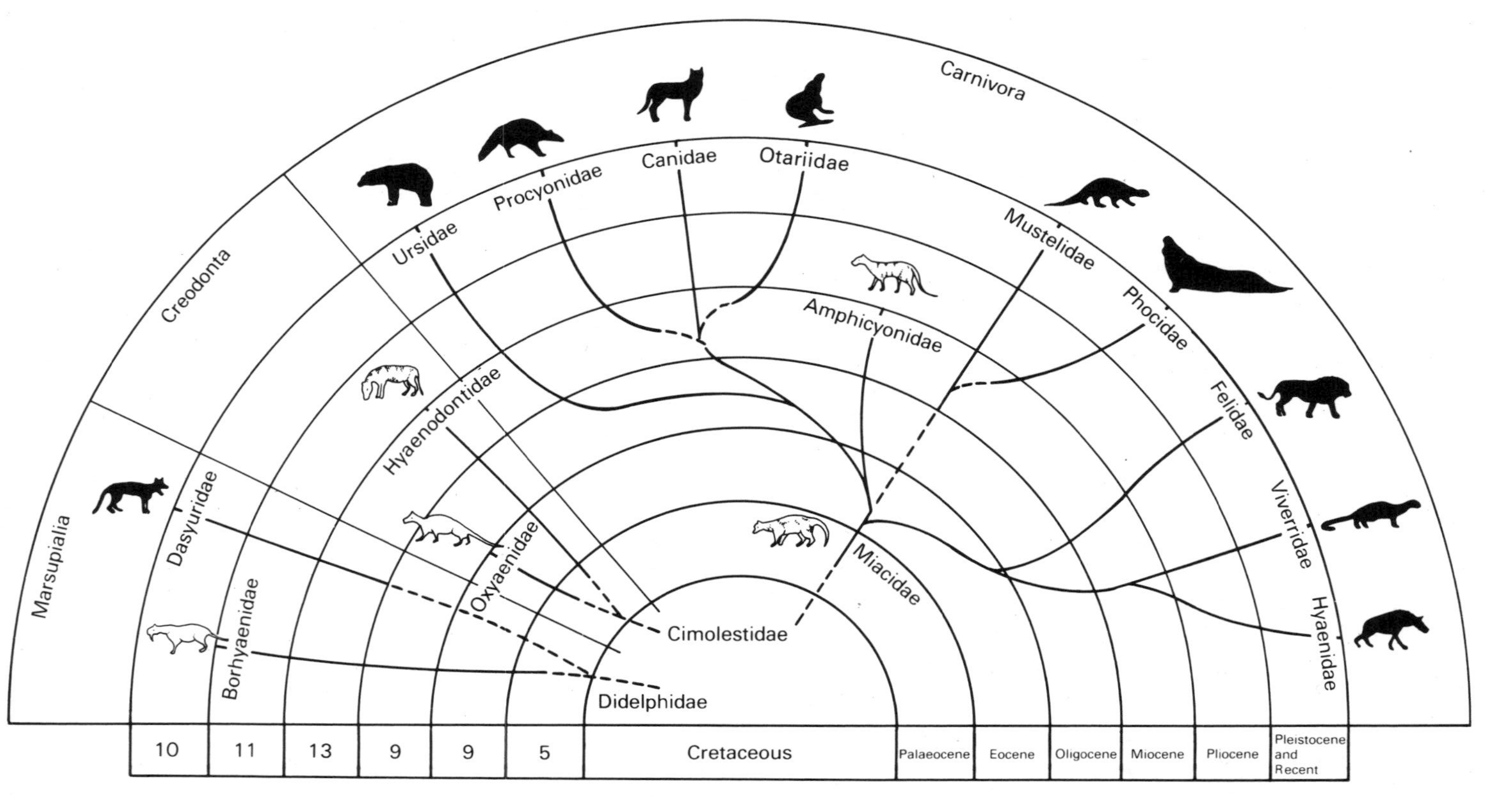

Figure 2.1 A phylogeny of carnivorous mammals (Savage, 1977). Recently, it has been suggested that the seals (Phocidae) and sealions (Otariidae) have a common ancestry (S.A.)

creodonts also showed a similar increase in cursoriality, although their vertebral columns were less flexible (Savage, 1977; Radinsky, 1982). The only obvious difference that Radinsky (1977) noticed between creodonts and carnivores was that different areas of the neocortex of the brain had expanded. However, the significance of this is unknown.

The miacids took advantage of the extinction of the creodonts and radiated into many different forms which were ancestral to the modern carnivore families (Figure 2.1). In fact, it has always been difficult to understand the hotchpotch of miacid forms and their evolutionary relationships. Recently, however, Flynn and Galiano (1982) have re-examined this problem and found that the miacids can be divided into two main groups: the miacines, which were ancestral to the dogs, bears, raccoons and weasels (Suborder Caniniformia), and the viverravines, which gave rise to the civets, genets, cats and hyaenas (Feliformia).

THE PALAEOFELIDS

The first cat-like carnivores to appear were the sabretooth palaeofelids (or nimravids) about 35 million years ago. These cats are usually classified as a separate subfamily of the Felidae, the Nimravinae, but recently Martin (1980) and Baskin (1981) suggested that they should occupy a separate, but closely related family, the Nimravidae (Figure 2.2). The evolutionary relationships within the Carnivora are based chiefly on the structure of the base of the cranium and, in particular, the structure of the auditory bulla, a bony case surrounding the middle ear (Hunt, 1974). In the aeluroid feliforms (civets and cats), the auditory bulla is formed from two bones which meet to form a bilaminar septum across the cavity, the functional significance of which is not known (Figure 2.3). Early true carnivores either did not have a bony auditory bulla, or it was only poorly ossified. Nimravids also had only a poorly ossified bulla, except for the most recent form, the enormous *Barbourofelis* from North America, which survived until about seven million years ago (Martin, 1980; Baskin, 1981).

Flynn and Galiano (1982) went one stage further and claimed that palaeofelids (nimravids) are more closely related to the Caniniformia. In other words, they are dog-like carnivores which have evolved a sabretooth cat-like form independently from the Felidae. Further investigation will be required to see just where the nimravids fit in. For the meantime, the nimravids will be excluded from our story of feline origins for the sake of simplicity.

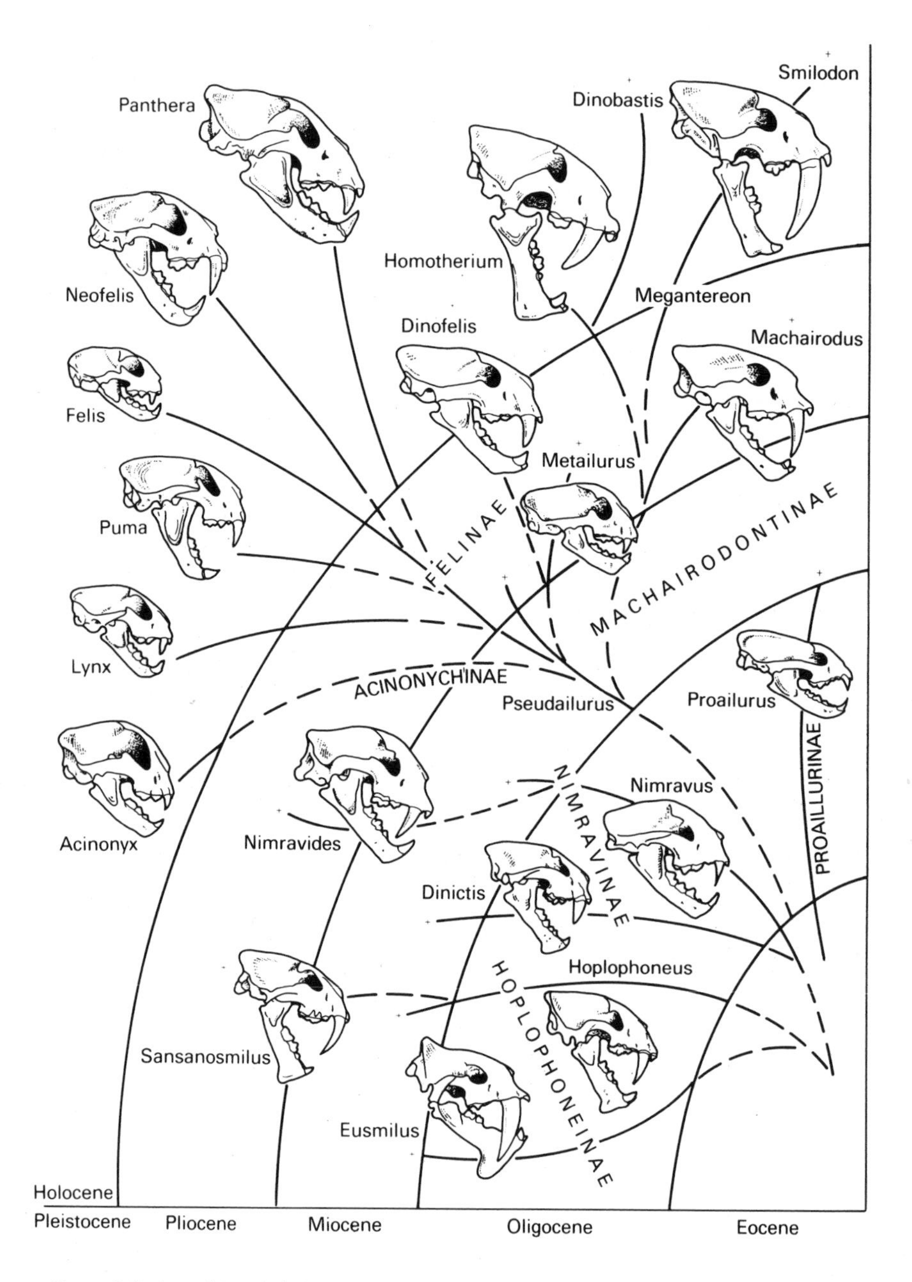

Figure 2.2 A traditional phylogeny of cats (after Thenius, 1969). The Nimravinae and Hoplophoneinae have recently been separated as the family Nimravidae (the palaeofelids) (S.A.)

THE NEOFELIDS

An example of the first true felid is *Proailurus* from France, which appeared about 25 million years ago (Figure 2.4) (de Beaumont, 1964; Thenius, 1967; Radinsky, 1977). Many of its features are civet-like, reflecting its early divergence from civet-like ancestors. In the past, *Proailurus* was classified with the civets (Viverridae) and it

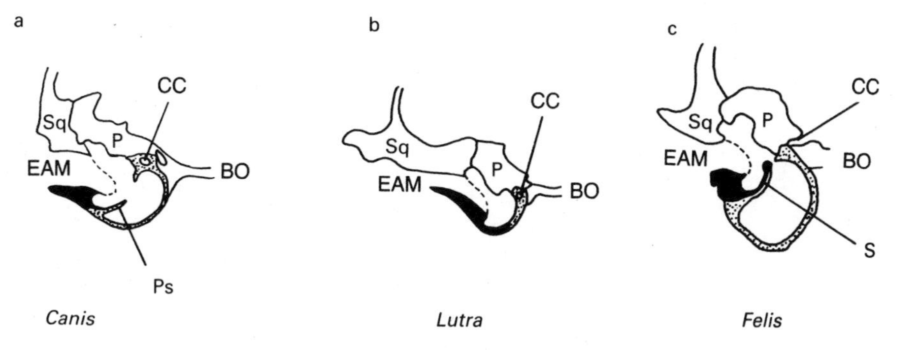

Figure 2.3 Transverse sections through the auditory bullae of (**a**) dog, (**b**) otter and (**c**) cat. Only the cat has a septum (S) that is made up of two layers of bone. The ectotympanic bone is shown in black and the entotympanic bone is stippled. The eardrum is shown as a dashed line. EAM, external auditory meatus leading to ear opening; Sq, squamosal bone of skull; P, petrosum bone; CC, carotid canal; BO, basioccipital bone; Ps, pseudoseptum (after Savage, 1978) (A.K.)

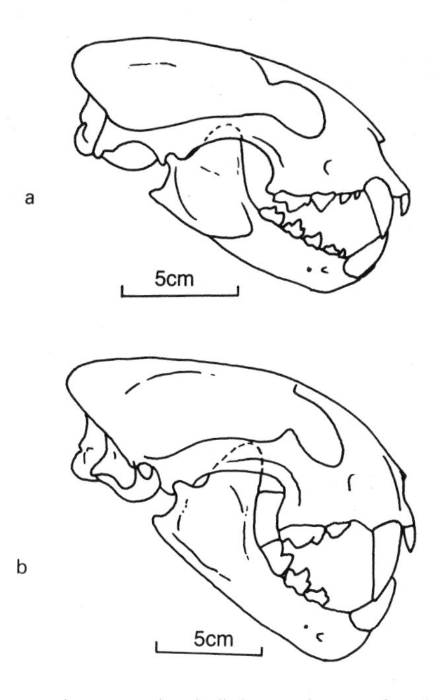

Figure 2.4 A comparison between the skull shape of (**a**) the first felid, *Proailurus* and **b**) the clouded leopard the closest living equivalent to a sabretooth (Radinsky, 1982) (A.K.)

has only fairly recently been recognised as a felid. *Proailurus* was probably plantigrade, unlike the cats of today.

About 20 million years ago the felids were represented by a generalised cat called *Pseudaelurus* which lived in North America and Europe (Figure 2.5). Several different forms of *Pseudaelurus* existed and these formed the basis for the later radiation and diversification of the neofelid cats of today. Modern felid genera begin to appear in the fossil record in the late Miocene (about ten million years ago) in Eurasia (*Felis*). The big cats, *Panthera*, appear in the Villafranchian (more than two million years ago) of Europe, but may have had an earlier African origin.

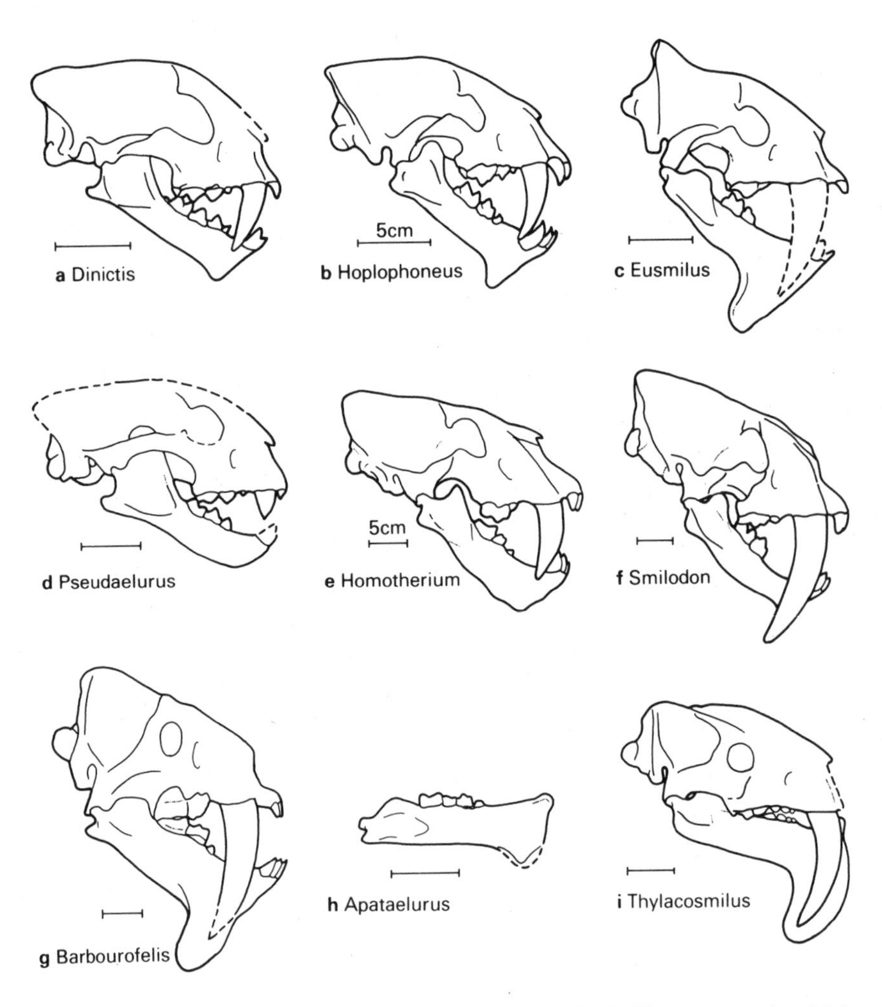

Figure 2.5 Sabretooths evolved at least four times independently. There were palaeofelids (*Dinictis, Hoplophoneus, Eusmilus, Barbourofelis*), neofelids (*Homotherium, Smilodon*), creodonts (*Apatelurus*) and marsupials (*Thylacosmilus*). *Pseudaelurus* was an ancestor of the modern felids and neofelid sabretooths (Emerson and Radinsky, 1980) (S.A.)

SABRETOOTHS

Sabretooths have evolved independently at least four different times. There were sabretooth creodonts (*Apatelurus*, *Machaerodus*), a sabretooth borhyaenid marsupial (*Thylacosmilus*), sabretooth palaeofelids (Nimravidae) and the sabretooth neofelids (Figure 2.5) (Emerson and Radinsky, 1980). There were two main functional types of sabretooth with equivalents both in the neofelids and palaeofelids, namely the dirk-toothed and scimitar-toothed cats. The dirk-toothed cats have long slender canines with fine serrations on the edges. They have short, strong limbs and rather resemble bears in morphology. They probably hunted by waiting in ambush (Figure 2.6) (Martin, 1980) and most were probably solitary (Gonyea,

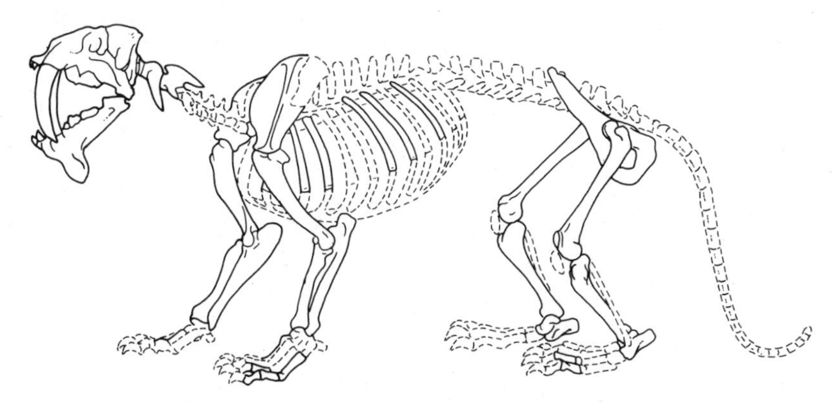

Barbourofelis, a bear-like, dirk-toothed palaeofelid

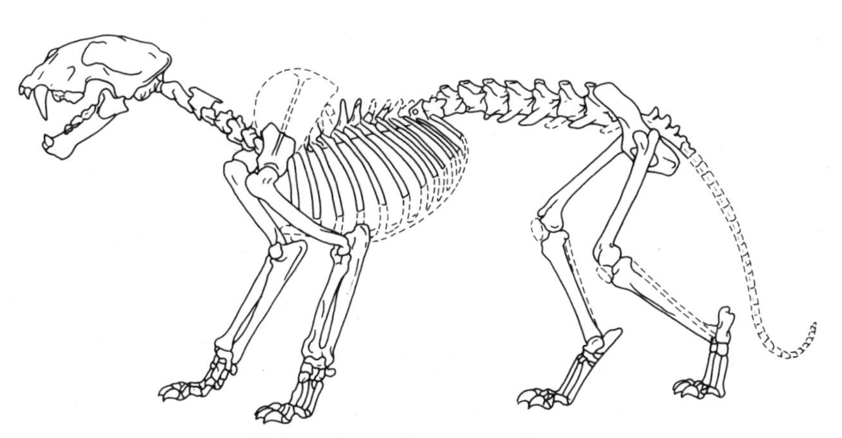

Machairodus, a scimitar-toothed cat

Figure 2.6 There were two main types of sabretooth: the bear-like dirk-toothed cats e.g. *Barbourofelis* (**a**) and the cat-like scimitar-toothed cats e.g. *Machairodus* (**b**) (Martin, 1980) (S.A.)

1976b). Examples of dirk-toothed cats include *Megantereon* and *Smilodon*. However, it seems that *Smilodon* lived in an open habitat, and may have lived in prides like lions (Akersten, 1985). The scimitar-toothed cats have short, broad canines with coarse serrations, and long cursorial limbs (Figure 2.6). They were probably pursuit predators living in open habitats. Examples of scimitar-toothed cats include *Machairodus* and *Homotherium*. Both types of sabretooth have retractile claws, which were probably very important for subduing the prey before biting with the delicate canines (Gonyea, 1976b).

Two mysteries continue to puzzle palaeontologists: firstly, how the sabretooths used their massive canine teeth to kill prey and, secondly, how they fed themselves. The functional constraints of being a sabretooth have brought about a most striking convergence in morphology between otherwise totally unrelated groups. Common sabretooth features include a narrow zygomatic arch width, small eyes, a compressed facial region, lack of a coronoid process on the lower jaw, a gape of more than 90° compared to the 65° of modern felids (Figure 2.7), massively developed head-depressing muscles, and carnassials set close to the jaw joint (Emerson and Radinsky, 1980). There is also often a bony flange at the end of the lower jaw just below the smaller lower canines (Emerson and Radinsky, 1980).

It was widely believed that because the sabretooths did not have a coronoid process, they could not bite powerfully with either their canines or their carnassials (Emerson and Radinsky, 1980; Akersten, 1985). To compensate, the carnassials are set closer to the jaw joint for a shorter lever arm and consequently a more powerful bite. In fact, sabretooth carnassials are just as large as they are in the other cats of similar size. Instead of using jaw muscles to push the canines into the prey, the head-depressing muscles were used instead. To allow them to use their enormous canines, sabretooths had a wider gape, which meant that the temporalis muscle had to stretch more. This was achieved by reducing the coronoid process and by changing its line of action to a more vertical direction (Emerson and Radinsky, 1980).

Many odd theories have been proposed to explain the functions of the elongated canines, including their use as a can-opener on glyptodont 'armadillos', a carrion slicer, a tree-climbing aid, a tool to grub for molluscs, like walrus tusks, and as a way of stabbing with a closed mouth (Akersten, 1985). Nowadays it is believed that sabretooths were active predators that slashed or stabbed with open mouths (Martin, 1980; Miller, 1980; Akersten, 1985).

However, the canines would have been very prone to breakage if they were subjected to moderate sideways forces (van Valkenburgh and Ruff, 1987), which means that they would have been used in a very controlled way. A sabretooth could not hurl itself at its prey

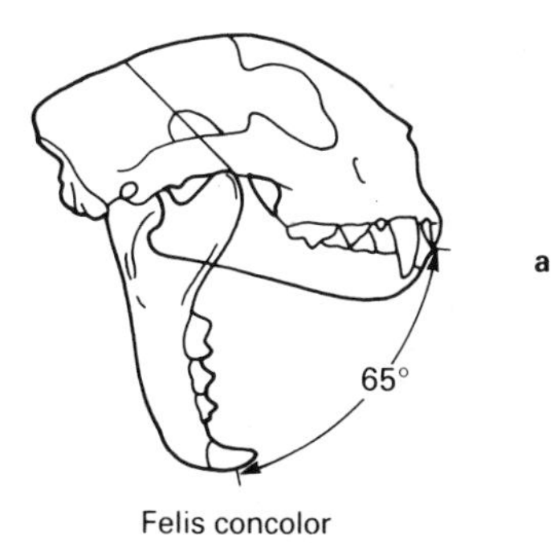

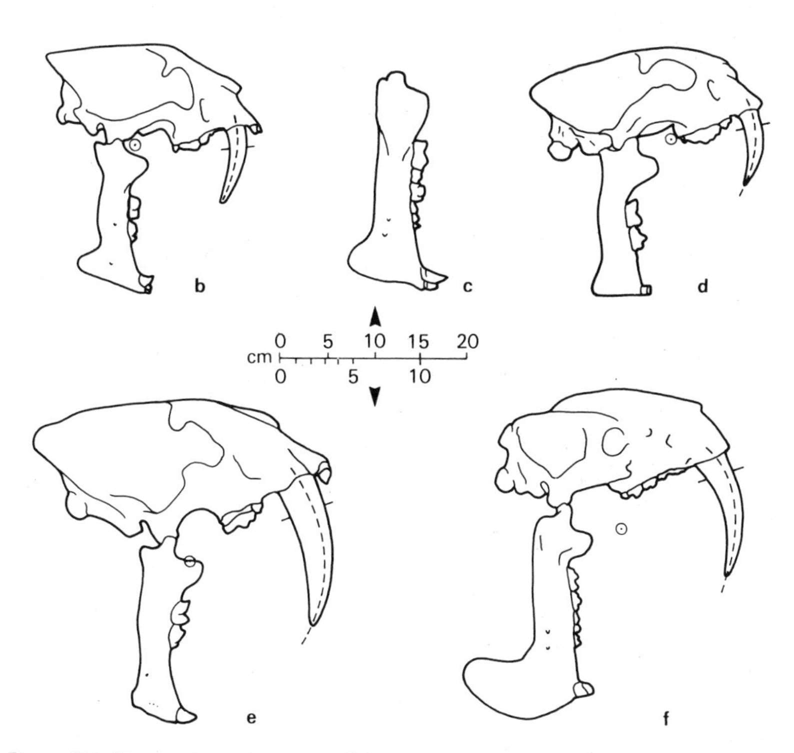

Figure 2.7 The maximum jaw gape of the puma (**a**) is only 65°, but sabretooths (**b–f**) could open their jaws to more than 90° (after Emerson and Radinsky, 1980; Savage, 1977). (**b**) *Hoplophoneus* (**c**) *Apataelurus* (**d**) *Nimravus* (**e**) *Smilodon* (**f**) *Thylacosmilus* (S.A.)

and use its massively developed head-depressing musculature to force its canines into its victim in a frenzied attack, because its teeth would probably just break off. Akersten (1985) has carefully analysed how the elongated canines of a sabretooth neofelid, *Smilodon*, could have been used to kill prey. He rejects the stabbing hypothesis as impossible, because the teeth are fragile and because there is insufficient clearance between the upper and lower canines

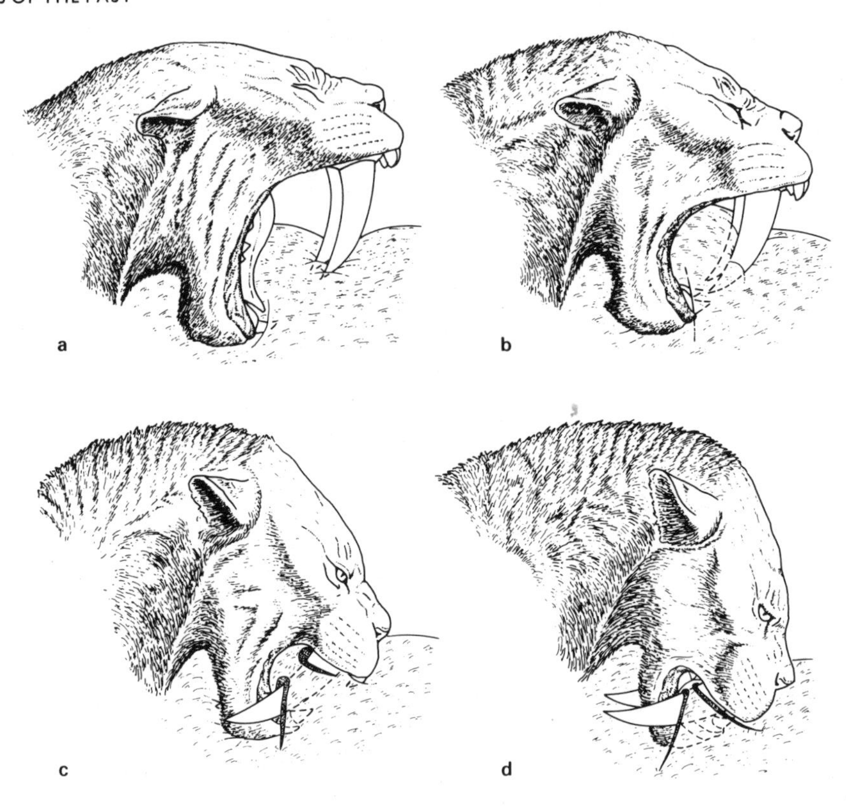

Figure 2.8 The successive stages in the canine shear bite of *Smilodon* (after Akersten, 1985) (S.A.)

to allow them to be used for stabbing. Instead, he proposes a canine shear bite, which is used once the prey has been subdued by the cat's forelimbs (Figure 2.8). Therefore, *Smilodon* would have used its forefeet to knock its prey off balance and then used canine shear bites to wound its victim's underside fatally (Figure 2.9). Neck bites either to the nape or the throat would be unlikely, because of the risk of delicate canines striking bones (Gonyea, 1976b; Akersten, 1985). The fine serrations on the tooth would have helped it to shear through the skin and flesh. The palaeofelid *Barbourofelis* even has blood channels along the length of the canines to aid penetration. The slight flange on the lower jaw would have helped the shear bite by acting as an anchoring point, so that the lower jaw could oppose the force of the shearing canines. Sabretooths probably used their highly developed incisors to pull meat off bones, or they used their tongues to position the meat correctly for the carnassials. Certainly, the canines would have been a hindrance during feeding. Recently, Martin (1989) has rejected Akersten's (1985) hypothesis that sabretooths attacked the belly of their prey. He instead suggests that sabretooths were adapted to killing large mammals using the typical throat bite (see Chapter 4).

Figure 2.9 The hunting technique of *Smilodon* (after Akersten, 1985) (S.A.)

The usual prey of sabretooths is thought to have been thick-skinned elephants and other pachyderms, but their skin may have been too thick for the canine teeth to penetrate. Also many sabretooths were quite small (e.g. the palaeofelid *Hoplophoneus* weighed between 12 and 19.5 kg), so it seems that relative prey size may have been large and included a variety of medium-sized and large mammals (Emerson and Radinsky, 1980). It is thought that both *Homotherium* and *Smilodon* preyed on young elephants (Figure 2.9) rather than adults, but not exclusively so (Akersten, 1985).

MORE FOSSIL FELIDS

Generally speaking, the fossil record of today's cats is rather poor, because most ancestral cats inhabited tropical forests, where conditions for preservation of fossils are also poor. However, our understanding of the evolution of some of the bigger cats is quite good, as they inhabited open ground where conditions for fossilisation of large animals were favourable. There have also been some significant advances in our knowledge of the evolution of some small cats, notably the lynxes. Most of this account is based on Hemmer (1976a), except where stated.

Present-day cats are usually divided into two genera: the big cats of the genus *Panthera* have a flexible hyoid allowing roaring, and the small cats of the genus *Felis* have an ossified hyoid so that they can only scream and purr. Owing to the relative completeness of their fossil record compared with the small cats, let us start with the evolution of the big cats, and in particular the lion. Here, the term 'lion' means the species *Panthera leo* and all its ancestors distinct from a common ancestor for one or more of all the big cats. The same goes for tigers (*P. tigris* and its distinct ancestors), leopards (*P. pardus* and its ancestors), etc.

The evolutionary centre for the lions is thought to be Africa (Hemmer, 1976a). The oldest remains are known from Bed II of Olduvai, East Africa, which are variously dated at between 500,000 and 700,000 years old (Hemmer, 1976a). Recently, however, Hemmer (1987) has begun to doubt the identity of these giant early lions which resembled other members of the genus *Panthera* more than they did modern lions.

By the Cromerian interglacial (approximately 250,000 years ago) the lion had spread to Eurasia. Remains are known from Greece, Germany and the Kolmya Region of Siberia. During the Riss/Illinoan glacial at the end of the Middle Pleistocene (about 100,000 years ago), populations of lions became isolated, resulting in speciation. The cave lion (*P. spelaea*) evolved in Europe and the American lion (*P. atrox*) probably evolved in eastern Asia and spread into North America. During the Late Pleistocene, lions spread throughout North and Central America as far south as Peru in South America. Warming of the climate at the end of the Pleistocene led to vegetational changes and the loss of open habitats, which seem to have triggered the decline of the lion to its former present day distribution in India, southwestern Asia and Africa.

The oldest known leopard remains are from the Indian Siwaliks

(Late Villafranchian, approximately two million years ago). Its teeth show it to have been a primitive leopard which resembled the jaguar and the now-extinct *P. gombaszoegensis*. Leopards are also known from the early Middle Pleistocene of Java (one million plus years ago) and the Early Pleistocene/early Middle Pleistocene of South Africa.

By the middle of the Pleistocene, the leopard had spread to eastern and southeastern Asia. Very large leopards inhabited Java, but local populations in southern China became isolated and evolved into forms which are said to resemble modern Chinese leopards.

Similar isolation occurred in Europe, where several different subspecies of fossil leopard are recognised. Leopards entered Europe earlier than the lion in the early Middle Pleistocene. These unspecialised leopards resembled the jaguars of the North American Pleistocene. A separate subspecies of leopard from southeastern Europe spread to central Europe at the end of the Riss glacial (100,000 years ago). Again, the repeated glaciation and isolation during the Pleistocene has resulted in a complicated pattern of speciation and subspeciation similar to that of the lion.

Jaguars evolved in Eurasia from a leopard ancestor and spread into North America via the Bering land bridge during the late Blancan (cf. late Villafranchian). They belonged to the subspecies *P. onca augusta*, which was larger and longer legged than modern jaguars (Kurten, 1973a). During the Pleistocene there was a decrease in size of jaguars and a shortening of legs. The latter probably occurred as open habitats were replaced by forests. Jaguars are known from Bolivia in the Middle Pleistocene, and giant jaguars inhabited Argentina during the Late Pleistocene.

Jaguar/leopard ancestors are thought to have spread over Africa, Eurasia and North America in the second half of the Early Pleistocene (one to two million years ago).

Fossil tigers are known from the Late Pliocene/Early Pleistocene of southeastern Asia. A small primitive tiger was living in North China during the Early Pleistocene. Between 1.3 and 2.1 million years ago, tigers were living in Java. In fact, they seem to have spread over eastern and southeastern Asia by two million years ago (Hemmer, 1987). The fossil record of tigers is well documented in Java, where throughout the early Middle Pleistocene, the local evolution of this species was continually interrupted by invasions from the mainland population. By the Middle/Late Pleistocene, tigers in Java were much larger, reflecting a cooling of the environment (Hemmer, 1987). Enormous tigers are aslo known from China during the Late Pleistocene, which have since reduced to the size of their modern day descendants. In summary, from about two million years ago, tigers spread from their evolutionary centre in eastern Asia in two directions. Tigers moving through the Central Asian woodlands to the west and southwest gave rise to the Caspian tiger. Secondly, tigers from China moved to the east of the central Asian mountains to southeastern Asia and the Indonesian islands, and thence westwards to India (Hemmer, 1987). There was a small tiger in Japan until the Late Pleistocene.

Little is known of the fossil records of the snow leopard and clouded leopard. Snow leopards are known from the Middle Pleistocene of northern China and the Late Pleistocene of the Altai in central Asia. Clouded leopards were found in Java in the Early Pleistocene, and southern China and northern Vietnam during the Middle Pleistocene. This ancestral clouded leopard (*Neofelis nebulosa primigenia*) had less specialised teeth than modern clouded leopards. Clouded leopards survived in Java until at least the Neolithic.

Most small species of cats are barely recorded as fossils, although very small leopard cats have been found in the Neolithic of Java, and also the early Middle Pleistocene in the Sunda Islands. The ancestor of the European wildcat is Martelli's wild cat (*F. silvestris lunensis*) from the Early Pleistocene. There is a transition to very large forms of the modern species during the early Middle Pleistocene, but at the end of the Pleistocene, wildcats reduced considerably in size. Hemmer (1976a) also refers to the limited fossil record of the jungle cat.

Recently a fossil cat from the Late Pliocene/Early Pleistocene of Argentina, *F. vorohuensis*, has been identified as the possible ancestor of the pampas cat (Berta, 1983). The ancestor of the margay, *F. wiedii amnicola*, is known from the middle and late Rancholabrean (about 100,000 years ago or less) of North America, giving rise to the modern species at the end of the Pleistocene (Werdelin, 1985).

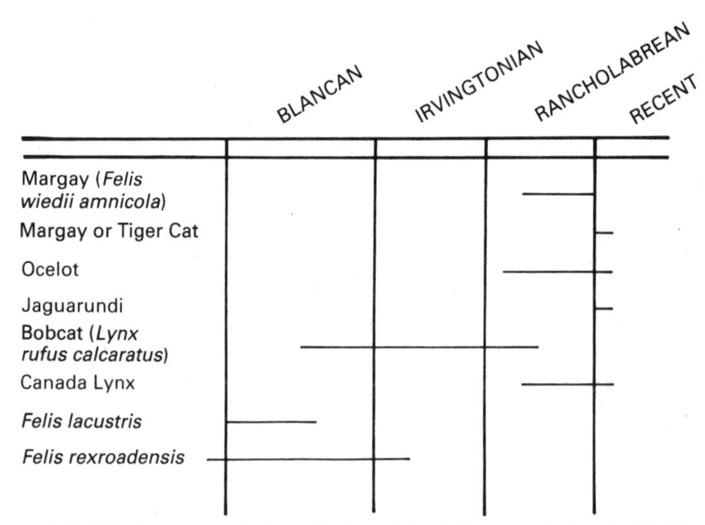

Figure 2.10 The fossil record of small cats in North America during the Pleistocene (after Werdelin, 1985)(A.K.)

Ancestral ocelots are known in Florida during the same time. The jaguarundi seems to have been a very recent arrival in North America and is known only certainly since the end of the Pleistocene (Werdelin, 1985). Figure 2.10 summarises our current knowledge of the fossil record of small cats in North America.

An exception to the poor fossil record of small cats is that of the lynxes, whose evolutionary centre appears to have been in the Late Pliocene (approximately four million years ago) of Africa (Werdelin, 1981). The ancestral lynx was the Issoire lynx (*Lynx issidorensis*), which is known from the Late Pliocene to the Cromerian (Werdelin, 1981). It has shorter legs than modern lynxes and is much more similar in appearance to cats of the genus *Felis*.

The Issoire lynx spread throughout the northern hemisphere by the middle of the Villafranchian. In Europe, Werdelin (1981) claims that it was the direct ancestor of the pardel lynx, which he recognises as a separate species (*L. pardina*). In China it was ancestral to the northern lynx which then spread westward to Europe by the Eemian interglacial (about 100,000 years ago). In the last 200,000 years lynx spread east to North America and gave rise to the Canada lynx.

But the Issoire lynx also reached North America in the Pliocene, where it gave rise to the bobcat. Competition with ancestral pumas probably resulted in a size reduction in bobcats, as did similar competition between lynx and *Panthera* in Europe. Kurten and Anderson (1980) summarise an alternative scheme for the evolution of the lynxes.

Pumas are known in North America from the Late Pliocene to the present day (Kurten, 1973b). Ancestral pumas (*Felis inexpectata*) occurred from the Blancan to the Irvingtonian (from about three to one million years ago). From the Rancholabrean, modern pumas are known. *F. inexpectata* was larger and more cursorial than modern pumas.

The oldest known cheetah is the giant cheetah, *Acinonyx pardinensis*, from the Late Pliocene of France. This species was larger and less cursorial than modern cheetahs and spread throughout the Old World. Like the evolution of most cat species, it showed a decrease in body size until the present-day *A. jubatus* appeared in the Late Pleistocene. Cheetahs apparently invaded North America where two species are known, *A. studeri* from 2.5 million years ago and *A. trumani* from about 12,000 years ago (Figure 2.11) (Martin, Gilbert and Adams, 1977; Adams, 1979). Alternatively, it has been suggested that these cheetah-like species are an example of convergent evolution, being derived from puma ancestors.

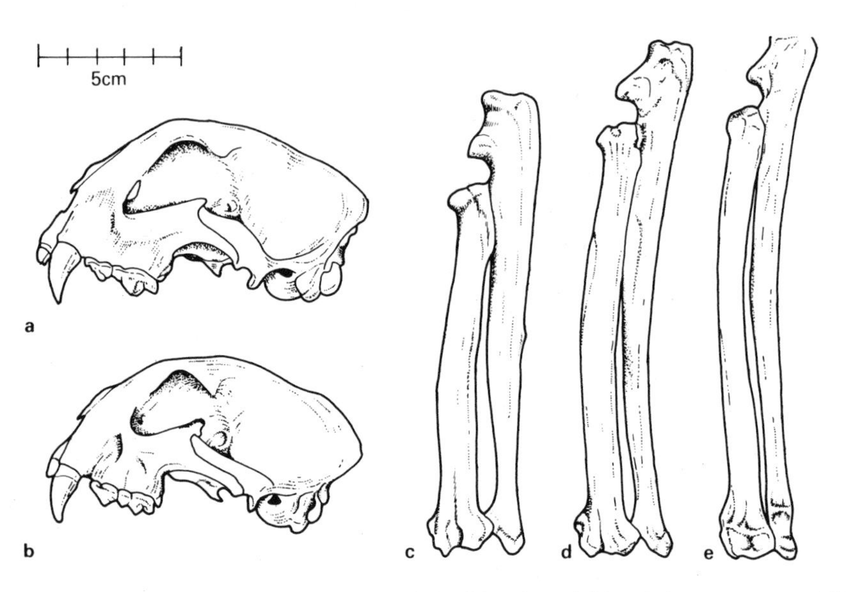

Figure 2.11 A comparison between the skull of the cheetah (**a**) and *Acinonyx trumani* (**b**) from North America, and also the radius and ulna of the puma (**c**), *A. trumani* (**d**) and cheetah (**e**) (Martin *et al.*, 1977) (S.A.)

THE PHYLOGENY OF MODERN CATS

One of the biggest problems facing systematists and evolutionary biologists is working out the relationships between living species of the Felidae, especially in the face of such a poor fossil record. Many different schemes have been proposed for classifying cats, and between two and twenty different genera have been used (Table 2.1). The various morphological and behavioural criteria seem to produce wildly different classifications. The problem is that most feline morphology is related to the function of prey capture, with the result that all cats are inevitably very similar.

In recent years more characters have been made available to the systematists, by the development of cellular and molecular techniques. For example, for a long time cytogenetics or the study of the karyotype (chromosomes) provided some hope of elucidating tangled feline relationships. Robinson (1976a, b) looked at the haploid number of chromosomes and the position of the centromere as important characters in determining evolutionary relationships. He found five basic karyotypes. The two main ones were the Old World cats, including the puma, jaguar and jaguarundi, with 19 pairs of chromosomes and the New World cats with only 18 pairs. Within this latter category he found evidence to put the jaguarundi (*Herpailurus*) (but see later), pampas cat, Geoffroy's cat, tiger cat (*Oncifelis*), margay and ocelot (*Leopardus*) in separate genera. Under

Table 2.1 Different classifications of the Felidae

Leyhausen (1979)	Hemmer (1978)	Ewer (1973)	Nowak and Paradiso (1983)
Genus *Acinonyx*	Genus *Acinonyx*	Genus *Felis*	Genus *Felis*
A. jubatus	*A. jubatus*	*F. silvestris*	Subgenus *Felis*
Genus *Prionailurus*	Genus *Puma*	*F. libyca*	*F. silvestris*
P. bengalensis	*P. concolor*	(including *ornata*)	(including *libyca* and ornata *ornata*)
P. rubiginosus	Genus *Herpailurus*	*F. chaus*	*F. bieti*
P. viverrinus	*H. yagouaroundi*	*F. bieti*	*F. chaus*
P. planiceps	Genus *Prionailurus*	*F. margarita*	*F. margarita*
P. iriomotensis	Subgenus *Prionailurus*	(including *thinobia*)	(including *thinobia*)
Genus *Leopardus*	*P. bengalensis*	*F. nigripes*	*F. nigripes*
L. pardalis	*P. rubiginosus*	Genus *Leptailurus*	Subgenus *Otocolobus*
L. wiedii	*P. viverrinus*	*L. serval*	*F. manul*
L. tigrinus	Subgenus *Mayailurus*	Genus *Prionailurus*	Subgenus *Lynx*
L. guigna	*P. iriomotensis*	*P. bengalensis*	*F. lynx*
L. geoffroyi	Subgenus *Ictailurus*	*P. rubiginosus*	(including *canadensis*
Genus *Lynchailurus*	*P. planiceps*	*P. viverrinus*	*F. pardina*
L. pajeros	Genus *Profelis*	Genus *Mayailurus*	*F. rufus*
(=*Felis colocolo*)	*P. aurata*	*M. iriomotensis*	Subgenus *Caracal*
Genus *Oreailurus*	Genus *Catopuma*	Genus *Ictailurus*	*F. caracal*
O. jacobita	*C. temmincki*	*I. planiceps*	Subgenus *Leptailurus*
Genus *Herpailurus*	(including *tristis*)	Genus *Otocolobus*	*F. serval*
H. yagouaroundi	*C. badia*	*O. manul*	Subgenus *Pardofelis*
Genus *Felis*	Genus *Leopardus*	Genus *Pardofelis*	*F. marmorata*
F. silvestris	*L. pardalis*	*P. marmorata*	*F. badia*
F. libyca	*L. wiedii*	*P. badia*	Subgenus *Profelis*
F. ornata	Genus *Oncifelis*	Genus *Profelis*	*F. temmincki*
F. thinobia	Subgenus *Oncifelis*	*P. temmincki*	(including *tristis*)
F. margarita	*O. tigrinus*	(including *tristis*)	*F. aurata*
F. bieti	*O. geoffroyi*	*P. aurata*	Subgenus *Prionailurus*
F. nigripes	*O. guigna*	Genus *Caracal*	*F. bengalensis*
F. manul	Subgenus *Lynchailurus*	*C. caracal*	*F. rubiginosus*
F. chaus	*O. pajeros*	Genus *Puma*	*F. viverrinus*
Genus *Leptailurus*	(=*Felis colocolo*)	*P. concolor*	*F. planiceps*
L. serval	Subgenus *Oreailurus*	Genus *Leopardus*	Subgenus *Mayailurus*
Genus *Lynx*	*O. jacobita*	*L. pardalis*	*F. iriomotensis*
L. lynx	Genus *Pardofelis*	*L. tigrinus*	Subgenus *Lynchailurus*
L. canadensis	*P. marmorata*	*L. wiedii*	*F. colocolo*
L. pardina	Genus *Neofelis*	*L. geoffroyi*	Subgenus *Leopardus*
L. rufus	*N. nebulosa*	Genus *Oncifelis*	*F. pardalis*
Genus *Pardofelis*	Genus *Uncia*	*O. guigna*	*F. wiedii*
P. marmorata	*U. uncia*	Genus *Lynchailurus*	*F. tigrinus*
P. badia	Genus *Panthera*	*L. colocolo*	*F. geoffroyi*
Genus *Profelis*	Subgenus *Tigris*	Genus *Oreailurus*	*F. guigna*
P. temmincki	*P. tigris*	*O. jacobita*	Subgenus *Oreailurus*
P. tristis	Subgenus *Panthera*	Genus *Herpailurus*	*F. jacobita*
P. aurata	*P. onca*	*H. yagouaroundi*	Subgenus *Herpailurus*
P. caracal	*P. pardus*	Genus *Lynx*	*F. yagouaroundi*
P. concolor	*P. leo*	*L. lynx*	Subgenus *Puma*
Genus *Uncia*	Genus *Leptailurus*	(including *canadensis*)	*F. concolor*
U. uncia	*L. serval*	*L. pardina*	Genus *Neofelis*
Genus *Neofelis*	Genus *Caracal*	*L. rufus*	*N. nebulosa*
N. nebulosa	*C. caracal*	Genus *Panthera*	Genus *Panthera*

Table 2.1 Continued

Leyhausen (1979)	Hemmer (1978)	Ewer (1973)	Nowak and Paradiso (1983)
N. tigris	Genus *Felis*	*P. leo*	Subgenus *Uncia*
Genus *Panthera*	Subgenus *Lynx*	*P. tigris*	*P. uncia*
P. pardus	*F. rufus*	*P. pardus*	Subgenus *Tigris*
P. onca	*F. canadensis*	*P. onca*	*P. tigris*
P. leo	*F. lynx*	*P. uncia*	Subgenus *Panthera*
	F. pardina	Genus *Neofelis*	*P. pardus*
	Subgenus *Otocolobus*	*N. nebulosa*	Subgenus *Jaguarius*
	F. manul	Genus *Acinonyx*	*P. onca*
	Subgenus *Felis*	*A. jubatus*	Subgenus *Leo*
	F. nigripes		*P. leo*
	F. margarita		Genus *Acinonyx*
	(including *thinobia*)		*A. jubatus*
	F. chaus		
	F. bieti		
	F. silvestris		
	(including *libyca* and *ornata*)		

his scheme, the clouded leopard and other big cats would not be placed in genera separate from *Felis*.

Kratochvil (1982) took this approach one stage further and combined karyology and baculum morphology to produce his own classification system. Many carnivores are classified according to the shape of their penis bones or bacula (see Chapter 7), but cats have very reduced bacula, so this character is not necessarily very useful. Kratochvil (1982) concluded that cats can be divided into two main groups on the basis of their karyology. The first group contains the big cats (Pantherinae), clouded leopard (Neofelinae), lynxes (Lynchinae) and cheetah (Acinonychinae). The second group contains all the other cats (Felinae). This latter group can be subdivided into the Old World cats (Felini) and New World cats (Leopardini) based on the number of chromosomes. Kratochvil (1982) suggested that *Felis* evolved from cats of the subgenus *Prionailurus* (i.e. leopard cat, etc.), which were also ancestral to the jaguarundi in North America. Study of baculum structure led Kratochvil (1982) to separate the caracal from the rest of the lynxes. In all he recognised 18 different genera!

New multivariate statistical methods (e.g. principal components analysis and correspondence analysis) have also been used to see if there are any patterns of skull shape within the Felidae (Werdelin, 1983). This analysis yielded few surprises. Large cats and small cats could be differentiated. Lynxes fell into a common group, but the caracal did not. The leopard cat, fishing cat, flat-headed cat and rusty-spotted cat (subgenus *Prionailurus*) all had similar-shaped skulls. The two oddities were the puma and clouded leopard which did not seem to fit in anywhere.

However, with the advent of molecular techniques it may now be possible not only to determine accurately the relationships between members of the Felidae but also to postulate an evolutionary timescale for this scheme. By looking at differences in either the amino acid sequences in proteins or the base sequences in DNA, and assuming a constant rate of change since two species split from a common ancestor, and setting it against known events in time, it is possible to calibrate a molecular clock for the evolution of members of the Felidae. For example, Collier and O'Brien (1985) looked at the blood serum albumin immunological distances (AID) between ten cat species to construct a phylogeny, which also included information from all but five of the rest of the living Felidae.

Collier and O'Brien (1985) concluded that cats radiated about twelve million years ago into two main groups—one which led to the South American cats, and the other, about eight to ten million years ago, to the Old World felines (Figure 2.12). About four to six million years ago, a pantherine lineage evolved from the Old World felids, which includes, to everybody's surprise, the cheetah, the serval, the jaguarundi, golden cats and puma. Despite its radically different morphology, the cheetah did not diverge early on in felid

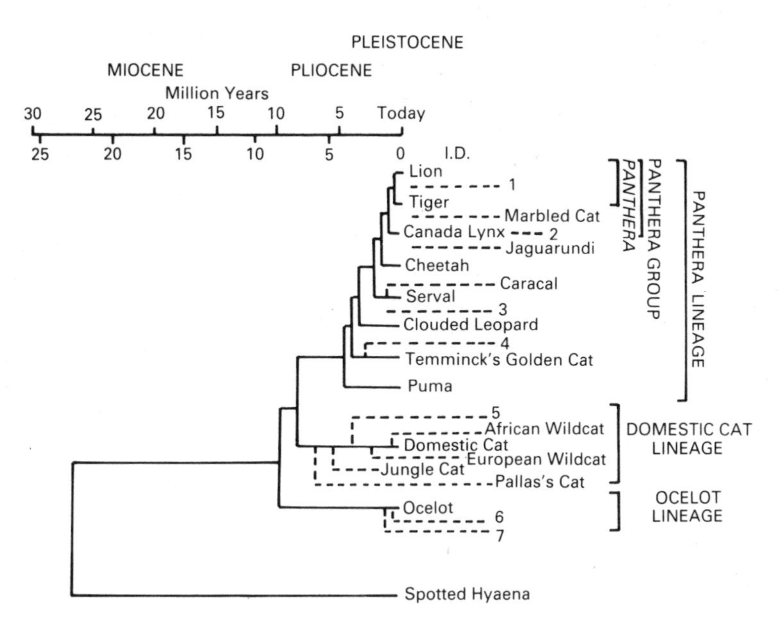

Figure 2.12 A new phylogeny of cats based on blood serum albumin immunological distances 1, Leopard, Jaguar, Snow Leopard; 2, Eurasian Lynx, Bobcat; 3, Flat-headed Cat, Fishing Cat, Leopard Cat; 4, African Golden Cat, Bay Cat; 5, Pallas's Cat, Black-footed Cat; 6, Pampas Cat, Margay, Tiger Cat; 7, Kodkod, Geoffroy's Cat. (after Collier and O'Brien, 1985) (A.K.)

history as suggested by traditional phylogenies. It is, in fact, closely related to the other pantherine cats. About two million years ago the big cats, including the marbled cat, split from the lynxes. Still more recently, O'Brien *et al.* (1987a) compared the results from albumin immunological distance studies, karyological studies, endogenous retroviral sequence studies, DNA hybridisation studies and isozyme genetic distance studies, and found that, in general, the results corresponded with Collier and O'Brien's (1985) earlier phylogeny (Figure 2.13). In the Pantherinae, it would seem that the clouded leopard diverged first about seven million years ago, followed by the snow leopard less than four million years ago and the rest of the big cats about two million years ago (Figure 2.13). The jaguar, leopard and lion seem to represent the final radiation of the Pantherinae.

What is remarkable is the close approximation between the known fossil record and the predictions about speciation from molecular techniques. But above all, it provides a lesson to the systematist that as many characters as possible ought to be used in constructing a phylogeny, if it is to be accurate.

THE FOSSA—AN ANALOGY OF A CAT

In 1833 Bennett described a curious cat-like carnivore from Madagascar. After careful study, he decided that this creature was a

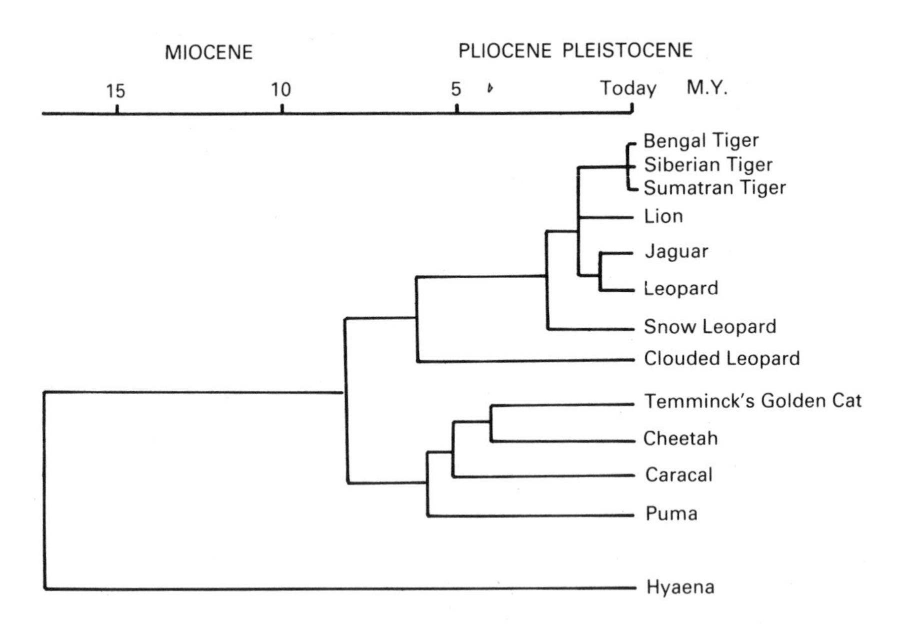

Figure 2.13 A phylogeny of cats based on isozyme genetic distances (after O'Brien *et al.*, 1987) (A.K.)

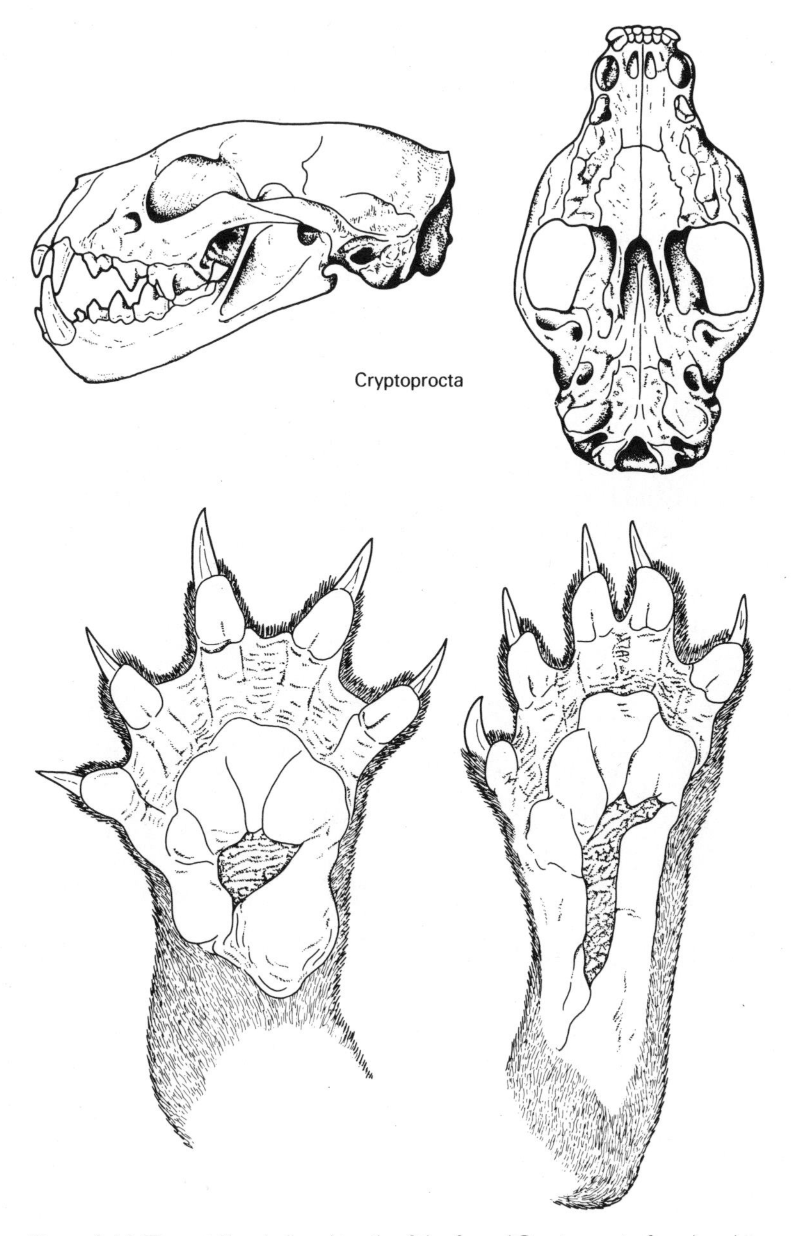

Figure 2.14 The cat-like skull and teeth of the fossa (*Cryptoprocta ferox*) and its un-cat-like feet (after Ewer, 1973; Pocock, 1916a) (S.A.)

specialised civet, which had some cat-like features. However, ever since its scientific discovery, the fossa (*Cryptoprocta ferox*) has been repeatedly classified as a primitive felid. This has continued until fairly recently. For example, de Beaumont (1964) and Thenius (1967) both considered the fossa to be closely related to *Proailurus* from the Oligocene, and hence a member of the Felidae.

It is true that the fossa has very specialised carnassial teeth similar to those of cats, and that it has retractile claws (Figure 2.14). But, as Pocock (1916a, 1951) emphasised, it shares many features with other carnivore families. For example, its paws are like those of palm civets, its anus is like that of a hyaena or mongoose, its genitalia are also like those of a palm civet or a hyaena. Moreover, its feet are semiplantigrade and not digitigrade, its head is too long for a cat, its claws are not withdrawn into cutaneous sheaths and it has an uncat-like inter-ramal tuft of whiskers.

Therefore, Bennett was quite correct to consider the fossa as a civet which has evolved cat-like features independently. However, it fills a similar ecological niche in Madagascar where the Felidae are not found (except recently as domestic cats introduced by humans). In other words, the fossa is an example of convergent evolution with the Felidae. The fossa is solitary and feeds on small mammals in forests and savanna woodlands, just like most species of small cats (Nowak and Paradiso, 1983).

A WHO'S WHO OF CATS

Thirty-six living species of wild cat are usually recognised today (Corbet and Hill, 1986). However, as with most families of mammals there is still some uncertainty about just how many species there actually are and also about the evolutionary relationships between the different members of the Felidae (see Chapter 2). It is possible that new species still await discovery, as a recent report of a new wild cat on Tsushima Island in Japan indicates (Jackson, 1989).

In this chapter all the cats of the world are introduced with a basic descrption of each species and its distribution and habitat preferences. The species accounts are based on my own observations, but also rely heavily on a number of general works including Guggisberg (1975), and Nowak and Paradiso (1983) in addition to more specialised accounts. The species classification follows Corbet and Hill (1986), but current differences of opinion are also discussed. Classification beyond species level follows Collier and O'Brien (1985) and Wozencraft (1989) with some modifications. (See Table 3.1 for classification list.) Common colour variations and distinctive subspecies or isolated populations are also described. Rare colour mutations and their genetics are discussed in Chapter 7. Measurements and body weights of all wild cat species taken from a variety of sources are given in the Appendix.

We start with the cats of the ocelot lineage (O'Brien, Wildt and Bush, 1986).

THE OCELOT LINEAGE

OCELOT *Felis pardalis* Linnaeus, 1758

This is one of the best-known South American cats, partly because recently it has been a mainstay of the fur trade, and also because in

Table 3.1 The classification of the Felidae used in this book

Lineage	Species
OCELOT LINEAGE	
	Ocelot *Felis pardalis*
	Margay *Felis wiedii*
	Tiger cat *Felis tigrina*
	Geoffroy's cat *Felis geoffroyi*
	Kodkod *Felis guigna*
	Pampas cat *Felis colocolo*
	Mountain cat *Felis jacobita*
DOMESTIC CAT LINEAGE	
	Domestic cat *Felis catus*
	Wildcat *Felis silvestris*
	Chinese desert cat *Felis bieti*
	Sand cat *Felis margarita*
	Black-footed cat *Felis nigripes*
	Jungle cat *Felis chaus*
	Pallas's cat *Felis manul*
PANTHERINE LINEAGE	
	Leopard cat *Felis bengalensis*
	Rusty-spotted cat *Felis rubiginosa*
	Iriomote cat *Felis iriomotensis*
PANTHERINE LINEAGE *contd.*	
	Fishing cat *Felis viverrina*
	Flat-headed cat *Felis planiceps*
	African golden cat *Felis aurata*
	Temminck's golden cat *Felis temminckii*
	Bay cat *Felis badia*
	Serval *Felis serval*
	Caracal *Felis caracal*
	Puma *Felis concolor*
	Jaguarundi *Felis yaguarondi*
	Cheetah *Acinonyx jubatus*
***PANTHERA* GROUP**	
	Eurasian lynx *Lynx lynx*
	Canada lynx *Lynx canadensis*
	Bobcat *Lynx rufus*
	Marbled cat *Pardofelis marmorata*
	Clouded leopard *Neofelis nebulosa*
	Snow leopard *Panthera uncia*
	Tiger *Panthera tigris*
	Leopard *Panthera pardus*
	Jaguar *Panthera onca*
	Lion *Panthera leo*

the past it was frequently kept as a pet. The ocelot is much larger than, although similar in appearance to, the margay, but it tends to be even more blotched than spotted compared with its smaller cousin. The chain-like blotches and spots, which run along the length of the animal, are bordered with black, but have a lighter-coloured centre. The ground colour of the coat varies from whitish or tawny yellow through reddish grey to grey. The underside is white. The backs of the ears are black wtih a central yellow spot (Guggisberg, 1975).

The ocelot's distribution and diversity in coat colour are very similar to the margay's. The ocelot used to range from Arizona and southwestern Texas to Paraguay and northern Argentina. Recently, an extension of its range into Uruguay has been confirmed by Ximinez (1988). Today it is almost extinct in the USA (only about 120 survive in Texas (Tewes and Everett, 1986)) and it is very rare in Mexico and many other parts of its range (Burton and Pearson, 1987). It is found in the mountains of Colombia, Ecuador and northern Peru, but not on the high plateaux of southern Peru and Bolivia. Pet ocelots have either been released or have escaped near Miami in Florida, where they survive and may one day form a self-sustaining population (Lever, 1985).

The habitats in which the ocelot is found are very diverse and include rain forest, montane forest, thick bush, semideserts, marsh and along river banks, but it is never found in open country. Like most cats, the ocelot is primarily terrestrial, although it climbs trees well.

TIGER CAT OR LITTLE SPOTTED CAT *Felis tigrina* Schreber, 1775

This is one of the smallest South American cats. It is covered all over with dark brown or black spots on a lighter ground colour. It is even more similar to the margay than the ocelot. Early naturalists often confused these two species. For example, Thomas Bewick (1807) describes perfectly what we know today as the margay. However, at that time only the tiger cat had received its formal scientific name. Indeed, Bewick considered the margay and tiger cat to be the same species. The tiger cat is not only smaller than the margay, but it has a relatively shorter tail, indicating its less arboreal habits (Figure 3.1).

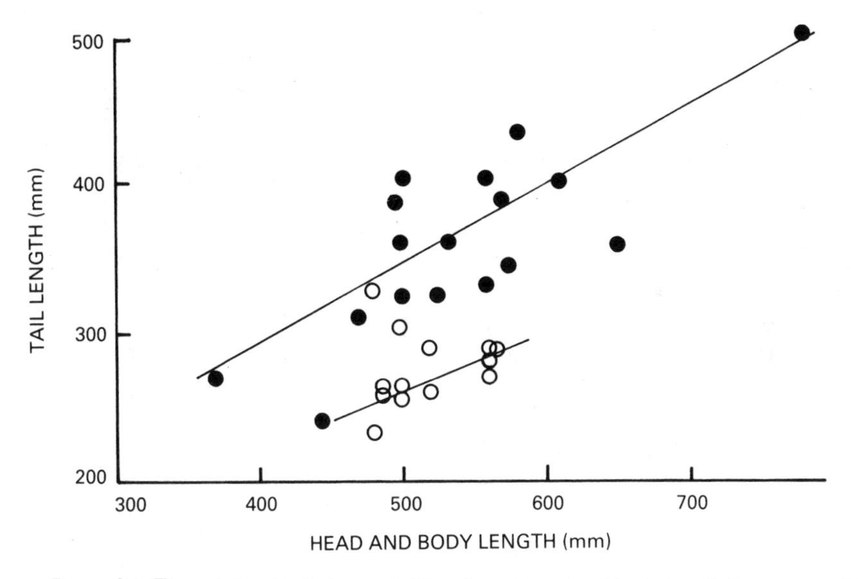

Figure 3.1 The relationship between tail length and head and body length for margays (●) and tiger cats (○). The lines show the least squares regression line for each species. Although the margay has only a slightly longer head and body length than the tiger cat on average, it has a relatively much longer tail indicating its more arboreal habits (A.K.)

Partly because of the confusion between these two species, they have received a variety of common names. The tiger cat is also known as the little spotted cat and the oncilla. It ranges throughout South America as far south as northern Argentina, but it is rare throughout its range (Burton and Pearson, 1987). It is also found in Costa Rica in Central America (Gardner, 1971). Tiger cats occur in a variety of habitats; those from mountains are often more heavily spotted and blotched than those from lowland rainforest. Melanistic tiger cats have been recorded, in particular, from southeastern Brazil.

MARGAY *Felis wiedii* Schinz, 1821

As has already been mentioned, the margay is slightly larger than the tiger cat and has a relatively longer tail. With its flexible ankle joint the margay is one of the most arboreal of cats.

The margay is more blotched than spotted, but is otherwise very similar in appearance to the tiger cat. Margays from the mountains are more heavily marked and thicker furred than lowland forest animals. The former distribution of the margay is almost identical to that of the tiger cat, except that it is said to have been found as far north as the southern USA. Unlike the ocelot, the margay is now extinct in the USA, and it is very rare in Central America (Burton and Pearson, 1987). It has suffered from habitat destruction and exploitation for its fur.

GEOFFROY'S CAT *Felis geoffroyi* d'Orbigny and Gervais, 1844

Since the decline of the ocelot and margay as providers of fur coats, the Geoffroy's cat has suffered the depredations of the fur trade. This small cat was named after the French naturalist Geoffroy St Hilaire and, like the other small South American cats described so far, is also very variable in coloration and markings (Ximinez, 1975).

The ground colour of the coat varies from brilliant ochre in the north of its range to silvery grey in the south. The body is covered in small black round spots. Spotting is more indistinct in the salt desert cat (*F. g. salinarum*) of northern Argentina, which used to be considered a separate species. The backs of the ears are black with a central white spot. The tail is spotted at the base, but ringed towards the tip. Four subspecies are currently recognised (Ximinez, 1975):

Felis geoffroyi geoffroyi—Argentina,
Felis g. salinarum—northern Argentina,
Felis g. paraguae—Paraguay, northeastern Argentina, southeastern Brazil, southern Uruguay,
Felis g. euxantha—northern Argentina, Bolivia.

Geoffroy's cat has a rather limited distribution, ranging from the Bolivian Andes and the mountains of northwestern Argentina to Uruguay and southern Brazil, and throughout Argentina to Patagonia where the largest cats are found. Geoffroy's cat is found in scrubby woodland and open bush in plains and foothills up to an altitude of 3,300 metres.

KODKOD *Felis guigna* Molina, 1782

The kodkod is the smallest South American cat and also has the most restricted distribution. It is very similar to Geoffroy's cat in appearance, but is much smaller.

The ground colour of the coat varies from grey brown to buff, marked with round black spots. The ears are black with a central white spot and the tail is ringed. There are two subspecies of kodkod; the small brightly coloured *guigna* from southern Chile and the larger, paler *tigrillo* from central Chile. The kodkod is found in forests, but very little else is known about this cat (Guggisberg, 1975).

PAMPAS CAT *Felis colocolo* Molina, 1782

The pampas cat is quite different from the other South American cats described so far, resembling more the European wild cat with its broad face and pointed ears. The fur of the pampas cat is long and may form a mane on its back up to 7 cm long. Its coloration is variable; from yellowish white and greyish yellow to brown, greyish brown, silvery grey and light grey (Guggisberg, 1975). There are bands of yellow or brown running obliquely from the back to the flanks, and a brown band running around the upper part of the forelegs. The ears are grey, but in northern animals they also bear a white central spot.

The pampas cat ranges from southern Patagonia through Argentina, Uruguay and Paraguay to southern and western Bolivia, Peru, Ecuador and southwestern Brazil. It occurs in a variety of habitats, ranging from open grasslands to humid forests.

MOUNTAIN CAT *Felis jacobita* Cornalia, 1865

This is a very rare and little known cat. It has long soft fur, which is a pale silvery grey in colour and is striped irregularly with brown or orangey markings. The ears are dark grey and the tail is bushy and ringed with black. The mountain cat is found in arid areas in the Andes from northern Chile and Argentina to southern Peru.

THE DOMESTIC CAT LINEAGE

WILDCAT *Felis silvestris* Schreber, 1775

The wildcat is one of the most widely distributed of the small cats, being found in a variety of habitats. Not surprisingly the coat colour of the wildcat varies with habitat to aid concealment, but this diversity in coat colour has led taxonomists to split this species into at least three different species groups which are described separately below. It is only recently that we have realised that these three distinct forms grade imperceptibly into one another where their ranges meet (e.g. Haltenorth, 1953; Ragni and Randi, 1986). This view is not, however, a unanimous one. Ansell and Dowsett (1988) and Smithers (1983) both uphold, for example, the specific distinctness of the African wildcat. Smithers (1983) points out that there seems to be a gap in the distribution between the *lybica* cats in the Middle East and the *silvestris* cats of the Caucasus as shown in Harrison (1968). The domestic cat has spread throughout the range of the wildcat and the subsequent hybridisation that has occurred over the last 4,000 years between mobile domestic cats and native wildcats has probably obscured the true relationships between the main wildcat subspecies forever. The closest wildcat to my home is the European wildcat (*Felis silvestris silvestris*):

EUROPEAN WILDCAT *Felis s. silvestris*

This is the wildcat which once roamed throughout the forests of Britain and Europe before it was extirpated by humans. The wildcat in Scotland is sometimes recognised as a separate subspecies and given the name *F. s. grampia* because of its darker coat colour compared with the wildcat of mainland Europe. However, Corbet (1978), for example, does not recognise this subspecies because these differences are not consistent or significant enough. After looking at the good series of wildcat skins in the Royal Museum of Scotland in Edinburgh, I can only agree, as there is a range of coat colour from the dark *grampia*-type of Scotland to the paler *silvestris* typical of continental Europe.

In Europe the wildcat is predominantly a forest cat, although in Scotland it can be found in open habitats such as rocky outcrops, heathland, etc. Wildcats in Germany prefer coniferous forest, whereas those of the Caucasus prefer broadleaved woodland.

The coat colour is grey-brown, but there is a wide variation in the ground colour. The coat is usually boldly marked with stripes which run along the neck and down the flanks just like a striped domestic tabby, although with fewer more widely spaced stripes. Wildcats usually have a white throat patch, and may have white patches on their abdomen and between their forelegs. Their ears are brown with no central, pale spot. The tail of the wildcat is bushy and blunt-ended compared with the thin tapering tail of domestic cats. Black wildcats (popularly known as Kellas cats after the Scottish village where the first one was found), which have become common in Scotland in the last ten years, are not a species new to science, but are merely hybrids between wildcats and domestic cats. They are not straightforward F1 hybrids, but represent introgressive hybrids which mostly contain wildcat genes. After considerable persecution until the beginning of the twentieth century, the wildcat has expanded its range in Scotland to occupy most of the suitable habitat available. The wildcat has recently been given full legal protection in Britain and it is hoped that one day it may be possible to reintroduce it into other suitable areas of Britain.

AFRICAN WILDCAT *F. s. lybica* Forster, 1780

The ancestor of the domestic cat is a little larger and more stocky than its tame descendant. It is basically a pale striped tabby. Its ground colour varies from sandy through yellowish grey to greyish brown and dark grey (Guggisberg, 1975). In some areas (e.g. West Africa) there are two colour phases, one greyish tan, the other steel grey. Its body is covered by reddish striped tabby markings. The ears are usually brownish, but may be grey. The ground colour of the coat varies with habitat, being darker in forests and paler in arid areas.

African wildcats are found in a variety of habitats, except for deserts and tropical rain forest. They are usually found in woodland, wooded grasslands and savanna. The wild cats of the Mediterranean islands appear to be more similar in appearance to the African wildcat than their burly north European cousin.

INDIAN DESERT CAT *Felis s. ornata* Gray, 1830

The Indian desert cat or steppe cat is adapted to the semideserts and steppes of southwestern Asia as far east as northern India. It is pale sand-coloured or grey and covered in distinct black spots. The Indian desert cat is thought to be the ancestor of Asian breeds of domestic cat (Kratochvil and Kratochvil, 1976).

DOMESTIC CAT *Felis catus* Linnaeus, 1758

The domestic cat, its colour varieties and ancestry will be dealt with fully in Chapter 7.

CHINESE DESERT CAT *Felis bieti* Milne Edwards, 1892

This is a very enigmatic cat. Almost nothing is known about it and if anybody bothers to study it, it may be found to be yet another, although larger, subspecies of the almost ubiquitous *Felis silvestris*. The Chinese desert cat is pale yellowish grey in colour. It has few body markings except for pale brownish streaks on its legs and haunches. Its tail has up to four rings towards its black tip. Its ears are yellowish brown and have short lynx-like tufts. Like the sand cat, it has hair-covered pads apparently as an adaptation to moving across hot and shifting desert sands. However, the Chinese desert cat is not a true desert animal, being found in steppes and mountains covered in bush and forest in North Central China and Mongolia up to 3,000 metres.

JUNGLE CAT *Felis chaus* Schreber, 1777

The jungle cat shares the distinction, with the African wildcats and domestic cat, of having been mummified and placed in tombs in Ancient Egypt. This fact, coupled with its sharing a very similar coloration to the Abyssinian breed of domestic cat, has led to speculation about the jungle cat's role in the ancestry of domestic cats. However, it is now believed that the much larger jungle cat was never an ancestor of the domestic cat.

The jungle cat's coat colour is basically pale or sandy brown, but varies from reddish through grey-brown to yellowish grey and sand-coloured. There are no body markings except for brown stripes on the legs. The ears are reddish and, like those of the Chinese desert cat, they have small lynx-like tufts. The tail is short, ringed faintly and has a black tip. Like the kittens of lions and pumas, jungle cat kittens are striped cryptically for safe concealment, although they lose these markings as they mature.

Jungle cats are found from Egypt through southwestern Asia to India. In Egypt they prefer swampy ground and reeds beds, hence the alternative common name of reed or swamp cat. In India they are found in woodlands, open plains, grasslands, agricultural crops and scrub. Jungle cats frequently use the disused burrows of other carnivores as dens.

SAND CAT *Felis margarita* Loche, 1858

The sand cat is one of the most widespread cats and yet it is one of the rarest and least known. The sand cat has large ears and a broad head. Its dense soft fur is pale sand or grey above and paler below. A reddish streak runs from its eyes across its cheeks, similar to the darker facial markings of tabby cats. The ears are reddish brown and black-tipped. There are faint stripes running down its flanks and black bands running around the tops of the front legs. Its tail has two or three black rings towards its black tip. To insulate it from and stop it slipping on the hot, loose sand, there is a dense layer of fur covering its sensitive pads (Harrison, 1968).

The sand cat is found in deserts and semi-deserts from the Sahara through the Middle East to Turkestan. Four subspecies are currently recognised (Harrison, 1968; Hemmer, 1978a):

Felis margarita margarita—the Sahara,
F. m. thinobia—Turkestan,
F. m. scheffeli—Pakistan,
F. m. harrisoni—Arabia, Jordan.

By day sand cats rest in burrows which they dig into the sand, from which they emerge to hunt at night. Their very acute hearing is used when hunting their prey.

BLACK-FOOTED CAT *Felis nigripes* Burchell, 1824

The black-footed cat is the sand cat of southern Africa. Its fur is dark to light yellowish brown, covered in black spots. Its front legs have transverse bars and there are cheek stripes as well. It has only a short tail, ringed and tipped with black. Its ears are reddish brown. As its name suggests, the undersides of its feet are black. However, domestic and wildcats have similar black feet, so that a new name, small spotted cat, has been proposed for this species (e.g. Smithers, 1983). This could be confused with the little spotted cat, *Felis tigrina*.

There are two subspecies:

F. n. nigripes—pale coat, restricted to East Cape Province,

F. n. thomasi—dark coat, Botswana.

The black-footed cat is found in deserts and grasslands. It often uses the burrows of springhares (*Pedetes capensis*) for dens.

PALLAS'S CAT *Felis manul* Pallas, 1776

This cat was named after Peter Pallas, the German naturalist, who described much of the Russian fauna (Gotch, 1975). Pallas's cats are very stocky and compact, and look like a ball of fluff with their long fur. They vary in colour from light grey through yellowish brown to russet. The tips of the hairs are white, giving the fur a frosted appearance. There are dark streaks on the side of the head, faint stripes across the flanks and four dark rings on the tail. The long, dense, soft fur not only gives a Pallas's cat its fluffy appearance, but provides excellent insulation from the cold. The ears are small and set wide apart on the broad head.

Pallas's cat is found in deserts, steppes and rocky country up to 4,000 metres. Daytime den sites include caves, crevices in rocks or burrows dug by other animals.

THE PANTHERINE LINEAGE

LEOPARD CAT *Felis bengalensis* Kerr, 1792

The leopard cat is one of the commonest and most widespread cats. It ranges throughout southern and eastern Asia, including the Philippines and Indonesia. Leopard cats are similar in size and shape to domestic cats, but are said to have longer legs.

A typical leopard cat from southern Asia is yellowish and covered with many black spots. Its underside is white and there are white streaks on its cheeks and running from its eyes to the top of its head. The spots sometimes coalesce to form bands running along the flanks of the body and along the back of the neck. However, the ground colour of its coat varies considerably throughout its range from pale tawny through yellow to reddish or grey. Sumatran

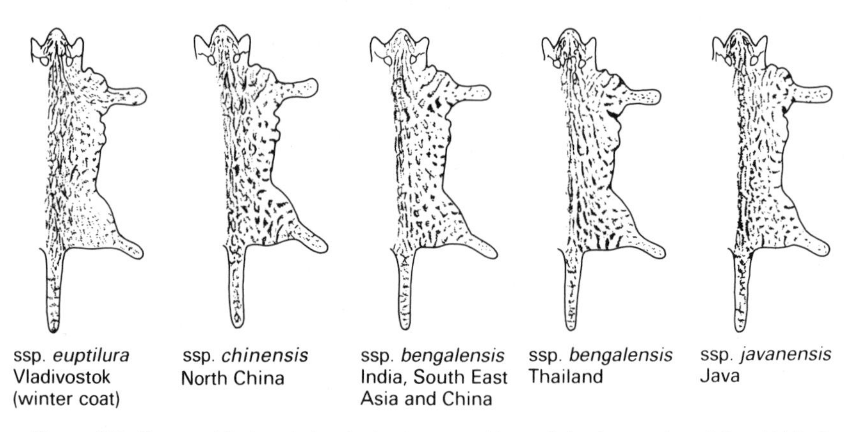

Figure 3.2 Geographical variation in the coat markings of the leopard cat (after Weigel, 1961) (S.A.)

leopard cats (*sumatranus*) are smaller and have fewer markings, while Javanese and Balinese leopard cats (*javanensis*) are duller. Bornean leopard cats (*borneoensis*) are brighter and redder, while Pakistani cats (*trevelyani*) are greyer (Roberts, 1977). The ears are dark with central white spots. Manchurian leopard cats (*F. b. euptilura*) are much larger than typical leopard cats. They have thicker fur and are greyer in colour with less distinct spotting. They look so different from their tropical cousins that they are sometimes recognised as a separate species. Variation in the markings on the fur of leopard cats throughout their geographic range is shown in Figure 3.2.

Leopard cats are found in a variety of habitats including woodlands, forests and scrub at all altitudes. The Manchurian race is found in taiga forests. Leopard cats may occupy dens in hollow trees, small caves or under tree roots.

IRIOMOTE CAT *Felis iromotensis* (Imaizumi, 1967)

Discovered in the early 1960s, this cat is confined to the 292-square-kilometre island of Iriomotejima in the Yaeyama Islands at the southern end of the Ryukyu Islands that run to the south of Japan. Barely 100 of these cats survive, due to the destruction of the subtropical rain forest there.

The Iriomote cat is similar to the leopard cat and may be a very distinct subspecies. It has a short tail, long body and short legs. It is brown with dark spots, which may coalesce into bands. Dark stripes run along its neck. The ears are dark with white central spots.

RUSTY-SPOTTED CAT *Felis rubiginosa* I. Geoffroy, 1831

The rusty-spotted cat is one of the smallest cats in the world. It is confined to India and Sri Lanka. The rusty-spotted cat is a smaller, 'washed-out' version of the leopard cat. Its greyish fur is marked with reddish spots.

The Indian race (*rubiginosa*) is found in relatively open country, including dry grassland and scrub. But the more brightly coloured Sri Lankan subspecies (*phillipsi*) is found in tropical forests.

FISHING CAT *Felis viverrina* Bennett, 1833

When Bennett (1833) first described the fishing cat scientifically, he thought that it looked similar to a civet, and hence, he dubbed it *viverrina*. This medium-sized, powerful cat is found throughout southern Asia including southern China, southeastern Asia and parts of Indonesia, but it is definitely a cat, not a civet. It has a robust build and a short tail. Its fur is greyish brown with dark round spots and its face is marked with white stripes running along its cheeks, and from its eyes to the crown of its head. The ears are dark with a central white spot.

It has often been stated that the fishing cat has webbed feet and that its claws do not retract completely (e.g. Roberts, 1977; Guggisberg, 1975). However, I have recently examined a live female fishing cat's paws in the flesh and have found that the webbing is no more than found, for example, in a bobcat (Figure 3.3).

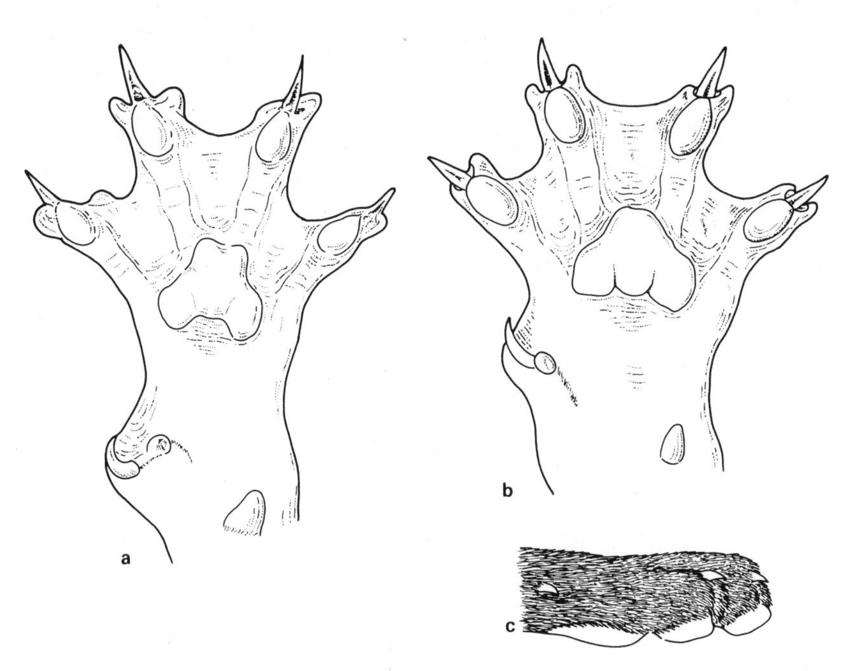

Figure 3.3 The undersides of the forepaws of cats with digits and claws extended to show the degree of webbing: (**a**) bobcat, (**b**) fishing cat. (**c**) Retracted claws of a fishing cat, showing that the cutaneous sheaths are not long enough to cover the claws (after Pocock, 1917) (S.A.)

In addition, I agree with Pocock (1917) that its claws do fully retract, but the cutaneous sheaths do not fully cover the claws. This condition has been found in other cats, including the flat-headed cat and Geoffroy's cat.

Fishing cats are found in a variety of watery habitats including mangrove swamps, marshy thickets and reed beds up to an altitude of 1,500 metres. As its name suggests, the fishing cat frequently enters water and hunts fish or frog prey, and even molluscs.

FLAT-HEADED CAT *Felis planiceps* Vigors and Horsfield, 1827

This is surely the oddest cat in the world. About the size of a domestic cat, its skull is strangely flattened. It has a long body, but short legs and tail. Its thick fur is reddish brown to dark brown. Its underparts are white spotted with brown. There are white streaks on its head. The tail is yellowish underneath. Its flat head and long muzzle are thought to be adaptations to fishing. Its curious characters have led Muul and Lim (1970) to suggest that it is the ecological equivalent of a semi-aquatic mustelid.

The flat-headed cat is found in Malaya, Borneo and Sumatra up to 700 metres altitude. It is found close to water in tropical forest, where it hunts fish and frogs. In fact, the flat-headed cat is probably more deserving of the name fishing cat than *Felis viverrina*.

AFRICAN GOLDEN CAT *Felis aurata* Temminck, 1827

This is another cat which is found in two different colour phases, originally described as separate species; the red phase was called the golden cat, while the grey phase was called the silver cat (Rosevear, 1974). An exhaustive study by van Mensch and van Bree (1969) showed that these two colour phases are found throughout the golden cat's range, but they did find that there are four different coat patterns which appear to separate into two geographical races:

(1) spotted all over;

(2) spots on the back and neck indistinct;

(3) no pattern on the neck and back; lower flanks distinctly spotted;

(4) virtually no pattern except on belly.

The first two colour patterns are found in West Africa, whereas the other two patterns are found in Central and East Africa. Van Mensch and van Bree (1969) recognised these two populations as subspecies which are geographically separated by the absence of any golden cats in Nigeria (Rosevear, 1974; Happold, 1987). The West African subspecies is called *celidogaster*, the unspotted sub-species from Central and East Africa is known as *aurata* (Figure 3.4).

The African golden cat is usually found in high rain forest, but it appears to be occasionally encountered in Guinea woodland as long

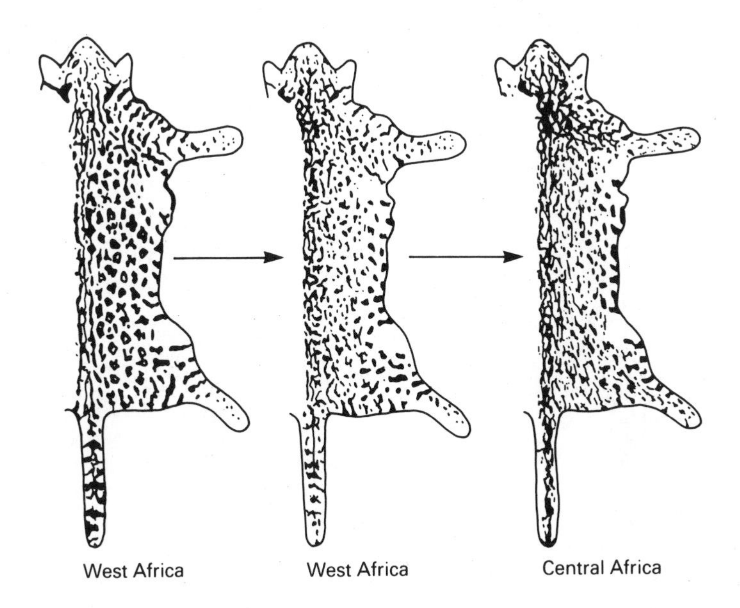

Figure 3.4 Variation in coat markings of the African golden cat (after Weigel, 1961) (S.A.)

as it is near water. It is nocturnal and said often to be found resting in the lower branches of trees during the day (Rosevear, 1974; Kingdon, 1977; Happold, 1987).

TEMMINCK'S GOLDEN CAT *Felis temminckii* Vigors and Horsfield, 1827

Temminck's golden cat, like its African relative, shows a variety of different colorations. In the south of its range, its dense fur varies from golden brown through dark brown to golden red and grey in colour. In the north of its range, it is usually heavily spotted and striped and resembles the leopard cat, although it is considerably larger. This distinct subspecies, Fontanier's cat (*F. t. tristis*) has been considered either as a subspecies of the leopard cat or as a separate species (Figure 3.5). The face of the golden cat is marked with white lines running across its cheeks and from the corners of its eyes up to the top of its head. Its ears are dark with a grizzled centre. The underside of the distal third of the tail, including its tip, is white.

Temminck's golden cat inhabits wooded and rocky areas of the Himalayas, China, southeastern Asia and through Malaya to Sumatra.

BAY CAT *Felis badia* Gray, 1874

The bay cat is a smaller island version of Temminck's golden cat. It occurs in two colour phases; one is bright reddish brown with faint spots on its paler underside and limbs. The other is dark, bluish, slatey grey. The face is marked with white stripes. There is a white

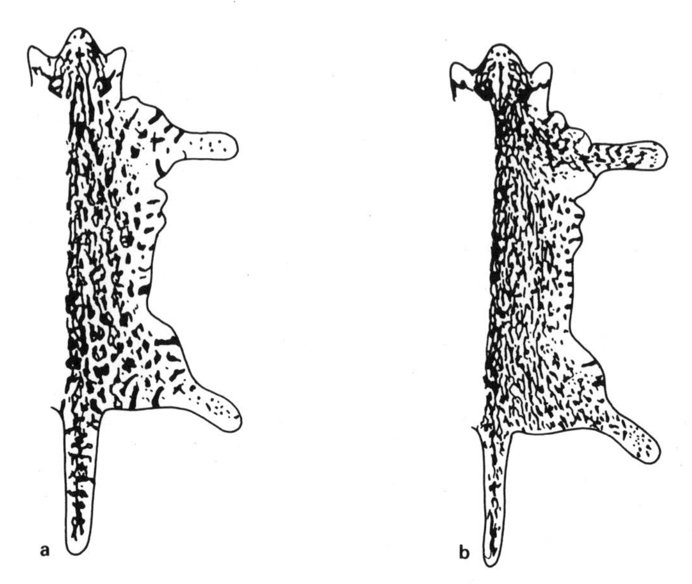

Figure 3.5 Variation in coat markings of Temminck's golden cat: (**a**) *Felis temminckii tristis* (**b**) *Felis temminckii temminckii* (after Weigel, 1961) (S.A.)

streak along the underside of the distal half of the tail. The tip of the tail is also white with a terminal black spot.

The bay cat is confined to Borneo, where it inhabits dense tropical forests and rocky limestone outcrops up to an altitude of 900 metres.

SERVAL *Felis serval* Schreber, 1776

The serval's name is derived from a Portuguese word meaning wolf-deer, which still does not really make it clear why it should have got this name in the first place (Gotch, 1975). The serval is a slender and graceful cat, with extremely long legs and large rounded ears. Although it is about the same size as the caracal, it is taller at the shoulder and lighter in body weight, in keeping with its more gracile form. The long legs of the serval are an adaptation to hunting in tall grass, so that it can use its large ears to pick up the low intensity sounds of its prey moving through the grass.

The serval has a pale sandy yellow coat which is typically marked with large black spots and stripes, but in the west of its African range, there is enormous variation in the spotting which led to the description of what was called the servaline cat, namely *F. servalina* or *F. brachyura*. The spotting on the coats of these servals is very fine or completely indistinct (Figure 3.6). In general, the more finely spotted servals are found in wetter habitats, whereas the boldly-marked servals are found in drier areas.

There is a relict population of servals in North Africa (*constantina*) which is endangered, but elsewhere south of the Sahara it is

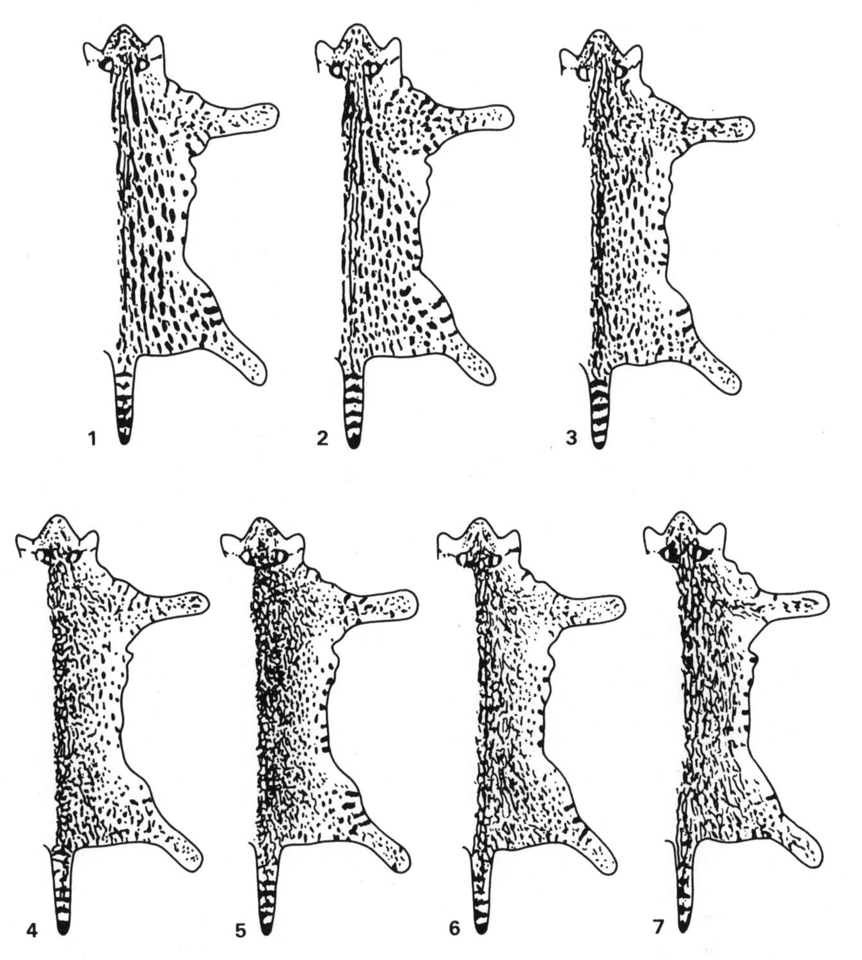

Figure 3.6 Variation in coat markings of the serval ranging from the spotted form (1) to the servaline form (7) (after Weigel, 1961) (S.A.)

common in all habitats except for rain forest, semi-desert and desert. It is more commonly associated with wetter habitats than the caracal, which has a similar distribution.

CARACAL *Felis caracal* Schreber, 1776

The caracal always used to be called a lynx, because of the long tufts of fur projecting from the tips of its ears, but we now know that the caracal is not closely related to the true lynxes (*Lynx* spp.). Its tufted ears have apparently evolved independently for use in communication (Collier and O'Brien, 1985; Werdelin, 1981). The caracal got its rather unusual name from a Turkish word, garah-gulak, which means 'black-ear', an entirely appropriate name for this cat (Rosevear, 1974; Kingdon, 1977; Smithers, 1983).

Caracals are about the same size as servals, but they are much more robust. Their fur is reddish brown, with white fur on the chin, throat and belly. There is a white ring around the eyes and a black line running from the anterior corner of the eyes to the nose (Guggisberg, 1975). The tail is reddish brown above and sandy below, lacking the black tip of the lynxes.

Caracals are wide ranging, being found in virtually every habitat except for rain forest from central India through southwestern Asia, Arabia, North Africa and throughout the rest of Africa south of the Sahara. Several subspecies have been described, but Corbet (1978) considers that only the North African race (*algira*) may be valid.

The caracal is repsonsible for the expression 'to put a cat amongst the pigeons'. In India, tame caracals were taken into an arena containing a flock of pigeons. Wagers were placed to see how many pigeons could be pulled down or disabled by each caracal as the flock took off (Rosevear, 1974).

PUMA *Felis concolor* Linnaeus, 1771

The puma probably has as many different colloquial names as it does alleged geographical races. It also has the greatest latitudinal distribution of any wild species of cat, being found in all habitats from the tip of Patagonia as far north as Canada. In all, 29 subspecies of puma are recognised by Hall (1981), but it is doubtful if many of these are true races. Most probably grade imperceptibly into each other from one end of the Americas to the other. Among the common names in the English language, the puma enjoys mountain lion, cougar, catamount, panther, painter, lion, Mexican lion, mountain demon, mountain devil, mountain screamer, brown tiger, red tiger, deer killer, Indian devil, purple feather (!), king cat, sneak cat and varmint (Guggisberg, 1975). I have always called this cat the puma.

Kurten (1973b) has looked at the geographical variation in puma body size with latitude and found that pumas from the equator are smaller than at the geographical extremes of north and south (Figure 3.7). In the past, this has been explained as an example of Bergmann's Rule, where large body size is expected in cooler climates (higher latitudes), because large bodies have a relatively lower body surface area to volume ratio, and so lose less heat to their surroundings. However, recently Geist (1987) has demonstrated that Bergmann's Rule does not hold for a similar variation in body size for wolves (*Canis lupus*) and various deer species, and that the latitudinal changes in body size are best explained by differences in seasonal productivity affecting body growth. In temperate climates there is a seasonal superabundance of food available to herbivores and carnivores that does not occur in the tropics. Therefore, temperate mammals achieve a greater propor-

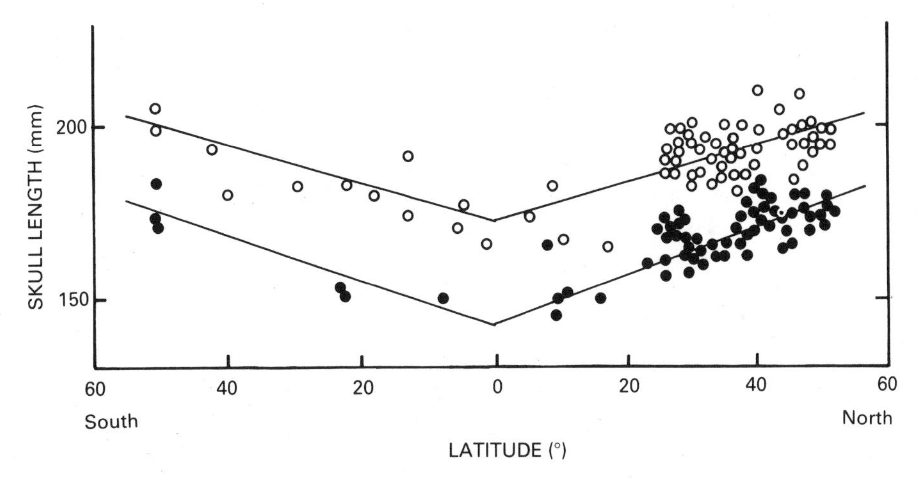

Figure 3.7 Geographic variation in the size of pumas as shown by a plot of skull length of male (○) and female (●) pumas against latitude (after Kurten, 1973b). The smallest pumas are found at the equator, the largest ones in the extreme north and south of their range (A.K.)

tion of their potential growth each year and on average grow larger than their tropical cousins. If the puma ranged further north into the Arctic Circle, we would expect a reduction in body size, as seasonal productivity falls away in the highest latitudes.

The puma is found in montane coniferous forests, lowland tropical rain forest, swamps, grasslands and bush up to an altitude of 4,500 metres. However, in mountainous regions bad winter weather forces the puma and its prey to lower elevations.

The puma varies in coat colour from brown/red through to grey/silver. Guggisberg (1975) states that red-coated pumas are most often found in the tropics, brown ones in arid areas, and sombre-coloured animals in the humid forests of the northern Pacific coast. However, more than one colour phase may be found together in most areas. Although adult pumas are not cryptically marked, their cubs are spotted to aid concealment in their first few weeks of life.

Recently, it has been suggested that a puma-like animal with cheetah-like cursorial adaptations is to be found in Mexico (Anonymous, 1986; Greenwell, 1987; Shuker, 1989). The onza, as it is called, was apparently known from Montezuma's zoo when the Spanish first began to exploit the New World. However, photographs of the alleged onza seem simply to show a young adult puma.

JAGUARUNDI *Felis yaguarondi* Lacépède, 1809

The jaguarundi does not look typically feline. Instead, it is almost weasel-like or otter-like in appearance. There are two colour phases of this otherwise uniformly-coloured cat which were at one time thought to be separate species: the reddish or chestnut eyra and the normal blackish or brownish grey phase.

Jaguarundis are inhabitants of lowland forests and thickets and are said to be crepuscular and more diurnal than other cats, in keeping with their plain unmarked coloration. Jaguarundis are found from the southern USA to northern Argentina in South America. However, they are now quite rare in the north of their range (Tewes and Everett, 1986). A feral population of jaguarundis is now well established in Florida, where they were introduced in the 1940s (Lever, 1985).

Jaguarundis are not closely related to the other small South American cats. Their ancestors probably evolved in the Old World before invading the Americas via the Bering land bridge at the same time as the puma's ancestors (Werdelin, 1985).

CHEETAH *Acinonyx jubatus* (Schreber, 1775)

The cheetah is the fastest-running carnivore, even if it can only manage this over a comparatively short distance. To achieve alleged speeds of 90 kilometres per hour, the cheetah shows many adaptations for fast running, most of which have been described in some detail in Chapter 1. Long legs and a flexible spine permit a long stride length for extra speed, exposed claws act like running spikes and wide nostrils help to pay back the oxygen debt and cool off after a sprint. The enlarged dew claw is used to pull down prey, and the long tail is essential as a counterbalance when turning. All these adaptations give the cheetah an almost greyhound-like form quite unlike any other cat.

The cheetah has fur that is pale yellowish, greyish or fawn and is covered with small round dark spots. There is a pronounced lacrymal or tear stripe running from the anterior corner of the eye down beside the muzzle. The ears are black, but tawny at the base and edges. The tail is spotted above and pale below with a white tip. Cheetah cubs are covered in a long woolly, bluish-grey mane which makes them less conspicuous to predators on the open savanna.

The cheetah used to have a very wide distribution from the open grasslands of central India through southwestern Asia, Arabia and throughout Africa where suitable open habitats were available. The cheetah (subspecies *venaticus*) is now extinct in most of its Asiatic range. It still just survives in Baluchistan and Iran. In Africa, numbers are declining rapidly in many areas. Only the East and South African populations exist in any numbers, but here they seem to be able to tolerate humans to some extent (Hamilton, 1986). Persecution, loss of habitat and loss of prey are all probably responsible for the cheetah's continuing demise.

Up to six subspecies of cheetah have been described in Africa (Wrogemann, 1975), but these are doubtfully distinct. Even the

Asian cheetah (*A. v. venaticus*) may not be a valid race. The king cheetah from southern Africa was described as a new species, *Acinonyx rex*, in 1927 (Pocock, 1927). Instead of spots, it is covered in blotches reminiscent of a blotched domestic tabby. Recent studies have shown that the king cheetah's unusual coat colour is due merely to a single genetic mutation (see Chapter 8) (van Aarde and van Dyk, 1986).

THE PANTHERA GROUP

LYNXES

Lynxes are medium-sized cats with short tails and tufted ears. Three or four species are found in the northern hemisphere, where they seem to be specialised for preying on rabbits and hares. The tail and the tufted ears are important in social signalling. Three species are recognised by Corbet and Hill (1986), although the Spanish race of the Eurasian lynx is sometimes considered a separate species (e.g. Werdelin, 1981).

EURASIAN LYNX *Lynx lynx* (Linnaeus, 1758)

The Eurasian lynx is the largest of the living species of lynx. Although it only occurs in relict and reintroduced populations in western Europe, it is still found widely throughout northern and central Asia. Eurasian lynxes have yellowish-brown fur which shows seasonal changes. In the summer, the coat is usually covered with dark spots, but these are barely visible in the winter. The distinctness of spotting seems also to vary with latitude: northern animals tend to be greyer and less spotted than southern animals. This reaches its extreme in the Spanish lynx (*L. l. pardinus*), which as a result is often recognised as a separate species. Lynxes have large furry paws as an adaptation for moving over snow; the fur insulates them from the cold and their large size spreads the weight of the lynx like a snowshoe. The tail is ringed towards its black tip. Lynxes have a characteristic facial ruff.

Lynxes are usually nocturnal terrestrial hunters, although in winter bad weather may force them to become diurnal hunters. Lynxes are found throughout north and central Asia. Four subspecies are currently recognised (Corbet, 1978; Tumlison, 1987):

L. l. lynx—Boreal forest of Eurasia and the Carpathians,
L. l. isabellinus—mountains of Central Asia,
L. l. pardinus—Iberian peninsula,
L. l. sardiniae—Sardinia.

The Eurasian lynx is usually found in high-timbered forests, but the

Spanish race is found in open forests of juniper, pine and pistachio scrub on old sand dunes (Guggisberg, 1975). Central Asian lynxes may be found in rocky areas with shrub cover (Guggisberg, 1975).

CANADA LYNX *Lynx canadensis* Kerr, 1792

The Canada lynx is smaller than its Eurasian counterpart. Its fur is usually white-tipped, which gives it a frosted appearance similar to the Pallas's cat. It is only indistinctly spotted. Several different colour varieties are recognised by the fur trade, including the very rare blue lynx, which is the result of a genetic mutation. The Canada lynx is probably conspecific with the Eurasian lynx. It is found throughout Canada and the north of the USA. The lynx on Newfoundland is sometimes recognised as a valid race, *L. c. subsolanus* (Tumlison, 1987).

BOBCAT *Lynx rufus* (Schreber, 1777)

The name bobcat is an abbreviation of bob-tailed cat, referring to this animal's short tail with its dark rings. Another name for the bobcat is the bay lynx, on account of its coloration. In fact, the coloration varies from buff to brown and the fur is spotted and striped with dark markings. The bobcat is smaller than the Eurasian and Canada lynxes and is confined almost exclusively to the USA and Mexico. Like the lynxes, the bobcat has a facial ruff of fur and tufted ears.

The bobcat has managed to colonise and successfully exploit a number of different habitats including pine forests, rocky mountainous regions, semideserts and scrub. For its den it uses crevices in rocks, hollow trees or the confines of a thicket. Bobcats are less dependent on lagomorphs than other lynxes. In recent years the bobcat and Canada lynx have become increasingly exploited for their fur as the tropical cats have become less available.

MARBLED CAT *Pardofelis marmorata* (Martin, 1837)

Looking like a miniature version of the clouded leopard, the diminutive size of the marbled cat belies its big cat affinities (Wozencraft, 1989). Study of its karyotype and its blood serum albumin show it to be very closely related to the big cats; perhaps it is similar in form to the forest ancestors of the big cats some ten million years ago (Collier and O'Brien, 1985). However, it may also have diminished in size more recently due to competition with other big cats.

The marbled cat has thick soft fur, which varies from brownish grey through yellow to reddish brown in colour and which is covered in large blotches, paler in their centres. There are black spots on its limbs and some black lines on its head and neck. The tail is long, confirming its arboreal habits. Indeed its adaptations to an arboreal lifestyle suggest that it is probably the Old World ecological counterpart of the margay.

The marbled cat is found in southern Asia from Nepal through southeastern Asia to Borneo and Sumatra. It is found in forests where it is thought to hunt some of its prey in the trees, although in Borneo it is said to be terrestrial. The marbled cat has always been rare, but is particularly vulnerable to the loss of tropical forests in southeastern Asia.

CLOUDED LEOPARD *Neofelis nebulosa* (Griffith, 1821)

The clouded leopard's long, bushy tail and flexible ankle joints are clear demonstrations of its arboreal ability. It is even said to be able to hang by one hind foot from a tree, waiting in ambush for potential prey. The clouded leopard is an animal of tropical forests, being found at altitudes of up to 2,500 metres. In Borneo, however, it was found to be mostly terrestrial (Rabinowitz, Andau and Chai, 1987).

The clouded leopard is greyish or yellowish and is covered by large blotchy cloud-like markings, which have a dark margin and a paler centre. The head, legs and tail are spotted. The clouded leopard is found from Nepal through southeastern China to Malaya, and on Taiwan (where it is now probably extinct (Rabinowitz, 1988)), Hainan, Borneo and Sumatra. Four subspecies are recognised:

N. n. nebulosa—south China and southeastern Asia;
N. n. brachyura—Taiwan (extinct?);
N. n. diardi—Borneo;
N. n. macrosceloides—Nepal to Burma.

Large black cats seen recently in Borneo may be melanistic clouded leopards (Rabinowitz *et al.*, 1987).

SNOW LEOPARD *Panthera uncia* (Schreber, 1775)

This inhabitant of the mountains of central Asia has thick soft fur for keeping out the cold weather. It uses its long bushy tail to wrap around itself to keep warm in the coldest of montane weather, and also as a counterbalance when jumping in its rocky habitat. Snow leopards are greyish above and white below. They are marked with dark spots, rings or rosettes. They have black ears marked with a central white spot. Snow leopards have relatively long hind limbs compared with their forelimbs, probably as an adaptation to leaping in a montane habitat (Gonyea, 1976a). They are said to be able to make leaps of up to 15 metres along the ground.

Snow leopards are found in alpine meadows and rocky areas between 2,700 and 6,000 metres during the summer months. In winter, bad weather and the migration of their prey means that snow leopards descend to forests below 1,800 metres.

Snow leopards are severely threatened in the wild from loss of prey and habitat, and also by humans hunting for their fur and to prevent loss of livestock. The recent introduction of a snow leopard hunt for tourists in the only remaining large population in Mongolia is, in particular, of great concern.

TIGER *Panthera tigris* (Linnaeus, 1758)

The tiger is one of the most widespread and familiar of wild cats. It is also one of the most endangered. Eight subspecies have been described:

(1) Bengal tiger *(tigris)*—Indian subcontinent,
(2) Siberian tiger *(altaica)*—Russian Far East and Manchuria,
(3) Caspian tiger *(virgata)*—Iran, Turkey,
(4) Chinese tiger (*amoyensis*)—south China,
(5) Sumatran tiger (*sumatrae*)—Sumatra,
(6) Javan tiger (*sondaica*)—Java,
(7) Bali tiger (*balica*)—Bali,
(8) Indochinese tiger (*corbetti*)—southeastern Asia.

Of these the Caspian, Javan and Bali tigers are extinct, the Chinese tiger is nearly so and the other subspecies have small wild populations. Only the Bengal and Siberian tigers are in any way secure in the wild.

Tigers are found in a variety of different habitats including rocky country, savannas, mangrove swamps, tropical rain forest, evergreen forests and riverine woodland. They are the largest living cats. They are so familiar that they hardly need describing. They have a reddish orange to reddish ochre coat marked with dark transverse stripes. The underside is white, the tail is ringed and the ears are black with a white central spot. Like the widespread puma, body size varies with latitude, the smallest occurring at low latitudes in Indonesia and the largest at high latitudes in Manchuria and Siberia.

LEOPARD *Panthera pardus* (Linnaeus, 1758)

The leopard is the most widespread big cat. It is found in virtually every habitat from tropical rain forest to desert and temperate forests, and from sea level up to more than 5,500 metres. In the humid rain forests of southeastern Asia melanistic leopards (black panthers) are common, but most leopards have fur which is pale, but of variable colour, marked with characteristic rosettes and spots. The underparts are white and the ears are black with a central white spot.

Many different subspecies have been described, based mainly on differences observed within the individual variation of one local population. However, there are variations in body size, coloration and markings which correlate with climate and habitat. As the leopard's former widespread distribution becomes fragmented, so it will become harder to assess the validity of these races. Among the races that are recognised, especially in captive populations, are the following:

Javan leopard (*melas*),
Amur leopard (*orientalis*),
Indian leopard (*fusca*),
North Chinese leopard (*japonensis*),
Somali leopard (*nanopardus*),
Zanzibar leopard (*adersi*),
Sinai leopard (*jarvisi*),
Sri Lankan leopard (*kotiya*),
Barbary leopard (*panthera*),
Persian leopard (*saxicolor*),
Arabian leopard (*nimr*),
Anatolian leopard (*tulliana*),
Caucasus leopard (*ciscaucasica*),
Indochinese leopard (*delacouri*).

JAGUAR *Panthera onca* (Linnaeus, 1758)

The jaguar is often confused with the leopard in zoos, but, of course, it would not be possible to mistake them in the wild. While the leopard is found in Eurasia and Africa, the jaguar is confined to South and Central America from northern Argentina to the very southern limits of the USA. It is now extinct in the USA, very rare in Central America (except Belize) and eliminated from many parts of its stronghold in South America. Both leopards and jaguars have similar coloration with spots and rosettes. However, the jaguar has additional spots inside its rosettes, which distinguish it from its close Old World relative (Figure 3.8). The jaguar is a much more stocky animal than the leopard, with shorter legs and tail. Gonyea (1976a) believes this to be an adaptation to living in a three-dimensional environment, i.e. forests. But jaguars are not restricted to forests and are also found in savannas, scrub and even deserts. The jaguar is a good swimmer and is usually found near water, where it hunts some of its prey.

LION *Panthera leo* (Linnaeus, 1758)

Like the tiger, the lion does not really require any introduction. It is the only cat that shows distinct intersexual differences which are probably related to its highly-developed social life. Male lions have shaggy manes which vary in colour from dark brown through to sandy. Lions from the extreme north and south of the range tend to have fuller manes which extend partially along the back and underside of the body. Females are maneless, but both sexes have a

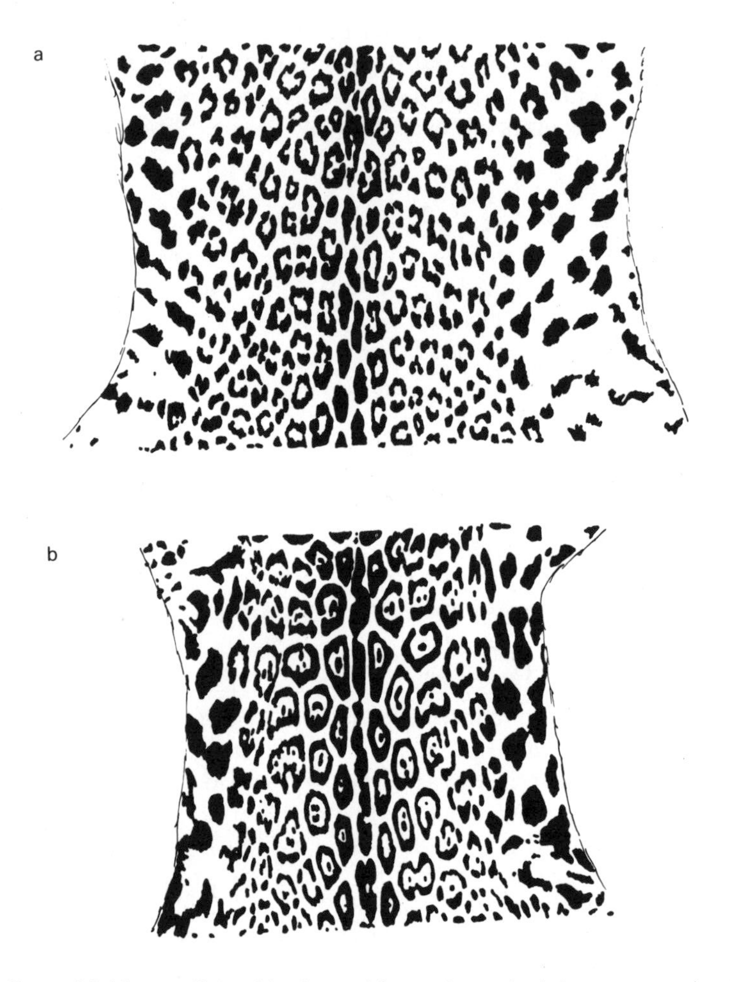

Figure 3.8 How to distinguish a leopard from a jaguar: both have rosettes, but jaguars (**b**) have extra spots inside their rosettes and leopards (**a**) do not (A.K.)

tuft of fur at the tip of the tail. It is often claimed that lions have a claw concealed in this tuft of fur. In fact, there is often a hard keratinous growth, but it is not a claw. Lions are usually sandy brown to reddish brown above and whitish below. There are no other body markings, although the cubs are spotted to help to conceal them from predators.

Lions were once common throughout Africa, southern Europe, and southwestern Asia as far east as India. Now, outside Africa, only a relict population of the Indian lion (*P. l. persica*) exists in the Gir Forest Reserve in India. In Africa the lion has been eradicated from North Africa (the Barbary lion (*leo*) may still survive in zoos) and South Africa (*melanochaita*) and it is becoming increasingly confined to national parks and game reserves elsewhere. Lions are principally animals of the savanna, but are found in a variety of habitats including woodlands and semideserts. They have even been recorded at altitudes of 5,000 metres in the mountains of Kenya. Lions may be active at night or day. They are the only cat to regularly hunt cooperatively.

4

KILLING AND EATING

Cats are the most specialised living carnivores and in this chapter the different ways in which they approach, capture and kill their prey will be described. The ways in which cats process their prey animals and feed on them will also be examined. However, it is all very well being specialised carnivores, but cats must be able to catch enough food to survive. Surprisingly the efficiency and effort that different cat species have to put into capturing enough food varies greatly. The highly efficient cheetah scavenges food only in exceptional circumstances, but the less efficient lion makes a regular habit of it. The factors that affect hunting success and behaviour will also be reviewed. Cats do not necessarily have to starve, if they are finding it hard to make a kill. Excess food from previous kills or multiple kills may be stored to be used later, thereby maximising the efficient usage of large prey animals.

Most small cats seem to be specialised for feeding on small rodents, but the largest cats have overcome the problems of feeding on prey animals sometimes very much larger than themselves. Cats generally catch prey up to a similar size to their own body size, and they tend to specialise in catching only a few different types of prey (Rosenzweig, 1966; Kruuk, 1986). For example, cats prey on an average of about four prey types, but dogs and civets prey on more than six (Kruuk, 1986). One cat, the cheetah, has chosen high speed pursuit as a very efficient way of catching food, and another, the lion, hunts cooperatively to catch very large prey, but most cats patiently stalk their prey using their cryptic coloration to good effect.

In the past predators were thought to have a detrimental effect on the populations of prey animals they exploit. This led to the wholesale destruction of predators to preserve game species for human hunters. In fact, predators have a wide variety of different effects on prey populations, ranging from total extinction when feral cats are introduced to islands lacking predators, to no effect at all such as lions preying on migratory species in the Serengeti (Schaller, 1972; Lever, 1985). In many cases the natural population regulation cycle of a prey species is in fact controlling the population size of the predator. The famous snowshoe hare and Canada lynx population cycles show how lynx populations are affected directly by the amount of available prey.

Most cats have virtually the same morphology and yet several species may coexist despite being limited to the same range of prey species. But how do they manage this without any apparent detriment to each other's populations? This chapter will describe some of the studies that have been carried out to try to explain why cats and other predators are able to coexist within the same community.

HUNTING STRATEGY

Most small cats have two main hunting strategies (Corbett, 1979; Geertsema, 1985; Kruuk, 1986). The mobile (M) strategy involves movement towards the prey and includes, for example, patrolling the home range until a potential prey animal is encountered. The stationary (S) strategy involves prey moving towards the predator and includes, for example, waiting in ambush until a potential prey animal comes along. For domestic cats, hunting rabbits on Hosta, North Uist, the M strategy is more successful than the S strategy—4.2 rabbits per hour compared with 1.1 rabbits per hour (Corbett, 1979). So why do cats have these two alternative stategies? Cats that use the M strategy are usually dominant cats, whereas subordinate cats usually use the S strategy. Also male cats use the M strategy more than females, and adult cats use the M strategy more than young cats (Corbett, 1979). It is thought, for example, that subordinate cats have no alternative but to use the less successful S strategy more often because, if they were to use the M strategy, they would be more conspicuous, and hence open to attack from dominant cats, which would probably result in disruption of their hunting and consequently an even lower hunting success than for the S strategy (Kruuk, 1986).

In Canada, lynxes often hunt snowshoe hares using the S strategy, by lying in wait on a 'hunting bed' in the snow. In Newfoundland, Saunders (1963a) found that 61 per cent of predated hares were caught in ambush, but Nellis and Keith (1968) found that only 12 per cent of hares were caught this way in Alberta. The difference in the use of the S strategy seems to be related to hare density. When the hare population is low, lynx are forced to search out their prey rather than waiting for it to pass by (Nellis and Keith, 1968).

Elliott, McTaggart Cowan and Holling, (1977) have defined three types of hunting strategy for lions in the Ngorongoro Crater in Tanzania. Type I hunting involves killing prey that a lion happens to encounter, type II hunting involves stalking and killing prey, but type III hunting involves searching out prey animals before stalking and killing them. Elliott *et al.* (1977) suggested that type III hunting was driven by hunger, but the other two types were more opportunistic and would be used more frequently at high densities of prey.

PREY CAPTURE BEHAVIOUR

Prey capture behaviour is very similar in all species of wild cats. Descriptions of prey capture behaviour and slight variations between species can be found in the references listed in Table 4.1.

Table 4.1 Descriptions of prey capture behaviour of cats

Species	Source
Ocelot	Emmons, 1988
Feral/domestic cat	Corbett, 1979; Leyhausen, 1979
Fishing cat	Breeden, 1989
Flat-headed cat	Muul and Lim, 1970
Iriomote cat	Yasuma, 1981
Serval	Geertsema, 1976, 1985
Caracal	Pringle and Pringle, 1979; Grobler, 1981
Puma	Hornocker, 1970; Wilson, 1984
Cheetah	Kruuk and Turner, 1967; Schaller, 1972; Eaton, 1974; Wrogemann, 1975
Eurasian lynx	Haglund, 1966
Canada lynx	Saunders, 1963a; Barash, 1971; Parker, Maxwell and Morton, 1983
Snow leopard	Schaller, 1977; Fox and Chundawat, 1988
Leopard	Kruuk and Turner, 1967; Schaller, 1972; Bothma and Le Riche, 1986
Jaguar	Schaller and Vasconcelos, 1978; Mondolfi and Hoogesteijn, 1986
Lion	Kruuk and Turner, 1967; Schaller, 1972; Elliott *et al.*, 1977
Tiger	Schaller, 1967; Sunquist, 1981

Most cats locate their prey primarily by sight and secondarily by hearing. However, some species specialise in detecting and capturing prey in other ways. For example, the serval specialises in hunting rodents in tall grass and has perfected hunting by hearing alone (Geertsema, 1985). Its widely-spaced, large ears help to pinpoint the slightest rustle in the grass (Figure 4.1). Servals hunt

Figure 4.1 The serval pounces on its rodent prey in the tall grasses of the savanna (after Geertsema, 1985) (S.A.)

different species of molerats in different ways according to their burrowing habits. Servals can listen for the position of the molerat *Tachyoryctes* in its shallow burrows and simply dig them up. However, they employ a different method for catching the molerat *Cryptomys* in its deep burrows. In this case the serval damages the entrance to the burrow and waits for the unfortunate molerat to come along to carry out repairs (Ewer, 1973; Kingdon, 1977).

Recently, Breeden (1989) had the good fortune of being able to watch a female fishing cat hunt in Bharatpur, India. The cat positioned herself on a log overhanging the water and held her nose just a short distance from the water surface, presumably to cut down on refraction. Then she launched herself head first into the water grabbing the fish in her mouth. This method of hunting is quite different to the way that domestic cats catch fish by dipping their paws in the water.

Cats rarely use their sense of smell to locate prey, although this has been observed for servals and leopards (Geertsema, 1985; Bothma and Le Riche, 1986).

It is often said that cats take account of wind direction when hunting, always approaching from downwind to avoid being detected (e.g. Tehsin, 1979). However, Schaller (1972) found that lions hunted randomly with respect to wind direction. Elliott *et al.* (1977) also looked at how wind direction affected hunting success of lions in the Ngorongoro Crater and found that it had no effect on hunting success and that lions ignored wind direction anyway. Feral cats and servals also take no account of wind direction when hunting (Corbett, 1979; Geertsema, 1985). Although prey animals are alerted by the scent of a predator, they only take flight when they see the direction from which the predator is coming (Schaller, 1972), since otherwise, if they panicked on detecting the smell of a predator, they might run towards it instead of away from it.

THE APPROACH

Once a potential prey item has been observed, most cats crouch and approach to the nearest available cover using a stalking or slinking run (Ewer, 1973; Leyhausen, 1979). When they are a certain distance from the prey animal (see Table 4.11), they stop and adopt a watching posture, lying as close to the ground as possible given the available cover. Whiskers are held widely spread and ears face forward while they wait to pounce. Further stalking runs may be necessary if a cat is still too far away, or it will just stalk very slowly towards the prey animal. To prepare for the final pounce, cats push back their hind legs gradually by treading each foot alternately. They then dart forward in a low run to attempt to catch the prey animal.

This hunting behaviour is typical for most of the smaller cats and is adapted for catching small rodents or other terrestrial mammals which retreat to a nearby burrow (Ewer, 1973). Although the stalk and capture hunting method is regarded as the classic felid hunting method, Bothma and Le Riche (1986) found that Kalahari leopards killed less than 5 per cent of their prey in this way. Perhaps prey density is so low in the Kalahari that virtually all kills are opportunistic. However, many cats kill much of their prey opportunistically. Stalking is not usually a good method for hunting birds which tend to flutter around before flying away (Ewer, 1973). Every time a cat gets closer, the bird flutters further away. This is why cats are usually very poor bird hunters, as indicated by the low proportion occurring in their diets (see Chapter 5). When cats hunt birds, they do not usually adopt a waiting posture before the attack, but rush straight at the bird, grabbing it with both paws. This has been observed for the jaguar, caracal, tiger cat and ocelot (Ewer, 1973; Schaller and Vasconcelos, 1978; Leyhausen, 1979; Emmons, 1988).

Lions tend to run at gazelles on sight from only a short distance away, but they prefer to stalk zebra and wildebeest. Gazelles are able to accelerate much faster and to a higher maximum velocity than lions, so that lions only enjoy a meal of gazelle is they catch them completely unawares (Figure 4.2) (Elliott *et al.*, 1977). The much slower zebra and wildebeest are more easily caught by stalking.

THE ATTACK

Prey animals are usually attacked from the rear or side except in long grass, where it may be necessary to pounce from above. It is

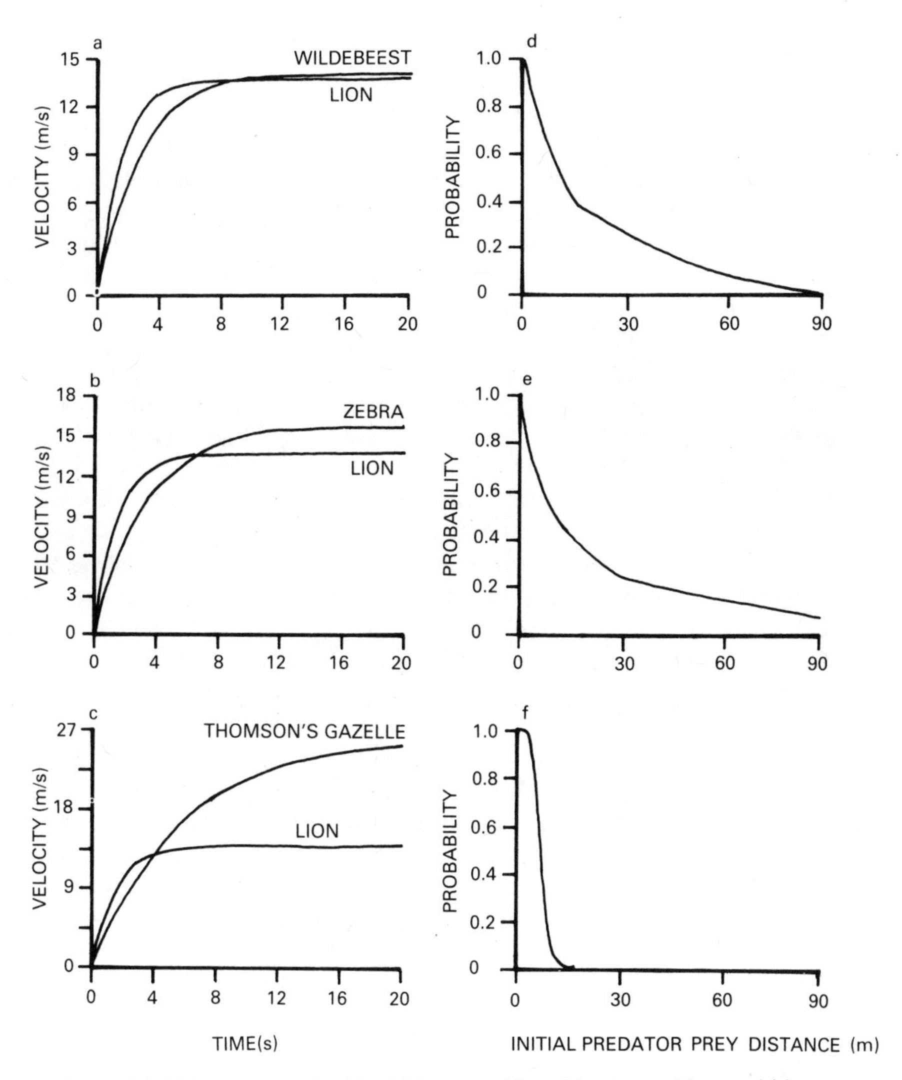

Figure 4.2 Velocity curves for (**a**) wildebeest and lion, (**b**) zebra and lion, and (**c**) Thomson's gazelle and lion. Probability of a successful attack by a lion on (**d**) wildebeest, (**e**) zebra and (**f**) Thomson's gazelle, over varying distances between lion and prey. To catch a gazelle, a lion must encounter it by surprise because of its faster acceleration and higher velocity (Elliott *et al.*, 1977) (A.K.)

very important for the cat, especially if hunting large prey, to keep its weight on its hind legs during the attack so that it can adjust its position, or if necessary retreat to avoid being injured (Leyhausen, 1979). Cats are quite often injured or even killed in prey capture, as recorded for pumas, lions and tigers (Schaller, 1972; Seidensticker *et al.*, 1973; Sunquist, 1981). Although leopards may wait in trees, they always leap to the ground before mounting an attack. However, it has been suggested that truly arboreal cats such as the margay, clouded leopard and marbled cat may well jump directly onto their prey (Leyhausen, 1979).

Cats usually grab small prey in their mouths, but larger prey are held by one paw before being bitten just in front of the paw hold. The paw hold gives the cat the opportunity, while capturing larger and more awkward prey, of readjusting the killing bite if necessary and knocking the prey off balance. The killing bite is usually directed at the nape of the neck/shoulder/back of cranium in relatively small prey. Long-necked prey such as birds are usually bitten in the shoulder or lower neck region at first, but the second bite is always a nape bite (Leyhausen, 1979). Grabbing by the mouth is an ancient behaviour found in many carnivorous mammals (Eisenberg and Leyhausen, 1972). Grabbing and holding the prey with forepaws is an advanced behaviour which is particularly associated with cats. Sometimes a cat may slap at small prey with its forepaws, as does the serval (Leyhausen, 1979), particularly if the prey is dangerous or defending itself. The blows raining down on the prey may be sufficient to stun or even kill it.

The cheetah, being the athletic felid, does things quite differently. It usually selects a victim from about 50–500 metres away and dashes after it with no hint of stalking at all. When it draws alongside the prey, it tries to knock it over with its forepaws, using the well-developed dewclaws to rake along the side of the body, both to injure and pull down the fleeing prey animal. The cheetah's pursuit distance is limited by its ability to store heat in its body. Taylor and Rowntree (1973) found that cheetahs store the heat produced by their body muscles until after a chase has ended, rather than dissipating it during the chase. They calculated that at its rate of heat storage, a cheetah could only chase its prey for up to 500 metres before its body temperature reached a lethal limit. Schaller (1972) rarely recorded cheetah pursuit distances of greater than 300–400 metres, well below the lethal distance. However, the distance at which cheetahs start their pursuit and the effort they have to put into it depend on the vulnerability of the prey. For example, adult gazelles are chased from as little as 80 metres away, but a vulnerable gazelle fawn may be chased from up to 500 metres away (Bertram, 1979). Schaller (1972) found that on average young gazelles are chased over 190 metres, but adults and older youngsters are chased 290 metres on average until captured.

THE KILL

Small prey are killed by a nape bite where the canines dislocate the cervical vertbrae (Figure 4.3a). The fast contraction time of the jaw muscles and the abundance of mechanoreceptors to the canines suggests that these killing teeth can 'feel' their way to the cervical vertebrae in a fraction of a second (Ewer, 1973). The nape bite may also crush the back of the skull. For example, jaguars may kill capybara in this way (Schaller and Vasconcelos, 1978). If the prey does not die instantly, the cat will often rake at it with its hind feet.

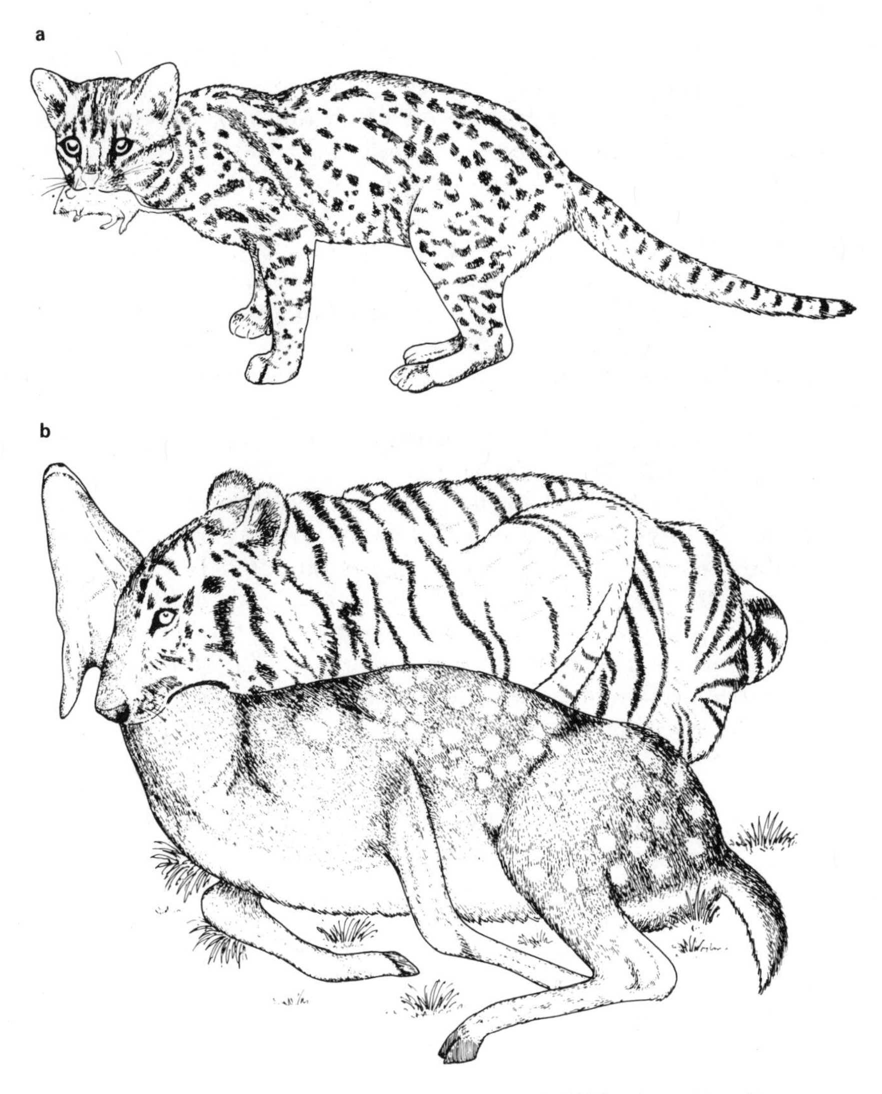

Figure 4.3 (**a**) The nape bite of smaller cats (e.g. tiger cat). (**b**) The throat bite of larger cats (e.g. tiger) (after Leyhausen, 1979) (S.A.)

Large prey are killed by a throat bite which occludes the trachea, leading to suffocation (Figure 4.3b). A nape bite would be ineffective against large prey, since the canines would not be long enough and the jaw gape wide enough to kill in this way. Suffocation may also be achieved by placing the cat's mouth over the mouth and nose of its prey. The very short canines of the cheetah make the throat bite essential for despatching medium-sized prey. It has been suggested that the throat bite of the cheetah may also result in damage to the central nervous system (Ewer, 1973).

Since smaller cats tend to hunt prey smaller than themselves, throat bites are usually only associated with larger cats. However, caracal kill mountain reedbuck with a throat bite, and Eurasian lynx kill reindeer and roe deer similarly (Haglund, 1966; Grobler, 1981). Sunquist (1981) found that tigers used the throat bite when the prey was more than just over half the tiger's body weight. To some extent the preferred killing bite will be up to the cat's individual preference, and can be altered if the prey is difficult to kill. For example, Fox and Chundawat (1988) describe a snow leopard attacking a goat with a nape bite, but switching to a throat bite as the prey struggled.

In some cases local traditions may develop in the killing of prey. Kalahari lions commonly prey on gemsbok (*Oryx gazella*) and kill them by breaking their lower vertebral column (sacro-lumbar joint) (Eloff, 1973). This behaviour is peculiar to some areas of the Kalahari, and is unknown in prides in the Etosha National Park, which use the normal throat bite.

After the kill has been made, a cat will often check out the surrounding area before beginning to feed, in order to see if other predators and scavengers have been attracted to the commotion. Many cats drag their prey into cover so as not to attract attention for this reason, but small prey are consumed on the spot. Schaller and Vasconcelos (1978) found that jaguars dragged their prey an average of 87 metres into cover before feeding.

PLAY

Another phenomenon which is often claimed to be only associated with domestic cats is playing with prey, either dead or alive. However, many species of wild cat play with their prey (Leyhausen, 1979). Indeed, it is an important part of a kitten's education when its mother brings back live prey to practise killing techniques and to recognise prey species (see Chapter 6). Even adult wild cats play with prey. For example, servals often play with rodent prey in the Serengeti, although it is mostly younger animals that do so (Geertsema, 1985).

Leyhausen (1979) has defined three types of prey play behaviour.

(a) Restrained play consists of normal catching actions which are very much reduced in intensity and modified playfully.

(b) Overflow play is the commonest form of play and there are two variants, 'chase' and 'catch and throw'. Leyhausen (1979) calls this overflow play, because it may be performed by very hungry cats and may be directed towards non-prey objects.

(c) Play of relief consists of high curving leaps over or around the dead prey. It usually occurs after having to deal with large or dangerous prey.

FEEDING

Small cats tend to start feeding at the head end and work down the carcase, but larger prey are started at the neck so that the head may become detached. Big cats and small cats feeding on very large prey start at the belly or the hind end. There are, however, some species differences which are given in Table 4.2. Cats use their whiskers to detect the direction of the fur in order to locate the appropriate end of the body to start feeding (Leyhausen, 1979). This is useful at night where vision may be impaired. Many cats will remove long fur or feathers, if they obstruct feeding. Lions and tigers are very adept at removing the long quills from porcupines, but may be severely injured as a result of tackling this prickly prey. Leyhausen (1979)

Table 4.2 Behaviour of cats with kills (after Sunquist, 1981)

Species	Initial feeding site	Stay with kills	Cover remains of kills	Reference
Tiger	Rump	+	+	1,2
Puma	Rump	+	+	3
Lion	Viscera	+	+	4
Leopard	Viscera, thigh, chest	+	+	4
Jaguar	Forequarters, chest, neck	+	?	5
Cheetah	Thigh	–	+	4
Snow leopard	Chest, lower abdomen, thigh	+	?	6
Canada lynx	Head	+	+	7,8
Wildcat	Head	+	+	9
Caracal	Head, forequarters	+	+	10
Ocelot	Head	+	+	11

References:
(1) Schaller, 1967
(2) Sunquist, 1981
(3) Hornocker, 1970
(4) Schaller, 1972
(5) Schaller and Vasconcelos, 1978
(6) Schaller, 1977
(7) Saunders, 1963a
(8) Parker *et al.*, 1983
(9) Corbett, 1979
(10) Grobler, 1981
(11) Emmons, 1988

Source: Sunquist (1981) (modified)

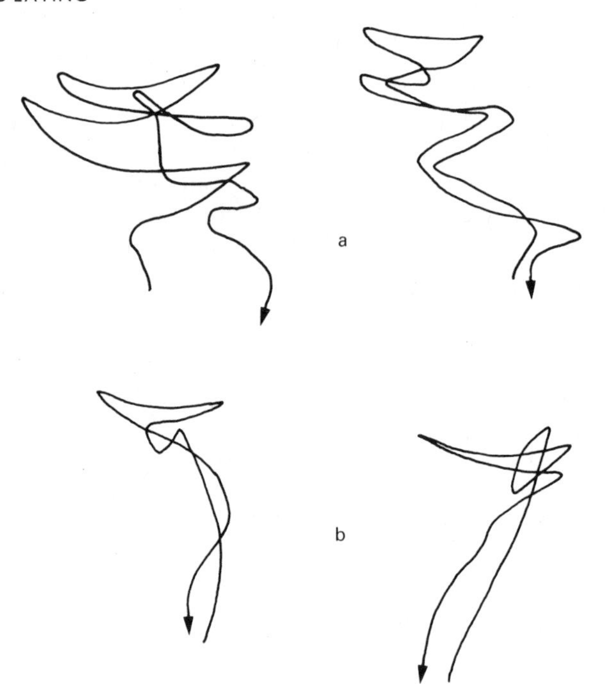

Figure 4.4 The plucking curves of **(a)** a female serval, **(b)** a tiger cat. Old World and New World cats have different patterns of plucking feathers. (Leyhausen, 1979) (A.K.)

describes in some detail the weaving pattern of the head while plucking fur or feathers from freshly killed prey (Figure 4.4). A captive-bred ocelot at the Cat Survival Trust was a very adept plucker when it came to chickens, and this behaviour has also been recorded for ocelots in the wild (Emmons, 1988). Jaguars often prey on river turtles and tortoises, which they feed on by scooping out the flesh from between the shells using their paws to leave a totally vacant chelonian home (Mondolfi and Hoogesteijn, 1986).

There is a difference between how big and small cats feed, although this probably reflects the relative size of prey on which they feed. Small cats crouch and eat their prey, but without using their forepaws to hold it (Leyhausen, 1979). However, big cats tend to hold their prey with their forepaws (Figure 4.5).

Cats can be quite selective when feeding. Small cats usually ignore the viscera (and bury them), larger bones and skin. This leaves curious remains as if the prey has been carefully skinned and all the meat removed from its bones. Wildcats feed on rabbits in this way, as do caracal on sheep (Corbett, 1979; Skinner, 1979). Big cats frequently eat the viscera of their prey or they will bury it. Lions and jaguars force the contents of the intestines out by squeezing them in their mouths before feeding (Schaller, 1972; Rabinowitz and Nottingham, 1986).

Cats can only use the carnassials on one side of their jaws at a time

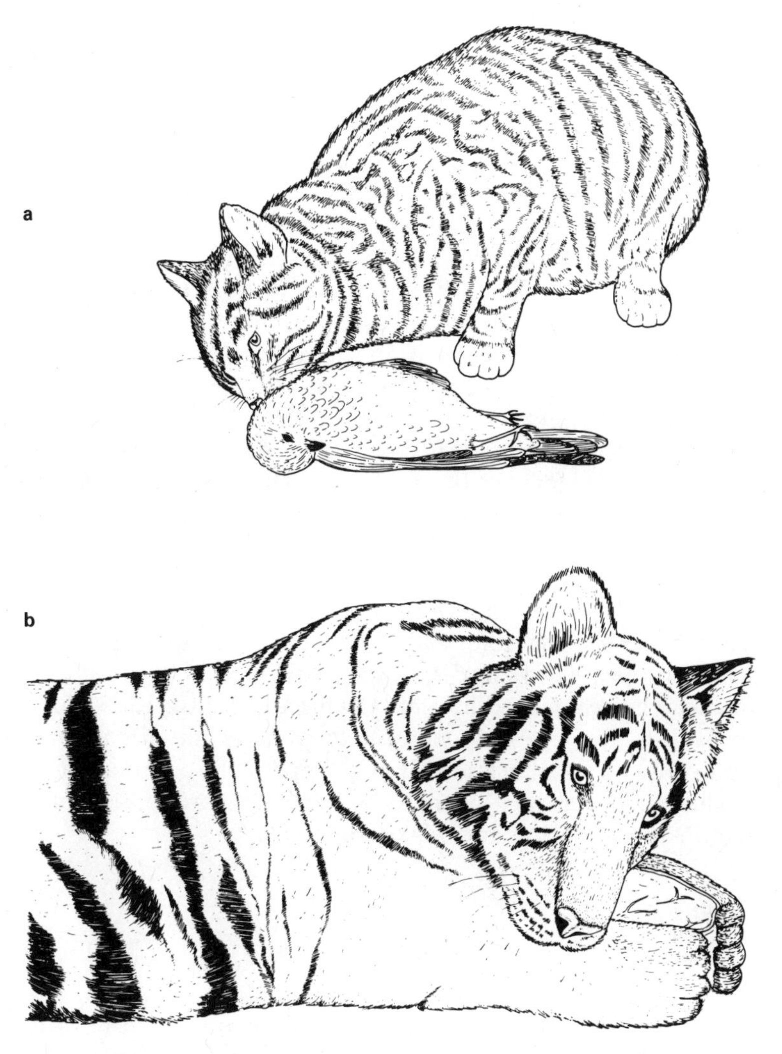

Figure 4.5 The typical feeding positions of (**a**) small cats and (**b**) big cats (after Leyhausen, 1979) (S.A.)

in order to slice up meat. The articulation of the mandible allows it to slide from side to side, bringing the carnassials on one side into alignment for use. Thus, cats characteristically hold their heads on one side when feeding. Big cats may tear at meat with their incisors by pulling back their heads and controlling the food with their forepaws.

CACHING

If too much food is caught for one sitting, it may be saved for later. Many cats are known to cache food (Haglund, 1966; Kruuk and Turner, 1967; Nellis and Keith, 1968; Hornocker, 1970; Berrie,

1973; Seidensticker, 1976; Corbett, 1979; Apps, 1983; Kruuk, 1986). It is obviously of great advantage to be able to kill large prey or multiple prey, leave it covered assuming it is safe from scavengers, and return later for another feed. There are, however, problems with caching. In hot climates, meat may become putrid very soon after the kill (e.g. caracal feeding on mountain reedbuck (Grobler, 1981)), although cats may feed on putrid meat if they are very hungry. In cold climates, the meat may freeze, making it impossible to chew through, e.g. Swedish lynx feeding on deer in the winter (Haglund, 1966). Scavengers may also steal food. One way to prevent this is to remain in the vicinity of the kill (e.g. pumas guard deer carcases until they are entirely consumed (Hornocker, 1970)). If scavengers are common, leopards drag their prey into trees. Strangely enough, this seems to deter vultures as well as terrestrial scavengers. Seidensticker (1976) calculated that leopards dragged about half of their kills into trees in the Royal Chitwan National Park. The Scottish wildcat has also been recorded using trees to cache rabbit prey (Kruuk, 1986).

Cats usually cover their prey with grass, earth, dead leaves or whatever else is available. Caracals cover their reedbuck kills with grass (Grobler, 1981) and Canada lynxes cache the rear ends of hares which are covered in snow (Nellis and Keith, 1968). However, if prey is scarce they do not cache prey at all. Feral cats cache cormorant chicks on Dassen Island (Apps, 1983). Pumas cache deer kills, and bobcats have been recorded hiding rabbits (Hornocker, 1970; Provost, Nelson and Marshall, 1973).

SCAVENGING

Consistent with their adeptness at saving food for later, cats are not at all adverse to scavenging food, if it is available. For feral cats in an urban environment, this skill is the key to survival, where natural prey may be rare or absent (Tabor, 1983; Dards, 1978, 1981). For the Canada lynx, scavenged deer meat (moose, caribou, etc.) may make up to 17 per cent of the diet (Nellis and Keith, 1968). This superabundance of food is available in the autumn and winter owing to deer carcases left after the hunting season, or which die due to disease and poor body condition after the rut.

Scavenging in lions is very common. Not only do they steal food from other predators such as cheetahs, leopards and hyaenas, but they also steal from each other (Schaller, 1972). Male lions in the Ruwenzori National Park in Uganda scavenge between 60 per cent and 75 per cent of their food requirements from lionesses in order to maintain their body weight (Van Orsdol, 1986).

The one exception to the scavenging rule is the cheetah. Its high hunting success makes scavenging unnecessary, even though it may lose up to 12 per cent of its kills to lions in the Serengeti (Schaller,

1972). Scavenging cheetahs have been recorded only exceptionally. In Kruger National Park in South Africa, anthrax wiped out the hoofstock and cheetahs were forced to scavenge from the carcases and succumbed to the disease as well (Wrogemann, 1975).

DRINKING

There is little mention in the literature about the water requirements of wild cats. Although cats do have access to free water, part of their water requirements is probably made up from their prey. Drinking frequency has been recorded in two desert cats—the lions and leopards of the Kalahari. Kalahari lions were seen to eat tsama melon and lick dew off the fur from other pride members to help maintain their water balance (Eloff, 1973). The lions were observed to drink every 3.3 days on average (Eloff, 1973). The Kalahari leopard drinks water at a similar frequency (Bothma and Le Riche, 1986).

SURPLUS KILLING

There are two types of surplus killing. One type should really be called multiple killing, and occurs when a cat is given the opportunity to make additional kills by chance. Lions, for example, may make multiple kills in cooperative hunts and lynxes are recorded making multiple kills of hares (Haglund, 1966; Schaller, 1972). Multiple killing is not necessarily a waste of food, because the cat can cache any excess food until it is hungry. Perhaps it should be called opportunistic multiple killing.

The other type of surplus killing occurs when large numbers of prey are killed and not utilised (Kruuk, 1972). The victims are usually domestic livestock which are penned and unable to escape from a predatory cat. Surplus killing has been recorded for caracal (on sheep), leopards (on sheep), snow leopards (on sheep and goats) and lions (on cattle) (Kruuk, 1972; Schaller, 1972; Stuart, 1986; Fox and Chundawat, 1988).

But why does surplus killing occur? It is partly because the domestic stock cannot escape, and also because the drive to kill is separate from the need for food (i.e. hunger) (Leyhausen, 1979). Cats continue to kill after they are satiated, thus being able to make multiple kills to maximise energy intake over time. This behaviour is seen as wasteful only in an artificial captive situation, where 'prey' animals are kept at very high densities and are unable to escape.

HUNTING SUCCESS

Hunting success is measured in different ways by different authors. Some utilise data from tracks found in the snow or sand (Saunders, 1963a; Haglund, 1966; Parker *et al.*, 1983; Bothma and Le Riche, 1986). Others observe their cats directly (e.g. Schaller, 1972; Geertsema, 1985), but it can often be difficult to tell whether a hunt is attempted or not. Sometimes lions run at ungulates to create confusion or see if they have been spotted by potential prey. They then may make a decision as to whether or not to continue the hunt, based on the reaction of the prey. Some authors calculate hunting success based on the final charge towards a prey item, others consider the beginning of the stalking phase. Despite this, it is useful to compare the hunting success of different cat species and the factors which affect it.

From over 2,000 observations Geertsema (1985) found that servals were successful in 49 per cent of their pounces and this did not vary with season or by day or moonlit night. Servals were, however, less successful in catching harlequin quails (23 per cent), but had greater success in catching insects (54 per cent). A female serval with kittens increased her hunting success to 62 per cent compared with 48 per cent when she had no kittens, so that she could provide sufficient food for lactation and also directly feeding her growing kittens.

Haglund (1966) carried out a detailed investigation of the hunting success of Eurasian lynx in Sweden during the winter by following tracks in the snow. He found that it varied with body condition, topography and the compactness of the snow. Over 70 per cent of the prey items were caught within a 20 metre chase. The success in very long chases of 100–300 metres was greater than for 20–100 metres, because the lynx could see that the body condition of the prey was poor and a long chase would be likely to be successful (Table 4.3) (Haglund, 1966).

In the north of Sweden, hunting success for hares (27 per cent) was about half the success for hunting them in the south (50 per cent). This was put down to the presence of snow, which, if soft,

Table 4.3 The hunting success of the Eurasian lynx over different chase lengths

Prey	Length of attack (m)				
	‹20	‹50	‹100	‹200	200
Reindeer	90%	71%	0%	50%	67%
Roe deer	80%	67%	40%	33%	0%
Hare	64%	11%	19%	33%	100%

Source: Haglund (1966)

thwarted the lynx's attempts. Success in hunting deer was high: 70 per cent for reindeer in the north and 66 per cent for hunting roe deer in the south—here snow thwarted the deer's movements. There was a similar effect on the hunting success of Canada lynx hunting snowshoe hare between winters when snow conditions differed (Table 4.4) (Brand, Keith and Fischer, 1976). Zheltukhin (1986) found that lynx hunting success was 23 per cent in the Upper Volga when hunting mountain hares, whereas in south-eastern Finland the success rate was 43 per cent (Table 4.5) (Pulliainen, 1981).

Table 4.4 The effect of snow conditions on the hunting success of Canada lynx

	Firm snow	Soft snow
Snowshoe hare	24	9
Ruffed grouse	19	8
Squirrel	12	0

Source: Brand *et al.* (1976)

Pumas were also very successful at hunting mule deer and wapiti in the Idaho Primitive Area. The presence of snow, and also the wapiti's habit of wintering in the rough bluff areas which afforded pumas maximum cover, gave a hunting success of 82 per cent (Hornocker, 1970). Lions were also more successful in hunting in areas with more cover, e.g. dense riverine habitats gave a hunting success of 41 per cent compared with 12 per cent in little or no cover (Schaller, 1972; see also van Orsdol, 1984).

Many factors affect hunting success in addition to weather conditions and cover. For example, hunting success for lions is greater on the plains at night than during the day (Schaller, 1972; van Orsdol, 1984), and increases again if it is a moonless night (van Orsdol, 1984). The darkness apparently makes up for any lack of cover on the open plains during daylight hours. Van Orsdol (1984) also noticed that lions would start hunting on moonlit nights, if it became cloudy. Although wind direction does not affect hunting strategy or hunting success in any cat, van Orsdol (1984) noticed also that lions would begin hunting if it got windier. Presumably, this would make it harder for prey animals to hear an approaching predator.

Corbett (1979) looked at the hunting success of domestic and feral cats. He had two measures of success: stalk effort (the number of stalks per hour) and stalk success (the percentage of successful stalks). Overall, combining the mobile and stationary strategies, the hunting success of domestic cats hunting rabbits was 17 per cent. He found that cats spent about the same amount of time using the

M and S strategies, but that more effort was put into the M strategy which was more successful (4.2 kills/hour for M, 2.1 kills/hour for S). Dominant and adult cats were more successful than subordinate and young cats.

Table 4.5 Hunting success in the Felidae

Predator	Prey	Hunting success (%)	Source
Feral/domestic cat	Rabbits	17	Corbett, 1979
Canada lynx	Snowshoe hare	12.5	Nellis & Keith, 1968
		9–36	Brand *et al.*, 1976
		42	Saunders, 1963a
		14	Parker *et al.*, 1983
		16	Nellis *et al.*, 1972
	Red squirrel	0–67	Brand *et al.*, 1976
	Ruffed grouse	19–67	Nellis *et al.*, 1972
Eurasian lynx	Mountain hare	23	Zheltukin, 1986
		27–50	Haglund, 1966
		43	Pulliainen, 1981
	Reindeer	70	Haglund, 1966
	Roe deer	66	Haglund, 1966
	Hazel hen, etc.	9	Pulliainen, 1981
Serval	Rodents	49	Geertsema, 1985
	Quail	23	
	Insects	54	
Cheetah	Thomson's gazelle	70	Schaller, 1972
		49	Bertram, 1978
Puma	Mule deer/wapiti	82	Hornocker, 1970
Leopard (m)	Various	13	Bothma & Le Riche, 1986
(f)		23	Bothma & Le Riche, 1986
Lion	Reedbuck	14	Schaller, 1972
	Topi	13	
	Warthog	47	
	Wildebeest	32	
	Zebra	27	
	Zebra/Wildebeest	22	Elliott *et al.*, 1977
	Thomson's gazelle	27	
Tiger	Various	8	Schaller, 1967

COOPERATIVE HUNTING

Cooperative hunting is usually only associated with the sociable lion, but it is probably important in all cats to some extent, especially when a mother takes her kittens/cubs with her when she hunts, and also with a consorting pair. But hunting together does not necessarily mean cooperative hunting. For example, servals may hunt together by moving down a trail and hunting on each side, but they do not work together to increase their chances of a kill, or to kill large prey (Geertsema, 1985).

Cooperative hunting among smaller cats is most commonly observed in the lynxes. Haglund (1966) recorded a pair of Eurasian lynx hunting mountain hares just like lions hunting wildebeest. One lynx lay in ambush beside a trail while the other drove a hare towards it. They shared the kill and went on their way. Canada lynx are also known to hunt cooperatively. A mother and her kittens spread out in a line and advanced through a fairly open area. Some of the snowshoe hares scared by one lynx were caught by one of the others (Parker *et al.*, 1983). This sort of hunting appears to be an important part of lynx education. In fact Canada lynx are more successful if they hunt together, but prey density probably limits this becoming a regular event (Table 4.6) (Parker *et al.*, 1983). Eurasian lynx mothers hunt in a similar way with their kittens (Haglund, 1966).

Table 4.6 The effect of group size on hunting success and distance between kills of Canada lynx preying on snowshoe hares

Group size	Hunting success (%)	Interkill distance (km)
1	14	7.6
2	17	4.9
3	38	2.9
4	55	0.5

Source: Parker *et al.* (1983)

There is still some dispute as to whether lions hunt together deliberately or accidentally (Kruuk and Turner, 1967; Schaller, 1972; Kruuk, 1986). Schaller (1972) considers that some hunts are examples of accidental cooperation, but that others, which are reminiscent in strategy to those of hunting lynxes, are examples of true cooperation. For example, lions do hunt in such a way that they may either completely surround potential prey or lie in ambush while other lions drive prey towards them (Figure 4.6).

In the Serengeti, lions tend to hunt large prey in larger groups

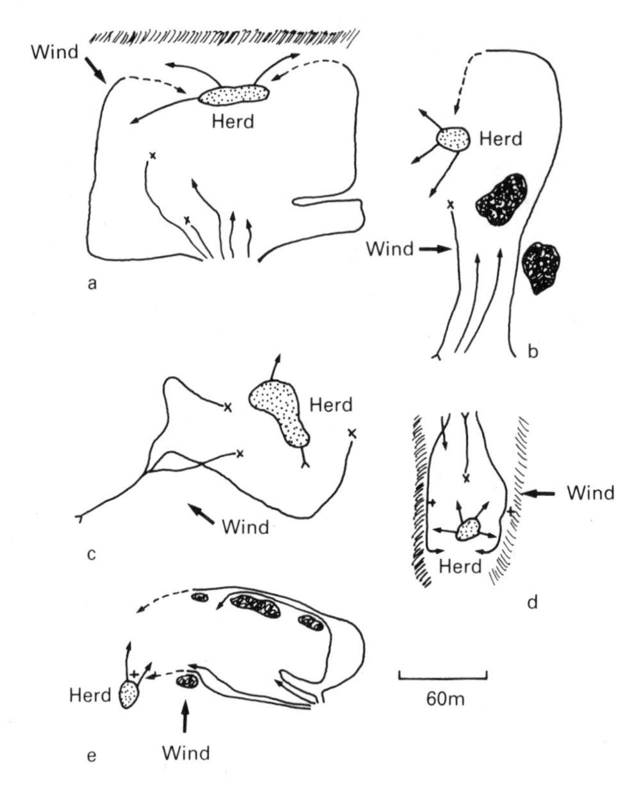

Figure 4.6 Examples of the routes taken by members of lion prides during cooperative hunts (Schaller, 1972). Continuous lines represent walking or stalking, dashed lines are running, crosses are crouches, + are kills, [thicket symbol] are thickets and the bold arrrows are wind direction (A.K.)

than smaller prey (Kruuk and Turner, 1967). Few single lions and no lionesses hunt very large prey (e.g. buffalo) alone, so it may be that cooperative hunting is essential for exploiting large prey animals. However, if a single lion were to hunt large prey exclusively, it could not eat all the meat before it rotted, or was scavenged by other predators, so that it would waste energy and risk injury unnecessarily. Therefore, cooperative hunting and sharing of prey may be a way of optimising the energetic return from the carcasses of large prey for lions that live together.

The data for hunting efficiency of lions hunting in groups seem to be contradictory. Schaller (1972) found that the optimum hunting group size was two; above this number, efficiency was not increased sufficiently and food intake per lion might actually be reduced (Table 4.7). Schaller's (1972) data were modelled by Caraco and Wolf (1975), who showed that it was not worth lions living together just for the sake of food. Elliott *et al.* (1977) found that hunting efficiency was hardly affected by hunting group size, so that it was even harder to see why lions hunt together (Table 4.7; Figure 4.7).

Table 4.7 The effect of group size on the hunting efficiency of lions

Group size	Prey	Hunting method	Time	Hunting success (%)
1	gnu/zebra	stalk	day	20
2				25
1	gazelle	stalk	day	29
2				25
1	gnu/zebra	run	night	11
2				30
1	gazelle	stalk/run		15
2				31
3				33
4/5				31
6+				33
1	gnu/zebra	stalk/run		15
2				35
3				12.5
4/5				37
6+				43

Source: Schaller (1972); Elliott *et al.* (1977)

Although feral cats may form social groups similar to lion prides, they do not hunt together. Perhaps lions associate for other reasons than cooperative hunting. These factors will be discussed more fully in Chapters 5 and 6. Kruuk (1986) has suggested that lions may hunt together because hunting sites may be limited in the open grasslands of East Africa. It is only by living together and scavenging off each other that lions can survive on the abundant ungulate food source. Packer (1986) considers that in the open savannas, it is likely that strange lions and other scavengers will steal a kill from a solitary lion. Therefore, it would be adaptive for lions to scavenge off close relatives rather than letting complete strangers share or steal their food.

Recently a dynamic model has been formulated, which looks at the food reserves of a lion over 30 days (Houston *et al.*, 1988). This model suggests that if only two lions within a group hunt large prey with fairly full stomachs and the remaining lions are hungry, it is an advantage for the lions to feed in a large group in order to use the food most efficiently (Houston *et al.*, 1988). However, smaller feeding groups are efficient at utilising smaller prey such as gazelles (Table 4.8a, b). The model also predicts that hungry lions should feed in smaller groups than well-fed lions. In the Kalahari Desert,

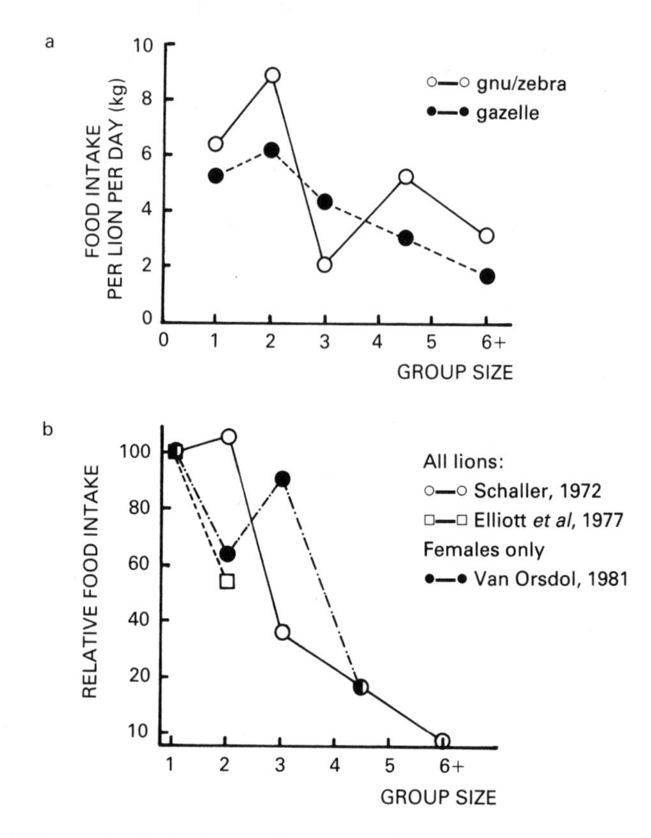

Figure 4.7 (**a**) Rates of daily food intake for individual lions in different group sizes while hunting wildebeest and zebra, and Thomson's gazelle. (**b**) Individual relative food intake in relation to group size compared with solitary hunters (100%) (Packer, 1986) (A.K.)

lions hunt in smaller groups during the dry season when prey is much scarcer. Packer and Ruttan (1988) have also produced models for social hunting based on different strategies in many animals. They found that lions should hunt cooperatively if they hunt large prey, and that lion prides should contain many cheaters, who benefit from a large kill.

ENERGETICS

Hunting energetics are measured in a variety of ways. Usually it is in terms of kills per hour/day/year, or kills per kilometre. From such calculations it is possible to work out how much food a cat needs per year, and hence its impact on prey populations.

For example, servals make 0.8 kills per hour during the day and 0.5 kills per hour at night (Geertsema, 1985). In terms of distance, this is 2.5 kills/km and 1.9 kills/km, respectively. These are average figures which mask a great deal of variation in killing frequency.

Table 4.8 (a) Maximum probability of survival over 30 days and hunting group size as a function of current stomach contents of lions preying on different prey species in the Serengeti

x (kg)	P(1)	n	P(2)	n	P(3)	n
5	0.86	3	0.26	2	0.06	1
10	0.90	4	0.43	2	0.13	1
20	0.97	4	0.64	2	0.26	2
30	0.99	6	0.74	2	0.35	2

Notes
x —stomach contents
P(1) —probability of survival hunting zebra in wet season.
P(2) —probability of survival hunting gazelle in wet season.
P(3) —probability of survival hunting gazelle in dry season.
n —optimal hunting group size.

(b) Maximum probability of survival, hunting group size and pride size for lions preying on zebras during the dry season

Pride size	Probability of survival	Hunting group size
2	0.18	2
4	0.75	2
6	0.95	2
8	0.993	2
10	0.999	2

Source: Houston *et al.* (1988)

Juvenile servals kill more frequently (4.2/km), but hunt smaller prey with a lower energetic return. Each serval in the Serengeti eats annually about 4,000 rodents, 260 snakes and 130 birds.

Similar data were collected for the lynx in Sweden. In the north, lynx make 1.2 kills/24 hours compared to 0.6 kills in southern Sweden and 0.3 kills in eastern Sweden, where there are no deer (Haglund, 1966). Of these kills, in northern Sweden 0.9 kills are of deer and 0.3 are hares and gamebirds compared with 0.4 and 0.2, respectively, in the south. Lynx need about 2 kg of meat per day (including wastage), which is equivalent to about one mountain hare, which just about balances the killing rate. In the Upper Volga lynx eat 90 mountain hare, 50 hazel hens, 20 capercaillie, 25 other birds and 10 red squirrels per year (Zheltukhin, 1986). The Spanish lynx eats about 74 g/kg body weight/day of food. Most of the prey is rabbits, amounting to 261 kills per lynx per year out of a total of 334 kills (Delibes, 1980).

In the Idaho Primitive Area, pumas prey almost exclusively on deer. On average the puma kills a deer every 10–14 days and needs 1.8–2.7 kg meat per day (Hornocker, 1970). Assuming a puma eats 70 per cent of a carcass, it would need 860–1,300 kg of deer per year, which is equivalent to 5–7 wapiti/puma/year or 14–20 mule deer/puma/year. In a year the puma population ate between 74–111 wapiti and 203–305 mule deer, which gives a predator:prey biomass ratio of 1 puma to 353 mule deer equivalents. In other words the pumas consumed just over 10 per cent of the standing deer population every year. Recently, Ackerman, Lindzey and Hemker (1986) have taken this approach one stage further by modelling the energetics of a puma population throughout the year, in order to aid the management of both predator and prey populations (Figure 4.8).

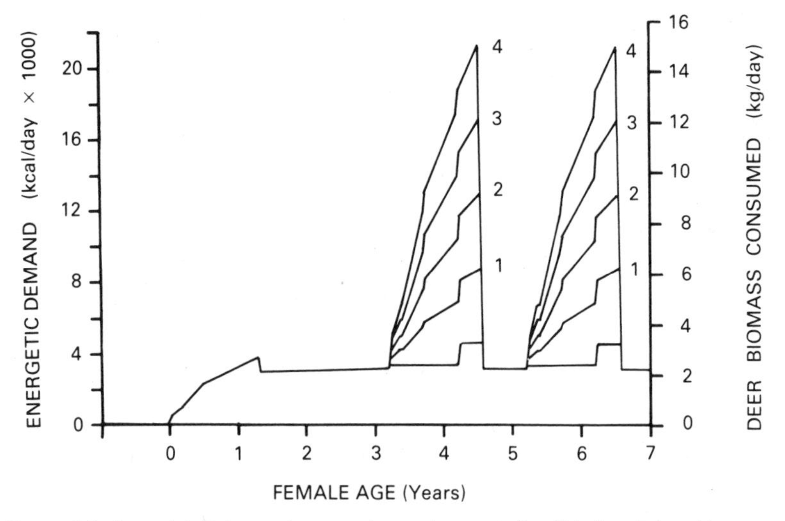

Figure 4.8 A model of the total energetic requirements (kcal/day) and deer biomass eaten (kg/day) over time by a female puma and a female puma with 1–4 cubs (Ackerman *et al.*, 1986) (A.K.)

THE BALANCE OF NATURE—EFFECTS OF PREDATORS ON PREY POPULATIONS

Mammal populations are regulated by a number of factors, of which the most important are availability of food or other resources such as denning sites, disease, migration and emigration, and predators. In the past it was believed that predators had a bad effect on prey populations. By eradicating the predators, the prey populations should rise, leaving more sport for the human hunter. As a result, the Scottish wildcat was very nearly eradicated by the beginning of the twentieth century and the large predators in African game parks were 'controlled' until relatively recently. The effect of predators on prey populations is not always clear. In the absence of predators and serious diseases, a potential prey population may increase until space or food become a limiting factor. The population may stabilise, if reproduction rates match decreases due to mortality and emigration. More usually, the population will go into a cycle of increase and decrease. As the limiting resource is exceeded the population will fall until that resource is no longer limiting. The amplitude and magnitude of the cycle and population numbers may vary considerably, depending on seasonal or stochastic changes in the limiting resource. It has been suggested that predators may weed out old or sick individuals, thereby improving the health of the prey population. Although the hunting techniques of cats mean that they may catch healthy prey as often as they do the infirm (Hornocker, 1970; Schaller, 1972), where the body size of a prey animal is approaching the upper limit for a cat species to prey on, infirm or younger animals are more likely to be caught. For example, Corbett (1979) found that feral cats feed most often on young rabbits and those afflicted with myxomatosis.

The effect of predators on prey populations varies considerably, ranging from extinction where feral cats have been introduced to islands on which endemic species have no anti-predator behaviours, to no effect in the Serengeti where lions make very little impact on migratory ungulate populations (Lever, 1985; Schaller, 1972). In some cases prey species are especially vulnerable to predation by their activities and this may limit populations. In several cases it has been suggested that predators may act to decrease the amplitude of population cycles, by preventing the prey population from exceeding the carrying capacity of the environment. In other cases, the predator population seems to be controlled by the prey population, apparently reducing the predator's role to that of a macroparasite. This variation in the effects of predator on prey will be reviewed below from the many field studies carried out throughout the world.

Pearson (1964; 1966) studied the effect of predators (mostly domestic and feral cats) on prey populations in a 14-hectare park in California. The main prey item was the vole *Microtus californicus*, which in the year of study had reached a population of over 4,000 by mid-summer, when breeding stopped until the next year. The predators (mostly cats) consumed more than 3,800 of these voles, but as the voles became rarer, they shifted their attention to harvest mice (*Reithrodontomys*), gophers and house mice. The increasing rarity of the voles should have reduced the predation pressure on their population, but the presence of alternative food meant that the cats could continue to hunt voles almost to extinction. Voles were the preferred prey probably because of their vulnerability to capture. The voles used persistent runs through the grass so that cats could wait by a run until a vole came by (Pearson, 1964). Pearson (1966) concluded that predators are very important in controlling the populations of voles and lemmings in the north of their range.

Sometimes prey become more vulnerable to predation, which can result in a rapid decrease in their numbers. For example, in Nairobi National Park, lions used to select strongly for wildebeest as prey. With the control of fires within the park, the increase in vegetation provided lions with more cover for more successful hunting of wildebeest (Figure 4.9) (Rudnai, 1974). The resident wildebeest population plummeted. Eventually the lions switched their prey preference away from wildebeest to kongoni, eland and zebra. This resulted in a recovery of the wildebeest population as fewer lions preyed upon them, even though they were still selected. It seems that cats have traditional prey preferences and that we have witnessed an example of cultural evolution among the lions of Nairobi National Park (Rudnai, 1974).

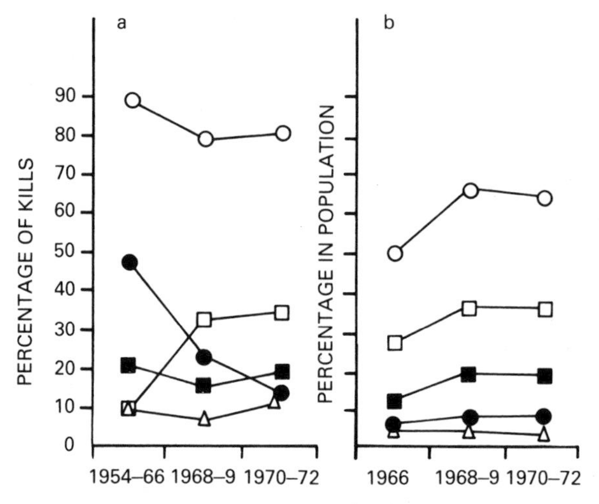

Figure 4.9 Percentages of four important prey species of lions (**a**) in their kills and (**b**) in the ungulate populations of Nairobi National Park. Wildebeest (●—●), kongoni (□—□), zebra (■—■), warthog (△—△) and all ungulates (○—○) (Rudnai, 1974) (A.K.)

Schaller and Vasconcelos (1978) documented the decline in a small, isolated population of capybara preyed upon by jaguars. The combination of disease and predation by jaguars resulted in a 20 to 30 per cent decline in the population in only two months.

Liberg (1984b) found that during a severe winter in southern Sweden, rabbits became weak and sick and were easy prey for cats and other predators (Figure 4.10). Normally domestic cats got up to 85 per cent of their food from their owners, but in this winter 93 per cent of their diet was made up of rabbit. The predators ate 23 per cent of the rabbit population, which led to a low population density and fewer rabbits to prey on the following year. Consequently, the predators switched their diets to include more small rodents. Normally the cats would eat up to 20 per cent of the vole and mouse population and 4 per cent of the rabbit population each year (Liberg, 1984b).

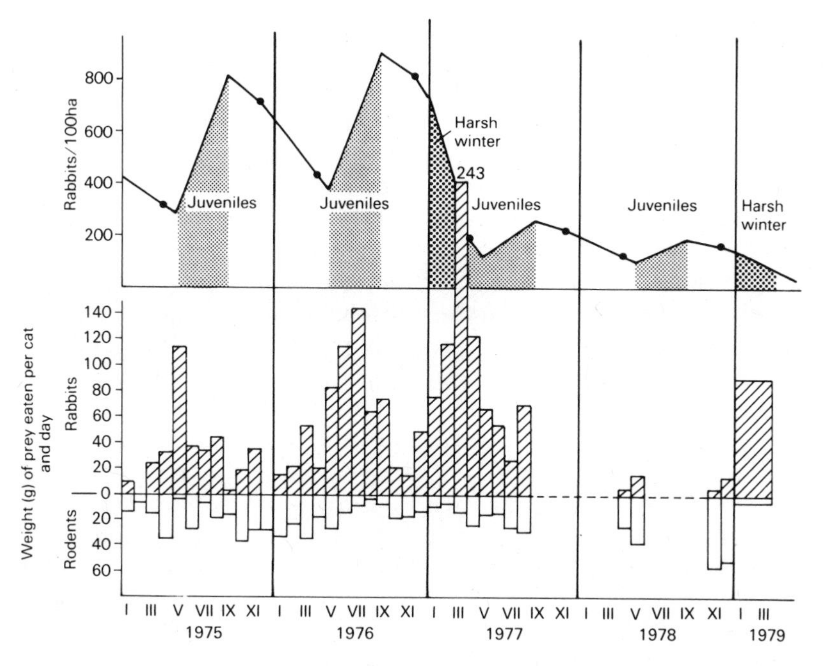

Figure 4.10 Average daily intake (g) per house cat of the two commonest prey, small rodents and rabbits, from 1975 to 1979. Also shown is the dynamics of the rabbit population (Liberg, 1984b) (S.A.)

The most serious effect that predators can have on prey populations is when they are introduced to oceanic islands, where the native fauna has evolved in the absence of predators and consequently has not developed or retained anti-predator behaviours. This can easily lead to extinction. Feral domestic cats have travelled the world with humans and have had a devastating effect on island faunas. One of the most bizarre examples was the discovery and extinction of the Stephen Island wren (*Xenicus lyalli*) by a single domestic cat. This tiny island in the Cook Strait of New Zealand was home to an almost flightless native species of wren. In 1894 a lighthouse keeper arrived on the island with his cat, which despatched the entire population. Subsequently sixteen specimens found their way into the museums of the world. After only a few months the Stephen Island wren was gone forever.

Another example of the devastation of island avifauna includes the eradication of the wedge-tailed shearwater (*Puffinus pacificus*) from Jarvis Island (Kirkpatrick and Rauzon, 1986). Jehl and Parkes (1983) also document the near extinction of the endemic dove

(*Zenaida graysoni*) and the endemic mockingbird (*Mimodes graysoni*) from Soccorro Island, Mexico between 1971 and 1978. The near extinction seems to have been caused by feral cats and has led to the replacement of the endemic forms by similar mainland species, the mourning dove (*Zenaida macroura*) and the northern mockingbird (*Mimus polyglottus*). Van Aarde (1980) found that feral cats on Marion Island were responsible for wiping out the diving petrel (*Pelecanoides urinatrix*). Iverson (1978) records the extinction of an iguana (*Cyclura carinata*) caused by feral cats and dogs. Between 1973 and 1978 the adult population on Pine Cay in the Caicos Islands fell from 5,500 iguanas to zero. Many other examples of the effect of feral domestic cats on island faunas are discussed by Lever (1985).

In contrast, several studies of the effects of cat predation on prey populations have concluded that there is little or no effect. In the Idaho Primitive Area, the puma population remained stable, but the mule deer and wapiti populations increased rapidly (Hornocker, 1970). It was concluded that although pumas do not regulate their prey populations, they probably dampened down any fluctuations in abundance during prey population cycles (Hornocker, 1970).

In the Kanha National Park in India, sambar and gaur populations were apparently not affected by tiger predation, but swamp deer declined during the study period (Schaller, 1967). In the Serengeti, lions had little impact on migratory wildebeest populations, but they may have checked the resident zebra population (Schaller, 1972). It is, of course, difficult to see how sedentary lions could have a dramatic effect on migratory species, whose populations are probably controlled by food availability and disease (Bertram, 1979). Rudnai (1974, 1979) also concluded that lions had little impact on most populations of prey animals in Nairobi National Park. A stable lion population of about 30 individuals saw a doubling of hoofstock between 1968 and 1972. However, it must be remembered that in cases where predators do not seem to have an effect on prey populations, other factors may be preventing this from happening. For example, factors may prevent the predator population building up to a level of effectiveness. Hunting by humans of pumas, lynxes and lions may limit their population increase in response to increasing prey populations. Also, it is important to look at all predators in a community to see the cumulative effect on prey populations. This has recently been achieved in southern Sweden, where the effect of various predators on vole, mice and rabbit populations was studied in detail (Erlinge *et al.*, 1984). The results were modelled and it was concluded that predators do affect the normal cyclic changes in abundance of prey animals, if the generalist predators (buzzard, fox and cat) maintained a high continuous predation rate and alternative prey were available in excess.

Swart *et al.* (1986) have modelled the effect of different predators, diseases and other causes of mortality on a population of rock hyrax in the Mountain Zebra National Park. This model predicts that the major factor limiting population increase is the level of juvenile mortality of hyrax, and that the numbers of hyrax may in fact be limiting the caracal population!

There are examples where predators have a significant, but not detrimental effect on prey populations. In the Bedfordshire village of Felmersham, cats preyed on at least 30 per cent of the house sparrow population present at the beginning of the breeding season (Churcher and Lawton, 1987). Although this seems to be heavy predation pressure on one species, the cats were preying mostly on small mammals.

In the Orongoro Valley of New Zealand, it was calculated that feral domestic cats had a significant effect on rat (*Rattus rattus*) and rabbit populations (Fitzgerald, 1978). The cats ate 1–2 times the standing crop of rats and 2–4.5 times the standing crop of rabbits each year. Although this appears to be a small effect given the reproductive potential of rats and rabbits, the climatic conditions in the Orongoro Valley seemingly reduced the reproductive success of the prey populations so that predation had a significant effect on them.

On Marion Island, feral cats have a significant predation effect on white-headed petrel and blue petrel populations, but hardly any effect on the huge prion population (van Aarde, 1980).

LYNX–HARE CYCLES

Records of fur returns from the Canada lynx over more than 200 years show that there is a constant 9.6 year population cycle of predator and prey, with the lynxes just lagging behind the hares (Figure 4.11) (Elton and Nicholson, 1942). The amplitude of these population changes is, however, erratic and in part reflects fluctuating demands for fur by humans.

Because of the economic importance of the lynx in the fur trade, much time and energy has gone into finding out what controls this cycle and how the population can be harvested to maximise fur production. Brand and Keith (1979) have reviewed the work done in Alberta, Canada, in elucidating how lynx respond to snowshoe hare population cycles.

The Canada lynx feeds almost exclusively on snowshoe hare, so that it is not surprising that its population should be affected by hare population changes (Saunders, 1963a; Nellis and Keith, 1968; Brand *et al.*, 1976). But how is the lynx population affected? It seems that the adult population is hardly affected at all, except that they have emptier stomachs when there are fewer hares around.

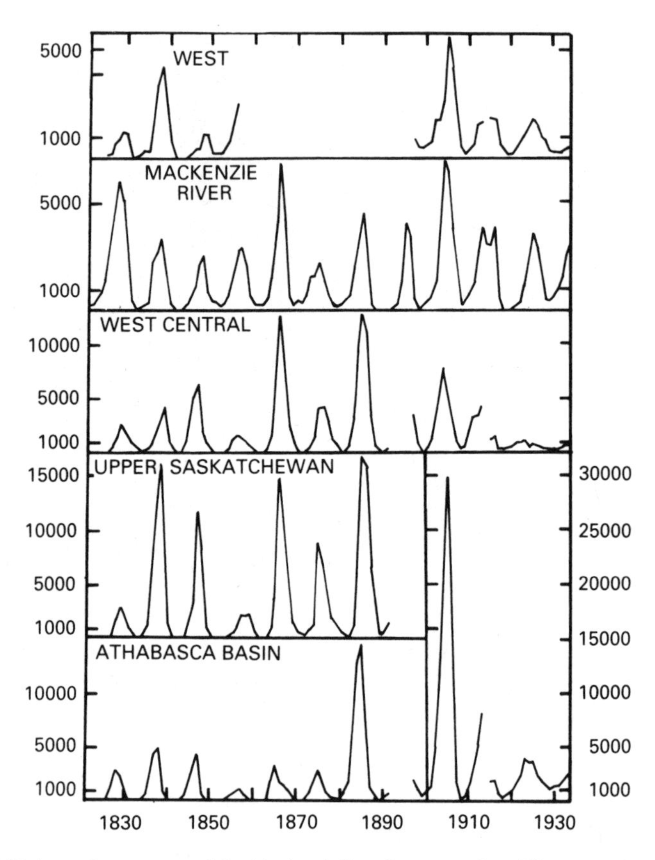

Figure 4.11 Lynx fur returns of the Hudson's Bay Company for different regions of Canada from 1820 to 1935 (Elton and Nicholson, 1942) (A.K.)

However, there is a profound effect on the breeding performance of the lynxes (Figure 4.12). Fewer young are conceived and fewer young survive when hare numbers are low. Also, young females tend not to breed. When hares are abundant, litters are larger, the young are more likely to survive to adulthood and young females breed. Studies of exploited lynx populations all recommend that trapping should be suspended for three years at the time of a hare low (Brand and Keith, 1979; Bailey *et al.*, 1986). This strategy would maximise fur production by allowing more animals to survive until the hare population had recovered. In some areas, trapping is so intensive that the lynx-hare population cycles are no longer linked and the lynx population remains low (Bergerud, 1983).

PREY SWITCHING

Prey populations often undergo cyclic changes in abundance, mostly influenced by food availability and factors other than preda-

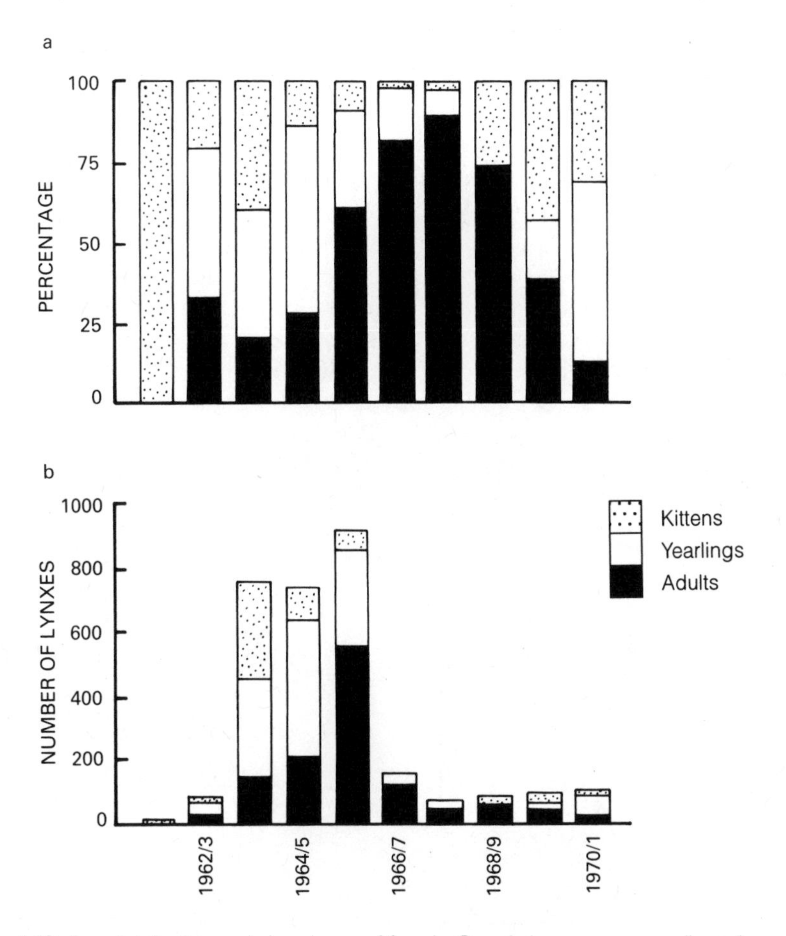

Figure 4.12 Age distribution and abundance of female Canada lynx carcasses collected from five areas of Alaska during one snowshoe hare cycle (O'Connor, 1986): (**a**) percentages, (**b**) numbers (A.K.)

tion. For example, Beasom and Moore (1977) looked at how bobcats in Florida responded to a fall in the populations of their main prey, the cotton rat (*Sigmodon hispidus*) and the cottontail rabbit. In a year of cotton rat and cottontail abundance, the bobcats ate only seven prey species. However, in the following year, when the rat and rabbit populations crashed, the bobcats switched their prey preference from seven to 21 different species.

Liberg (1984b) observed the effect of a severe winter on a rabbit population in southern Sweden. The cat population ate 23 per cent of the weakened rabbit population during the winter so that, in the following year, the cats switched their prey preference from now-scarce rabbits to more abundant voles and mice. We have also already mentioned the prey switching observed in the lion population of Nairobi National Park during the early 1970s (Rudnai, 1974).

We have had a unique opportunity to observe a prey switching cycle evolve. In Newfoundland, the Canada lynx was very rare until 1860, when the snowshoe hare was introduced to provide food for humans (Bergerud, 1983). Until that time, the very small lynx population preyed on the endemic caribou (calves only) and arctic hare (Figure 4.13). With the introduction of the snowshoe hare, there was a rapid increase in the numbers of hares until they ate themselves out of food and the population went into decline, to start a typical snowshoe hare ten-year population cycle. The lynx population had increased in response to the abundance of food, but at the time of the decline they switched their preference to caribou calves and arctic hares, both of which declined due to this increased

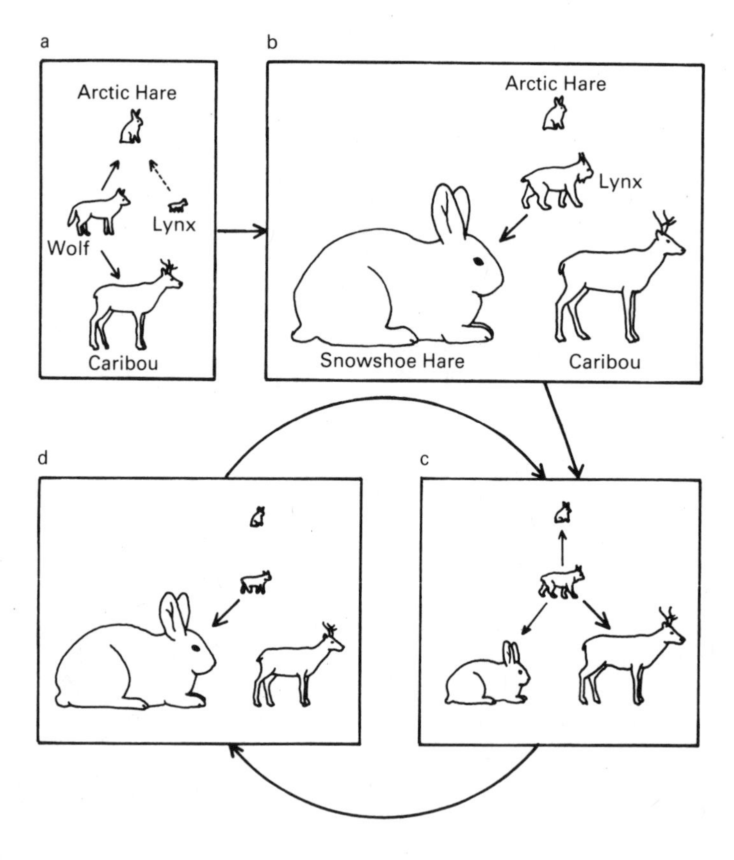

Figure 4.13 The establishment of a prey-switching cycle. (**a**) Until 1864 timber wolves were the main predator of caribou in Newfoundland. (**b**) In 1864 snowshoe hares were introduced and increased rapidly, peaking in 1900. The rare lynx population also multiplied, but the wolf became extinct in 1911. (**c**) In 1915 the snowshoe hare population crashed and the lynx began to feed on caribou calves and arctic hares, resulting in a reduction of their populations. (**d**) When the snowshoe hare population recovered, the lynx switched back. Ever since then there has been a typical ten-year prey switching cycle (Bergerud, 1983) (A.K.)

predation pressure. The caribou population dropped from 40,000 to 200 between 1900 and 1925, and the arctic hare became much rarer.

The caribou population declined due to lynx predation on the calves, which led to a 70 per cent mortality in the first year, coupled with humans hunting the adult population. Arctic hares were not outcompeted by snowshoe hares, but were less well adapted to running on snow and so were easily caught by lynxes with their furry 'snowshoes'. The lynx switched back to snowshoe hares as their population recovered, which allowed the caribou and arctic hare to recover slightly. The large lynx population now supported a new fur trade, so that as the lynx population has become increasingly exploited, the amplitude and persistence of the ten-yearly population cycle has declined in recent decades.

INTERSPECIFIC COMPETITION

In the rainforests of South America, up to six species of wild cat coexist without any apparent competition. These include, in order of increasing body size, the tiger cat, jaguarundi, margay, ocelot, puma and jaguar. But how do all these cat species coexist without direct competition, when they are all restricted to the same range of prey animals? Theoretically, there ought to be an ecological separation between these species so that they do not directly compete for the same food resources. If there is competition between sympatric species, the superior competitor will supersede the inferior in a particular area. This is clearly not the case in a South American rainforest.

Kiltie (1984) looked at functional aspects of skull morphology and body size of these sympatric cats, to see if there was either a constant difference in maximum bite force or maximum gape between species, or if the minimum ratio between species of nearest body size was significantly different from chance, which might lead to a difference in the way that these cats exploit prey populations. Kiltie (1984) was only able to find a significant difference in maximum gape, and then only between the four largest species (Figure 4.14). Margays and jaguarundis appear to be ecologically identical. However, they probably avoid competition because the margay is arboreal and nocturnal, whereas the jaguarundi is mainly terrestrial and diurnal. Therefore, there does seem to be a definite ecological separation between rainforest felids based on maximum gape size. But what does this mean to a living cat? Presumably a larger gape size allows a cat to prey on animals with a larger neck size. In other words, it sets an upper limit to the most profitable prey size a cat can catch. Cats will tend to prey on the largest possible prey animal in order to get the maximum energetic return

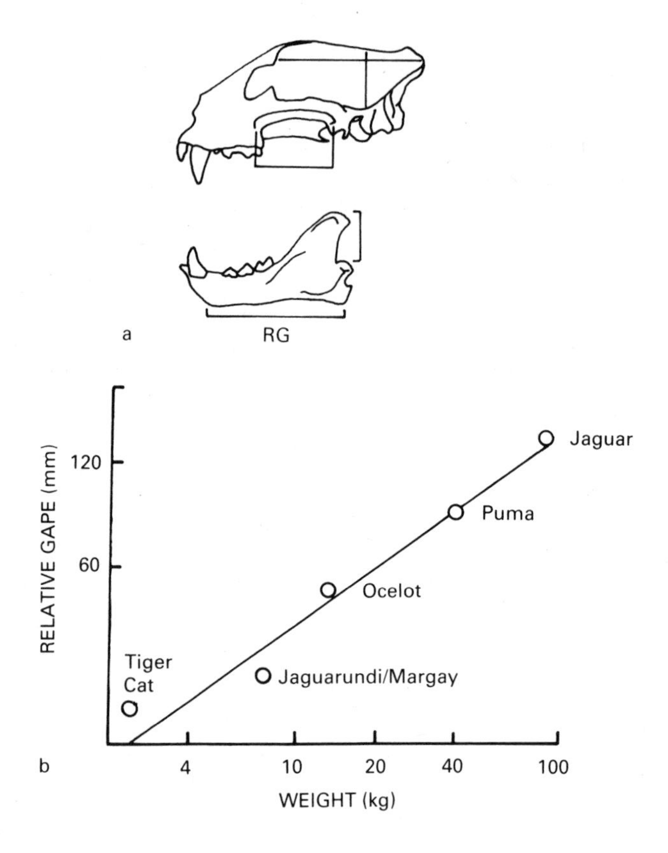

Figure 4.14 (**a**) Kiltie (1985) measured several different features of cat skulls to find out if there was any apparent morphological feature which allowed the six species of cats in South and Central American rain forests to survive without competing with each other. Only relative gape (RG) was found to be significantly different. (**b**) The relationship between the mean relative gape of South and Central American rainforest cats and their mean body weights. The margay and jaguarundi appear to be identical, but may avoid competition by hunting at different times of the day and night and in different parts of the same habitat (A.K.)

from the kill. Therefore, even though different-sized cats may prey on the same range of prey animals, each has a different preference for prey of the appropriate size.

Recently, a field study in Peru has gone some way to confirm Kiltie's (1984) theoretical study. Emmons (1987) looked at prey preferences of ocelots, pumas and jaguars. Her results show that most ocelot prey (92 per cent) is below 1 kg in weight, but most jaguar prey (85 per cent) is more than 10 kg in weight. The few data for the puma show that most of its prey occur in the 1–10 kg size range (Emmons, 1987). Eisenberg (1990) has developed this further by looking at the smaller species, but the data are rather limited so that his conclusions should be treated with caution. However, the data do show that the smallest species, the tiger cat, feeds on more small prey items than the margay, which in turn feeds on more small prey than the jaguarundi, etc. (Table 4.9).

It is usually claimed that the tiger and the leopard do not coexist in the wild. For example, leopards are found in Sri Lanka in the absence of tigers, whereas in the Kanha National Park in India, tigers are common and leopards are very rare and usually transitory.

Table 4.9 Percentage composition of stomach contents of four small South American cats. Prey are arranged in roughly ascending order of size

	Ocelot	Jaguarundi	Margay	Tiger cat
Small rodents	20	25	66	85
Birds	10	20	20	15
Reptiles and amphibians	20	35	6	–
Medium-sized mammals	20	20	8	–
Caviomorph rodents	30	–	–	–

Source: Eisenberg (1990)

However, a study of the leopard and tiger in the Royal Chitwan National Park shows that where they do exist together, they use different habitats and prey on different animals (Seidensticker, 1976). The presence of an abundant population of medium-sized animals in the Royal Chitwan National Park and the different habitat usage allows tigers and leopards to coexist (Table 4.10) (Seidensticker, 1976). In the Kanha National Park, there are few medium-sized prey available to the smaller leopard.

In the Big Bend National Park in Texas, three predators coexist, the puma, the bobcat and the coyote (Leopold and Krausman, 1986). But here again there are differences in prey preferences despite the fact that all three feed on the same range of food. Pumas feed mainly on white-tailed deer and collared peccaries, but bobcats feed on rabbits and rodents. Coyotes have a very similar diet to bobcats, but also feed on fruit and seeds (Leopold and Krausman, 1986). In fact, there may be competition between bobcats and coyotes in some areas during the winter. Major *et al.* (1986) noticed that bobcats in Maine often starved to death in the winter, when bobcats and coyotes both exploit snowshoe hare as their major winter food.

On a larger scale, Bertram (1979) has examined why several large predators, all of which seem to utilise the same food animals, can coexist in the Serengeti National Park. Lions, leopards, cheetahs, hunting dogs and spotted hyaenas each hunt in a different way, in different sections of the habitat and at different times of the day, so that they each utilise different portions of the same populations of gazelle, wildebeest and zebra etc. (Table 4.11) (Bertram, 1979).

Table 4.10 Kills made by tigers and leopards near Sauraha, Royal Chitwan National Park, Nepal

	Size classes of kills (kg)					Kill weight/predator body weight				
	‹25	25–50	50–100	100–200	200–400	‹0.25	0.25–0.50	0.50–1.0	1.0–2.0	2.0–4.0
Tiger: Wild species[a]	–	3	5	1	3	3	5	–	4	–
Domestic livestock[b]	–	–	1	–	13	–	1	–	2	11
Leopard: Wild species[c]	1	9	2	–	–	1	2	5	4	–
Domestic livestock[d]	–	–	2	–	–	–	–	2	–	–

Notes

a. Species killed (No.): *Rhinoceros unicornis* (1), *Cervus unicolor* (4), *Sus scrofa* (2), *Cervus axis* (2), *Cervus porcinus* (3).
b. Species killed (No.): Domestic cattle (2), domestic water buffalo (12).
c. Species killed (No.): *Sus scrofa* (+), *Cervus unicolor* (3), *Cervus axis* (4), *Cervus porcinus* (4), *Muntiacus muntjac* (1).
d. Species killed (No.): Domestic cattle (2).

Source: Seidensticker (1976)

Table 4.11 Predation by the main predators in the Serengeti

Species	Cheetah	Leopard	Lion	Hyaena	Wild dog
Adult body wt. (kg)	40–60	35–60	100–200	45–60	17–20
Population	220–500	800–1200	2000–2400	300–4500	150–300
Habitat	plains, woodland	woodland	woodland, plains	plains	plains, woodland
Activity	day	night	night	night, dawn	day
Hunting group	1	1	1–5	1–20	2–19
Method	stalk, long chase	stalk short spring	spread out, stalk, short spring	long distance pursuit	long distance pursuit
Initial chase distance (m)	10–70	5–20	10–50	20–100	50–200
Speed (km/h)	95	60	50–60	65	70
Distance of pursuit (m)	350	50	200	0.2–3 km	0.5–2.5 km
Hunting success %	37–70	5	15–30	35	50–70
Prey	Thommie, hare, Grant's gazelle, impala	Impala, Thommie, dik-dik, reedbuck, many others	Zebra, gnu, buffalo, Thommie, warthog, others	Gnu, Thommie zebra	Thommie, gnu, zebra others
Health of prey	Good	Good	Good	Poor and good	Poor and good
Age and sex	small fawns	all ages; young topi	all ages more young	male and young gnu	male
% Kills lost to scavengers	10–12	5–10	~0	5–20	50
% Food scavenged	0	5–10	10–15	33	3
Important competitors	hyaenas, possibly lions	possibly lions	none	possibly lions	hyaenas

Source: Bertram (1979) (modified)

5

WHAT DO CATS EAT?

The diets of wild cats are usually determined in one of two ways. Either scats are analysed for the presence of different prey species, or the contents of stomachs are sifted through. Results are expressed either as the frequency of occurrence of a particular species in the scats or stomachs, or as the percentage volume or weight of the total amount of food in the stomach. The need for the two approaches is clear. Caracals, for example, kill rock hyraxes most frequently in the Mountain Zebra National Park, but the larger mountain reedbuck is the most important prey by weight because a caracal may feed more than once from the kill and thereby consume more meat at each sitting (Grobler, 1981). Both methods have their advantages and disadvantages. If scats alone are analysed, this will tend to underestimate the incidence of highly digestible morsels of food, but it does not harm the cats producing them. If stomachs alone are analysed, they may be empty, so providing no data from the sacrifice of a valuable life. The latter approach may not be acceptable in the study of a highly endangered species.

In exceptional cases it is possible to observe directly what cats have preyed on. However, this is usually restricted to the larger cats in open habitats, e.g. lion, cheetah, leopard and serval (Schaller, 1972; Geertsema, 1985). This may not provide an accurate analysis. Geertsema (1985) underestimated the proportion of frogs in the diet of servals because it was not possible to follow them into the wetter areas of their range.

AN OVERVIEW OF FELID DIETS

The maximum prey size that a cat species can kill is related to its body size. In other words, the bigger the cat, the bigger its maximum prey size. Therefore, ocelots (approx. 10 kg) prey on mammals up to the size of a paca (6–10 kg) and the Scottish wildcat (approx. 4–5 kg) can tackle a rabbit (1.2–2 kg), but the caracal (approx. 12 kg) preys on the mountain reedbuck (25–30 kg) and the lion (150–250 kg) can manage a Cape buffalo (500 kg) (Schaller, 1972; Corbett, 1979; Grobler, 1981; Emmons, 1988). It is advantageous for cats to prey on the largest possible prey, in order to get the maximum energetic return, but the potential costs of tackling large prey are possible failure, wastage if too much food is caught and injuries caused in any struggle. Adult servals prefer to catch the large vlei rat (*Otomys*) which provides a much better energetic return than the smaller prey that young adults usually kill. In some cases a kill can be stored, if excess meat is left after one sitting. Pumas will stay near a deer kill for several days until it is all consumed (Hornocker, 1970). However, the cheetah never attempts to cache its food, because it would never be able to defend its kill from other predators in the Serengeti (Schaller, 1972).

Sometimes the population of the optimum prey size declines, so that cats are forced to switch their diet. This occurred in the Big Bend National Park in Texas, where pumas and male bobcats usually feed on deer (Leopold and Krausman, 1986). In 1980–81 the deer population crashed and both cats were forced to switch their preferences to lagomorphs and peccaries, the next biggest prey. Other examples of prey switching are given in Chapter 4.

Many cats are opportunistic in their prey preferences. In particular, tropical forest cats seem to take what they can find up to their maximum prey size. Ocelots in Peru and jaguars in Belize have both been recorded as taking prey in the same proportion as occurs in the habitat (Rabinowitz and Nottingham, 1986; Emmons, 1987). Leopards in Le Parc National de Tai in the Ivory Coast feed on over 30 different mammal species, indicating how unfussy they are about what they eat (Hoppe-Dominik, 1984).

Other cats are, however, much more specific in their choice of food. The Canada lynx preys almost exclusively on the snowshoe hare on Cape Breton Island off Newfoundland, and the Scottish wildcat and Spanish lynx eat mostly rabbits (Corbett, 1979; Delibes, 1980; Parker *et al.*, 1983). Other cats are probably specialised in their dietary requirements, including the fishing cat and flat-headed cat, which both show adaptations to piscivory. However, to date no detailed study of their diets has been undertaken.

Often there are substantial differences in diet between the sexes, and between adults and young, which means that different-sized prey are consumed by males and females to avoid intraspecific competition. Adult male bobcats can tackle white-tailed deer, but females and young are restricted to rabbits and rodents (Litvaitis *et al.*, 1986). Lions can tackle prey up to the size of the Cape buffalo on their own, but lionesses are unable to exploit this huge prey alone (Schaller, 1972). Feral cats on Heisker, in the Monach Islands, feed birds to their kittens, while continuing to prey on rabbits themselves (Corbett, 1979).

Diets can change with season depending on the availability of prey. On oceanic islands feral cats are, in particular, dependent on the nesting seasons of seabirds to provide their food. On Macquarie Island, feral cats feed on prions during the summer, but have to switch to white-headed petrels in the winter (Jones, 1977). In the Serengeti, resident lion prides suffer badly from hunger, having to turn their attention to smaller prey when migratory species leave their area (Schaller, 1972). Sometimes availability is not subject to migration, but to the changing vulnerability of a potential prey. Therefore, in Scotland wildcats show two peaks of predation on rabbits: the spring peak corresponds to the presence of young rabbits naive of predators, while the autumn peak is due to adult rabbits afflicted with myxomatosis (Corbett, 1979). Winter snow makes deer more vulnerable to predation by pumas, Eurasian lynx and bobcats (Haglund, 1966; Hornocker, 1970; Litvaitis *et al.*, 1986). However, winter snow may also reduce predation by Canada lynxes on snowshoe hares for the same reason. In contrast, arctic hares are much more vulnerable than snowshoe hares to predation by lynxes because they are less well adapted to walking on snow. Voles may use the same runs through grasslands, making them very vulnerable to predation by waiting cats and leading to almost total eradication of the population in some areas (Pearson, 1964, 1966).

Humans can also affect prey availability indirectly. Canada lynxes and bobcats may both benefit from deer carcasses left by human hunters in the autumn (Saunders, 1963a; Litvaitis *et al.*, 1986). Lynxes may also benefit from being able to kill deer that are exhausted or injured in the autumn rut. Adult pumas in the Idaho Primitive Area seem to be unable to kill healthy adult male wapiti, and so their diet comprises mostly old and young animals (Hornocker, 1970). Female wapiti are less likely to be predated upon because they are usually in herds, and occupy areas where the pumas have little cover and are thus less successful at hunting. Unlike some other predators, cats do not usually select unhealthy animals preferentially. Lions and cheetahs both kill their fair share of healthy prey, and pumas in Idaho kill the same proportion of unhealthy and healthy deer as occurs in the population. This is because cat hunting methods do not rely on testing the stamina of the prey in pursuit, when any weaknesses are most likely to show up.

Prey availability can affect body condition dramatically. Whereas adult male bobcats can hunt deer all winter and remain in good condition, females cannot maintain their body fat reserves in the face of depletion during the summer due to lactation, and the lack of rodents and lagomorphs in the winter (Litvaitis *et al.*, 1986).

Snowshoe hares are also scarcer in the winter, so that Canada lynxes may deplete their fat reserves greatly (Parker *et al.*, 1983). However, the larger Eurasian lynx easily hunts roe deer and reindeer in the Swedish winter and stays in good body condition at this usually harsh time of year (Haglund, 1966).

Some species of wild cat have very wide geographical distributions which encompass a diversity of different habitats and potential prey. These include the wildcat, leopard, caracal, bobcat, Eurasian lynx and lion. For example, in Scotland wildcats feed mainly on rabbits, in Europe their favourite food is rodents, in Botswana they prefer gerbils and in Turkmenia it is the fat sand rat (*Rhombomys*) that is most frequently killed (Sapozhenkov, 1961; Corbett, 1979; Schauenberg, 1981; Smithers, 1983). In Turkmenia the caracal preys mostly on the tolai hare, but in South Africa its most frequent prey is the rock hyrax (Sapozhenkov, 1962; Grobler, 1981).

Feral domestic cats have had a dramatic influence on island faunas with endemic species that have poor or undeveloped anti-predator behaviours. In Australia and New Zealand, feral cats probably took their toll of the native fauna when they were introduced initially. Ground-nesting birds in Australia and New Zealand were probably the most vulnerable to their depredations. Now that the native fauna is more or less confined to National Parks, it is interesting to see that outside these areas the cats feed mainly on other introduced species such as rabbits and rats, but inside the parks native species are much more commonly taken as prey (e.g. Coman and Brunner, 1972; Jones and Coman, 1981).

The destruction of natural prey and its replacement with domestic livestock has brought many of the big cats into conflict with humans. In extreme cases big cats have even chosen humans as their preferred prey (see Chapter 8). An example of this conflict occurs in Patagonia, where the puma feeds mainly on introduced European hares and domestic livestock rather than native species (Yanez *et al.*, 1986). Lions, tigers, jaguars, leopards, snow leopards and many other cat species have all subsequently suffered because of this conflict. However, the cats have very little choice, but to continue to exploit whatever prey is available until they are driven to extinction.

This brief discussion has highlighted some of the more interesting aspects of cat diets. Most of the studies of the diets of wild cats have been concentrated on the big cats and those of economic importance in the northern hemisphere. Until recently, very few studies of the diets of tropical forest cats had been carried out. Even today, many tropical cats' food preferences are only sketchily known or altogether unknown. The remainder of this chapter is intended as a reference section for easy access to information about the diets of the different species of cats, where known. It will take into account seasonal and intersexual variations in diet, and any positive selection for particular prey types. Not all cat species are discussed

—it is to be hoped that these gaps in our most basic knowledge of the world's wild cats will be filled in the not too distant future.

OCELOT

Of all the small South American cats, only the ocelot's diet has been studied in any detail. Emmons (1987, 1988) analysed 177 scats of the ocelot in Manu National Park in Peru. The most important prey items were spiny rats (*Proechimys*), which made up 32 per cent by frequency of the diet. Cricetine rodents (e.g. *Oryzomys*) and birds were also important prey items (Table 5.1).

Table 5.1 Percentage of total numbers of individual prey items found in ocelot scats in the wet and dry seasons in Manu National Park

Prey	Percentage total numbers	
	Wet	Dry
Proechimys	28	33
Cricetine rodents	26	27
Opossums	9	5
Large mammals (›1 kg)	6	7
Arboreal mammals and bats	9	5
Birds	13	10
Snakes	6	4
Lizards	0	6
Crocodilians	0	1.5
Fish	0	3

Source: Emmons (1987, 1988)

The ocelot is an opportunistic hunter, taking prey less than 1 kg in the same proportion as found in the wild. However, larger mammals such as agoutis, pacas and acouchis are at the limit of prey size for the ocelot and occur less frequently as prey body size increases. Relatively fewer larger mammals are taken than occur in the wild, and when these are predated there is selection for the smaller, and hence younger individuals. Aquatic prey was only eaten during the dry season as pools and rivers began to dry up, so that aquatic animals were more easily captured.

Bisbal (1986) examined the stomach contents of ten ocelots from Venezuela. Mammals were found in 90 per cent of stomachs and occupied 88 per cent of the volume. Opossums, bats (*Sturnira*), sloths, armadillos and various rodents (*Proechimys*, *Heteromys*, *Sigmodon*, *Oryzomys* and *Dasyprocta*) were the commonest mammalian prey. Birds were found in 20 per cent of stomachs, but occupied only 1 per cent of the volume and included passerines and woodcreepers (*Dryocopus*). Frogs and insects were minor components of the diet.

Mondolfi (1986) also examined stomach contents from 16 ocelots in Venezuela. Most of the prey were mammals (81 per cent) (opossums, agoutis, armadillos, spiny rats, cotton rats, pocket mice and bats), but reptiles (25 per cent) were also common and included tortoises, iguanas and teiid lizards. Again, frogs and birds were minor components of the diet.

The stomach contents of a few margays and tiger cats are also mentioned in Mondolfi (1986), but there are too few data to draw any meaningful conclusions about the diets of these two cats, except that they mainly consume small rodents.

WILDCAT

EUROPEAN WILDCAT

In Europe the predominant prey of the wildcat is rodents (Schauenberg, 1981). For example, in the central Apennines of Italy the commonest prey species are the pine vole (*Pitymys savii*) (37.3 per cent of stomachs) and wood mouse (*Apodemus sylvaticus*) (14.2 per cent) (Table 5.2) (Ragni, 1978). Arthropods (19.4 per cent) were also quite important, but birds (7.5 per cent) and reptiles (1.5 per cent) were comparatively rare food items.

In France, the stomach contents of 89 wildcats revealed that 97 per cent contained rodents, 8 per cent birds, 6 per cent shrews, 5 per cent amphibians, 2 per cent insects and only 1 per cent hares (Conde *et al.*, 1972). In the West Carpathians, the main food item

Table 5.2 An analysis of the stomach contents of wildcats in the central Apennines of Italy

Species	Frequency (%)	Weight (%)
Pine vole	37	28
Wood mouse	14	16
Bank vole	7	6
Vole (*Microtus nivalis*)	2	3
Mole	3	8
Hare	1	10
Hedgehog	1	8
Birds	7	8
Reptiles	2	1
Arthropods	19	1

Source: Ragni (1978)

was rodents (88 per cent) followed by birds (13 per cent) and hares (4 per cent) (Sladek, 1962). In the East Carpathians, wood mice and voles were the main prey (65 per cent) in the stomachs of 28 wildcats, followed by birds (14 per cent), squirrels (12 per cent), hares and marmots (5 per cent) (Lindemann, 1953). In the Iberian Peninsula, the main food item is wood mice (42 per cent), but rabbits were also an important prey (21 per cent). Other rodents made up most of the rest of the diet (Aymerich, 1982). In Azerbaijan, rodents are again the most important prey (Table 5.3) (Nasilov, 1972).

Table 5.3 An analysis of the contents of 124 stomachs and scats of the wildcat in Azerbaijan

Prey	Frequency %	Prey	Frequency %
Shrews	2	Social vole	10
Hare	3	Chicken	2
House mouse	13	Pheasant	1
Wood mouse	21	Egret	2
Water vole	6	Other birds	26
Pine vole	30		

Source: Nasilov (1972)

However, in Scotland the main prey seems to be rabbit. In 546 scats from eastern Scotland, 92 per cent contained rabbit, 18 per cent rodents and shrews and 14 per cent birds (Corbett, 1979). Other rare items included sheep, roe deer (both scavenged), frogs and toads (Corbett, 1979). Rabbits were the major prey in the spring, and in the autumn and winter. Young rabbits were easily caught in the first part of the year because of their inexperience in dealing with predators, but adult rabbits suffering from myxomatosis were vulnerable to predation in the latter part of the year when this disease has its greatest effect. In North East Scotland the main prey item in 18 wildcat stomachs was also rabbit and mountain hare (14 stomachs), followed by wood mice and voles (5), and birds (4) (Hewson and Kolb in Kolb, 1977). However, in a study of wildcat scats in western Scotland, the main prey was rodents (51 per cent) just as with the rest of the European population (Hewson, 1983). Among the prey eaten were rodents (mainly field voles), very few rabbits, but a substantial proportion of birds (30 per cent) (Hewson, 1983). Contrary to popular belief the wildcat is not an important predator on red grouse. In a study of the mortality of grouse in both upland and lowland areas, the fox was a major predator, but the wildcat was regarded as insignificant (Jenkins, Watson and Miller, 1964).

AFRICAN WILDCAT

An analysis of nine stomachs of the African wildcat in Natal revealed that the most common item of prey is small rodents (70 per cent frequency), including *Otomys*, *Mastomys*, *Rhabdomys*, and *Dendromus* (Rowe-Rowe, 1978). Other items included chickens and other birds (20 per cent), and insects (10 per cent).

The diet of the African wildcat has been studied in some detail in Botswana and Zimbabwe (Smithers, 1971; Smithers and Wilson, 1979). In both places, rodents were the most important prey, but the species that were eaten depended on what was available in the different habitats in each area (Table 5.4).

Table 5.4 The food items found in the stomachs of African wildcats from Zimbabwe (n = 58) and Botswana (n = 80)

Prey	Frequency (%)		Prey	Frequency (%)	
	Zimbabwe	Botswana		Zimbabwe	Botswana
Murid rodents	72	74	Frogs	2	1
Birds	21	10	Spiders	2	1
Insects	19	19	Myriapods	2	1
Reptiles	10	13	Scorpions	–	2
Other mammals	9	6	Fruit	2	1
Solifuges	7	18			

Sources: Smithers (1971); Smithers and Wilson (1979)

In Zimbabwe, the main rodent prey were multimammate rats (*Praomys*), vlei rats (*Otomys*) (14 per cent) and gerbils (*Tatera*), whereas in Botswana the main rodent prey were two species of gerbil *(Tatera)* (40 per cent) and pygmy mice *(Mus minutoides)*, with a few multimammate rats, pouched mice *(Saccostomus)* and hairy-footed gerbils *(Gerbillurus paeba)*. Other mammals included scrub hare *(Lepus saxatilis)*, rock rabbits *(Pronolagus)*, elephant shrews, juvenile grysbok and springhare. Birds included domestic fowl, button quails, harlequin quails, bustards and quelea. Reptiles included various lizards, geckos, skinks and snakes. Most of the insects were grasshoppers and crickets.

INDIAN DESERT CAT

In the Eastern Kara Kum in Turkmenia, the main food of the Indian desert cat is the fat sand rat (*Rhombomys opimus*) followed by the mid-day gerbil (*Meriones meridianus*), tolai hare (*Lepus tolai*) and ground squirrels (*Spermophilopsis leptodactylus*) (Table 5.5) (Sapozhenkov, 1961).

Table 5.5 An analysis of the main prey items of the Indian desert cat in Kara Kum, Turkmenia

Prey	Frequency %	Prey	Frequency %
Tolai hare	9.6	Jerboas	3.1
Ground squirrel	4.2	Birds	5.5
Fat sand rat	38.2	Reptiles	5.2
Mid-day gerbil	19.1		

Source: Sapozhenkov (1961)

In Western Rajasthan, the main prey item of the Indian desert cat is the desert gerbil (*Meriones hurrianae*), but hares (10 per cent) are also an important component (Sharma, 1979). Indian desert cats also eat turtle doves, francolins, sandgrouse, peafowl, bulbuls, sparrows, cobras, saw-scaled vipers, sand boas, geckos, scorpions and beetles.

FERAL DOMESTIC CAT

Feral cats are worldwide in distribution and as such have the most varied diets of any species of cat. There have been a considerable number of studies of the diets of feral domestic cats. A very detailed discussion of feral cat diets worldwide has been completed by Fitzgerald (1988). Its scope is beyond what can reasonably be covered here, so below is given a selection of the diets of feral cats from localities worldwide. Populations of feral cats on small oceanic islands have had catastrophic effects on seabird populations and endemic faunas.

AUSTRALIA

The food habits of feral cats have been studied in several parts of Australia. In Victoria, a study of stomach contents revealed that mammals were the most important prey (88 per cent volume; 67 per cent frequency) followed by insects (0.5 per cent volume;

15 per cent frequency) and birds (3.5 per cent volume; 5 per cent frequency) (Coman and Brunner, 1972). Cats were taken from two study sites; the primary site contained native fauna unaffected by humans, the secondary site contained a fauna that had been modified by human activities.

Most of the prey at both sites included mammals. In the primary site the cat was an important predator on native mammals, but in the secondary site the cats preyed almost exclusively on introduced rabbits and house mice (Table 5.6) (Coman and Brunner, 1972).

Feral cats are very common in all habitats in Australia. This study shows that they could have a significant impact on the native fauna in primary habitats. Jones and Coman (1981) carried out a similar study at three sites: the Mallee of Victoria, Kinchega National Park and the eastern highlands of Victoria. Introduced mammals made

Table 5.6 Food items found in the stomachs of 80 feral cats by percentage volume and percentage occurrence in primary and secondary habitats in Victoria, Australia

Prey	Primary		Secondary	
	Volume (%)	Frequency (%)	Volume (%)	Frequency (%)
Rabbit	20	11	62	30
Mouse	12	8	27	41
Ringtail possum	11	6	0	0
Other	7	6	1	7
Phalanger	5	6	0	0
Rattus fuscipes	5	6	0	0
R. lutreolus	7	4	0	0
Murids	2	4	0	0
Macropus spp.	10	2	0	0
Schoinobates	4	2	0	0
Trichosurus	2	2	0	0
Sheep	0	0	2	4
Antechinus	1	2	0	0
Sminthopsis	0.3	2	0	0
Perameles	+	2	0	0
Insects	1	15	+	15
Birds	6	6	+	11
Lizards	1	2	+	4
Frogs	0	0	1	7
Household scraps	4	6	0	0
Carrion	0	0	3	15

Source: Coman and Brunner (1972)

up 85 per cent, 64 per cent and 45 per cent of the total prey weight in each habitat, respectively. Birds made up 9 per cent, 18 per cent and 13 per cent of the total prey weight at each site. Native mammals made up only 2 per cent and 4 per cent of the prey weight in the Mallee and Kinchega National Park, and included brushtail possums (*Trichosurus*), planigales and bats. However, in the eastern highlands the native mammal fauna made up 40 per cent of the diet. The most important native species consumed included the southern bush rat (16 per cent), common ringtail possum (8 per cent), brushtail possum (5 per cent), brown antechinus (4 per cent) and sugar glider (4 per cent). Seasonal variation in diet of feral cats in the Mallee is shown in Figure 5.1(c).

In the arid regions of Australia, the feral cat feeds mainly on reptiles (100 per cent occurrence; 32 per cent humid volume) including skinks, monitors, geckos, agamas and snakes (Bayly, 1976). Mammals (36 per cent occurrence; 43 per cent volume) included rabbits and house mice. Birds (7 per cent occurrence; 1 per cent volume) included the galah cockatoo. The feral cats from these arid regions also ate many arthropods (100 per cent occurrence; 14 per cent volume) including scorpions, chafers and grasshoppers, although they contributed little to the total diet by volume.

NEW ZEALAND

The diet of feral cats in New Zealand has been studied by Fitzgerald and Karl (1979) in the Orongoro Valley. Black rats were the most important prey item (42 per cent by weight), followed by rabbits (23 per cent), brush tail possums (18 per cent), house mice (8 per cent) and stoats (1 per cent). Birds made up about 4.5 per cent by weight of the diet and insects (mainly wetas) made up 2 per cent of the diet. Seasonal variation in diet of feral cats in the Orongoro Valley is shown in Figure 5.1(d). The cats do not seem to eat many birds, although they may have been responsible for reducing and eliminating the most vulnerable species in the past when other introduced fauna may have not been so common as it is today.

EUROPE AND NORTH AMERICA

Corbett (1979) studied the diets of domestic cats and feral cats in Scotland. He found that the main dietary item was rabbits, just like the sympatric wildcat. On Heisker in the Monach Islands to the west of the Outer Hebrides, the feral cats have a similar social and spatial organisation to wildcats. In this marginal habitat, rabbits make up 88 per cent of the diet. Passerine birds were the second most important component of the diet. The adult cats preferentially feed their kittens with birds, so that they make up 65 per cent of the youngsters' diet. As with wildcat predation in eastern Scotland,

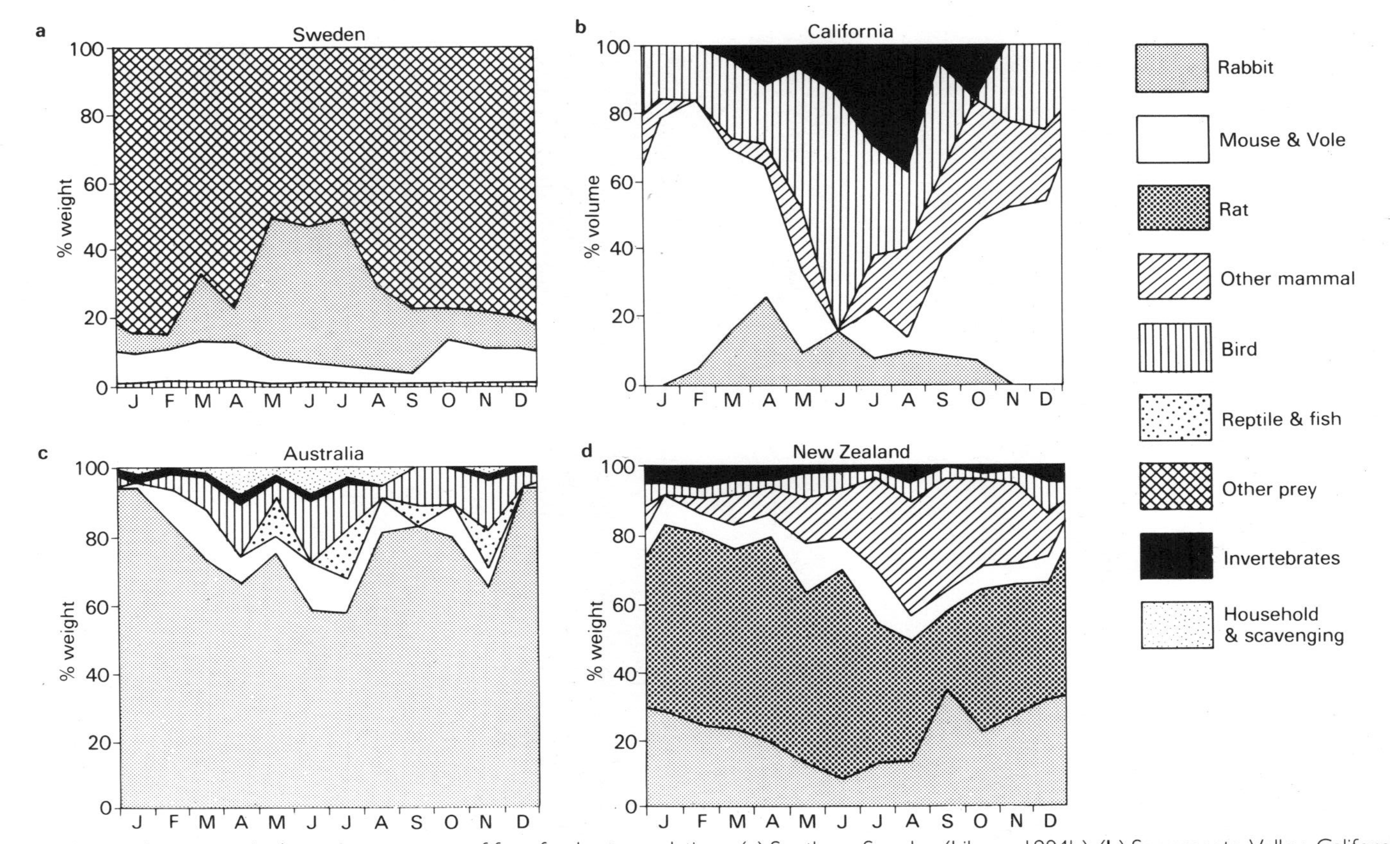

Figure 5.1 Seasonal variations in the main prey types of four feral cat populations. (**a**) Southern Sweden (Liberg, 1984b). (**b**) Sacramento Valley, California (Hubbs, 1951). (**c**) Mallee county, Australia (Jones and Coman, 1981; Jones in Liberg and Sandell, 1988). (**d**) Orongoro Valley, New Zealand (Fitzgerald and Karl, 1979) (after Liberg and Sandell, 1988) (S.A.)

young rabbits are more vulnerable to predation in the first part of the year, and adult rabbits with myxomatosis towards the end of the year.

In the Bedfordshire village of Felmersham, a study of the diet of house cats found that mainly rodents and shrews were consumed (Table 5.7) (Churcher and Lawton, 1987)

Table 5.7 The prey items of the domestic cat in the village of Felmersham, Bedfordshire (n = 832 kills)

Prey	Frequency %	Prey	Frequency %
Wood mouse	17	Other mammals	8
Bank vole	7	House sparrow	16
Field vole	14	Song thrush	4
Common shrew	12	Blackbird	3
Pygmy shrew	4	Robin	3
Rabbit	3	Other birds	10

Source: Churcher and Lawton (1987)

In contrast to many other studies on cat diets, these cats killed a high proportion of birds. In general, the more urban the environment the fewer rodents and other small mammals are available, so that cats probably have no option but to concentrate on birds as prey. This appears to be the case in Portsmouth dockyards, where feral cats subsist mainly on scavenged food, but because rodents are controlled by poison, the cats prey only on birds (Dards, 1978).

In southern Sweden, house cats and feral cats were found to prey mainly on rabbits and, to a lesser extent, voles (Liberg, 1984b). However, the main food for house cats was that scavenged from or provided by households, which made up to 85 per cent of the diet during the winter. Seasonal variation in house cat diet in Sweden is shown in Figure 5.1(a).

Several studies of feral and domestic cat diets have been carried out in the USA. In California, voles were the main component of their diet (Pearson, 1964, 1966). Seasonal variation in the diet of house cats in the Sacramento Valley in California is shown in Figure 5.1(b). In Wisconsin, voles were also the main prey, but house mice, deer mice, cottontail rabbits and brown rats were also important items. Although cats do prey on rats, an adult rat seems to present a formidable adversary to overcome. In Baltimore, the rats killed by domestic cats in an urban environment were analysed (Childs, 1986). It was found that the cats preyed invariably on juvenile or

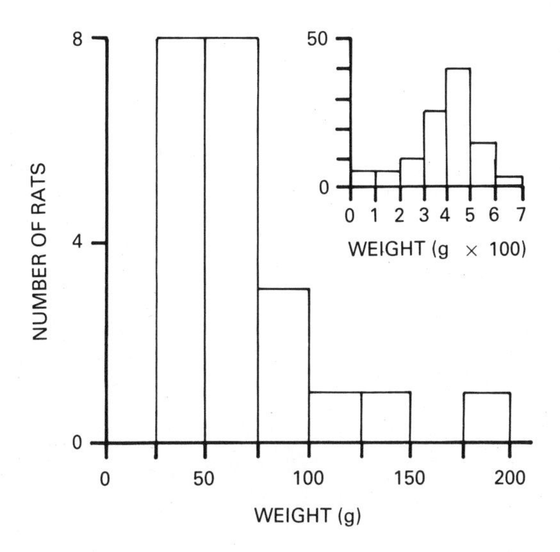

Figure 5.2 Size distribution of the rats caught by domestic cats in Baltimore. The inset shows the size distribution of the free living population of rats. It is evident that cats only prey on small rats (after Childs, 1986) (A.K.)

subadult rats, adult rats seemingly being immune to any cat predation (Figure 5.2).

ISLANDS

Feral cats have had a devastating effect on island faunas, which are vulnerable to introduced predators. Many studies of the diet of feral cats on islands have now been carried out. Some of these will be reviewed below.

An analysis of the stomach contents of 73 feral cats on Jarvis Island in the Central Pacific showed that sooty terns were the most important prey (Table 5.8) (Kirkpatrick and Rauzon, 1986). Other birds, house mice, skinks, geckos and insects were minor prey items.

Domestic cats were first introduced to Macquarie Island at the beginning of the nineteenth century. Jones (1977) analysed 756 scats from these cats and found that rabbits (mostly young), prions and white-headed petrels were the most important prey items (Table 5.9). To look at seasonal variations in diet, the gut contents of 41 feral cats were examined at different times of the year. Rabbits (71 per cent frequency) were the most important prey item throughout the year. Prions (24 per cent) were fed on during the summer and white-headed petrels (24 per cent) during the winter. Weka rail (12 per cent), elephant seal (10 per cent) (scavenged only!) and rats (2 per cent) were also important winter food items.

On Dassen Island off the western coast of South Africa, feral cats are a recent introduction. Important prey items include rabbits, house mice, jackass penguins (adults scavenged, chicks killed), Cape

Table 5.8 Prey items from the stomachs of feral cats on Jarvis and Howland Islands in the central Pacific

Prey	Jarvis Island		Howland Island	
	Frequency (%)	Volume (%)	Frequency (%)	Volume (%)
Sooty tern	67	91	3	44
Wedge-tailed shearwater	0	0	2	25
Brown noddy	0	0	1	23
Red-tailed tropic bird	3	6	0	0
Other bird	2	+	0	0
Snake-eyed skink	0	0	1	8
Gecko	1	+	0	0
Mouse	6	2	0	0
Insects	19	+	0	0
no. of stomachs	73		5	

Source: Kirkpatrick and Rauzon (1986)

Table 5.9 Occurrence of prey species in 756 scats from feral cats on Macquarie Island

Species	Frequency %	Species	Frequency %
Rabbit	82	Rat	3
Prion	29	Weka	2
White-headed petrel	16	Other bird	1.5
House mouse	4	Elephant seal	0.3
Penguin	3		

Source: Jones (1977)

cormorants (mainly chicks), gulls and terns (Apps, 1983).

On Marion Island in the southern Indian Ocean, remains of prey (1,224) and stomach contents (124) of feral cats were examined by van Aarde (1980). The cats were introduced in the early 1950s and a large population of over 2,000 cats was present at the time of the study in 1975/6. The main prey items were prions, but house mice and other petrels were also important food items (Table 5.10).

Prions were eaten throughout the year, but peaked in August/September. White-chinned petrel chicks (*Procellaria aequinoctialis*) were taken in December, but the Kerguelen and soft-plumaged petrels (*Pterodroma brevirostris* and *P. mollis*) were taken through-

Table 5.10 The percentage frequency of occurrence of prey items in stomach contents and remains of kills of feral domestic cats on Marion Island

Prey	Stomachs (%)	Remains (%
Prions:		
Pachyptila vittata	30	60
Petrels:		
Pterodroma mollis	6	9
P. brevirostris	5	13
P. macroptera	5	10
Procellaria aequinoctialis	5	1
Halobaena caerulea	1	2
Other birds:		
Pelecanoides sp.	–	0.5
Fregatta sp.	1	0.1
Chionis minor	–	1
Penguin remains	7	–
Other petrel remains	44	3
Mammals:		
House mice	16	–

Source: van Aarde (1980)

out the year. Great-winged petrels (*P. macroptera*) were found in stomachs only between 22 June and 21 August, when they were nesting on the island.

Cook and Yalden (1980) looked at the diet of feral cats on Deserta Grande Island in Madeira. They found that scats contained 62 per cent bird remains, which included Bulwer's petrel, Cory's shearwater and Madeiran storm petrel. Mammals made up 38 per cent of the diet and included rabbits and house mice. Cook and Yalden (1980) have suggested that cat predation may have caused the extinction of manx shearwater and soft-plumaged petrel on Deserta Grande.

SAND CAT

There is very little information concerning the diet of sand cats. Lay, Anderson and Hassinger, (1970) have reviewed the Russian literature. A study of 182 stomachs and scats in Turkmenia showed that 65 per cent of stomachs contained rodents (34 per cent *Rhombomys opimus*; 19 per cent *Meriones meridianus*), but the diet also included birds, reptiles and arthropods. *Rhombomys opimus* was also the most common prey species in the Kyzlkum Desert and the Ust Urt plateau in Uzbekistan, where rodents, including jerboas, made up 88 per cent of the diet.

BLACK-FOOTED CAT

Smithers (1971) analysed the stomach contents of seven black-footed cats. Murid rodents were found in 57 per cent of stomachs and included *Dendromus*, pouched mouse, hairy-footed gerbil and other gerbils (*Tatera* spp.). Arachnids were the second most important prey item (43 per cent) and consisted mainly of solifuges and spiders. Elephant shrews, double-banded coursers (*Rhinoptilus africanus*), agamid lizards and beetles were also found.

JUNGLE CAT

An analysis of 30 scats in central Punjab, Pakistan, revealed that jungle cats feed primarily on small mammals (70 per cent), but their diet also includes a few birds (10 per cent), reptiles and amphibians (10 per cent) (Khan and Beg, 1986). Most of the small mammal bones were unidentifiable to species, but belonged to rats and mice, including the bandicoot rat which was successfully preyed upon despite its subterranean lifestyle.

In Uzbekistan, the summer and winter diets of the jungle cat have been investigated by Ishunin (1965). The main component in summer and winter was rodents, particularly gerbils (*Meriones*). More birds were eaten in the winter, whereas more reptiles and insects were eaten in the summer (Table 5.11).

Table 5.11 A seasonal analysis of prey animals in the diet of jungle cats from Uzbekistan

Prey	Summer (%)	Winter (%)	Prey	Summer (%)	Winter (%)
Hare	10	17	Birds	12	44
Rodents	85	92	Reptiles	16	0
(Gerbils	75	26)	Beetles	12	0
(House mouse	4	13)			

Source: Ishunin (1965)

LEOPARD CAT

To date there has only been one detailed study of the diet of the leopard cat. Inoue (1972) has meticulously analysed the prey items found in the scats of the leopard cats on Tsushima Island. (However, it has been recently suggested that the Tsushima Island leopard cat may actually represent a new species of cat, closely related to the Iriomote cat (Jackson, 1989).) The main prey items were moles, murids, birds and a huge number of insects (Table 5.12) (Inoue, 1972).

Alcala and Brown (1969) recorded rats in the stomachs of leopard cats from the Philippines.

Table 5.12 Seasonal variation in the percentage frequency of occurrence of prey items in 230 scats of the leopard cat from Tsushima Island, Japan

Prey	March	August	December	Total
Moles	3	21	13	16
Murids	35	92	40	73
Passerines	13	27	8	21
Ducks	0	0	1	0.4
Pheasants	0	0	1	0.4
Fish	4	10	15	8
Amphibians	13	14	33	14
Snakes	0	6	8	3
Insects	18	69	14	44

Source: Inoue (1972)

IRIOMOTE CAT

Although only described in 1967, the Iriomote cat has been subjected to detailed investigation since its discovery (Imaizumi, 1967; Yasuma, 1988 and pers. comm.). Yasuma (pers. comm.) has looked at the diet of the Iriomote cat for every month of the year by analysing scats, remains of kills, stomach contents and direct observations. The total annual frequency of occurrence of different prey items is shown in Table 5.13.

Among the mammal species eaten were fruit bats (*Pteropus dasymallus*), black rats and wild pig. Birds included night herons, quails, rails, pigeons and doves, scops owls, kingfishers, robins, thrushes and crows. Yamaya and Yasuma (1986) have studied in very great detail the 39 species of beetle that the Iriomote cat regularly consumes. Although insects make up about one-third of

Table 5.13 The frequency of occurrence of prey items in 849 scats of the Iriomote cat

Prey	Frequency (%)	Prey	Frequency (%)
Mammals	57	Amphibians	12.8
Birds	69.4	Fish	2.5
Reptiles	65.3	Insects	100

Source: Yasuma (pers. comm.)

the diet in terms of total number of items found in scats, they contribute very little in terms of mass. The most important prey items by mass are mammals (50 per cent), birds (25 per cent) and reptiles (20 per cent) (Yasuma, pers. comm.). Looking at seasonal variations, slightly fewer mammals are eaten during the summer, but more birds and reptiles are consumed instead. The total prey mass is more or less constant throughout the year.

CARACAL

The caracal is found throughout sub-Saharan Africa and south-western Asia. Consequently, its diet varies throughout its distribution. In the Karakum of Turkmenia, the main food item is the tolai hare (*Lepus tolai*), but rodents including fat sand rats (*Rhombomys*), jerboas and ground squirrels (*Spermophilopsis*) were also important components (Table 5.14) (Sapozhenkov, 1962).

Table 5.14 Percentage frequency of prey animals in the diet of the caracal in Kara Kum, Turkmenia

Prey	Frequency (%)	Prey	Frequency (%)
Tolai hare	48	Jerboas	11.8
Fat sand rat	17.6	Ground squirrel	6.9

Source: Sapozhenkov (1962)

In the Mountain Zebra National Park in South Africa, the caracal's diet is completely different. The most commonly preyed upon species is the rock hyrax (*Procavia capensis*) which was found in 53 per cent of 200 scats (Grobler, 1981). However, the most important prey item by mass is the mountain reedbuck (*Redunca fulvorufula*), which occurred in only 20 per cent of the scats, but contributed to 70 per cent of the mass of food (Table 5,15). Even though a reedbuck contains too much food for a caracal to eat in one sitting, it can be cached and returned to subsequently. Birds in the caracal's diet included guineafowl and francolins.

Table 5.15 An analysis of 200 caracal scats from the Mountain Zebra National Park

Prey	% Occurrence	% Mass	Prey	% Occurrence	% Mass
Hyrax	53	17	Grysbok	1	2
Reedbuck	20	70	Mongoose	1	0.2
Rock rabbit	5	1	Springbok	0.4	2
Scrub hare	5	3	Springhare	0.4	0.2
Rodents	5	0.04	Birds	5	–
Duiker	3	5	Reptiles	1	–

Source: Grobler (1981)

SERVAL

In the Aberdares of Kenya, servals feed mainly on hares which graze along roadsides, and rodents including mole rats, crested rats (*Lophiomys imhausi*), ground squirrels and giant rats (York, 1973).

In the Serengeti National Park, servals hunt primarily nile rats (*Arvicanthis niloticus*) (Geertsema, 1976). In the Ngorongoro Crater of Tanzania, their diet includes more vlei rats (*Otomys*) and frogs (Geertsema, 1985). Servals are mainly nocturnal in the Ngorongoro Crater, but diurnal in the Serengeti, which reflects the different activity peaks of their main prey species in these two areas.

The most detailed study of the serval's diet to date was carried out by Geertsema (1985) in the Ngorongoro Crater. In almost 1,000 direct observations of what servals were feeding on, 89 per cent of the kills were mammals, 6 per cent snakes, 2 per cent frogs, 1.4 per cent insects and 1.3 per cent birds. However, Geertsema was unable to follow the servals into the wetter areas, and so an analysis of 56 scats revealed that the number of frogs in the serval's diet is much greater than direct observations would otherwise suggest (Table 5.16).

The commonest rodent prey was the vlei rat (43 per cent), followed by the pygmy mouse (18 per cent), nile rat (11 per cent), pocket mouse (*Pelomys*) (7 per cent), multimammate rats (6 per cent) and shrews (6 per cent). Birds included quails, flamingoes, quelea and teal. Although vlei rats and pygmy mice were the main prey

Table 5.16 Incidence of prey animals in 56 serval scats in the Ngorongoro Crater in Tanzania

Prey	% Occurrence	Prey	% Occurrence
Rodents, shrews	98	Grasshoppers, beetles	20
Frogs	77	Snakes	14
Birds	21		

Source: Geertsema (1985)

throughout the year, half as many vlei rats were eaten during the wet season compared with the dry season. The numbers of nile rats (doubled) and pocket mice (trebled) that were eaten increased during the wet season to make up for this shortfall. Vlei rats are the preferred prey of the serval because of their larger body size compared with other rodents available. In a sample of 65 serval stomachs from Zimbabwe, the commonest prey animals, vlei rats and multimammate rats, accounted for more than 80 per cent of the prey (Smithers, 1971).

Although the caracal and serval have a similar body size and African distribution, the more heavily-built caracal feeds on larger mammals than the specialist rodent-killing serval.

JAGUARUNDI

The stomach contents of ten jaguarundis from Venezuela have recently been analysed by Bisbal (1986). Mammals included rabbits (*Sylvilagus floridanus*) and rodents (*Zygodontomys* and *Sigmomys*) and were found in 40 per cent of the stomachs, occupying 46 per cent of the volume. Birds were an important component, being found in 70 per cent of stomachs and taking up 26 per cent of the volume. Passerines and domestic fowl were among the birds eaten. Reptiles were equally important (50 per cent occurrence; 29 per cent volume) and included teiid lizards and iguanas.

Mondolfi (1986) looked at 13 stomachs of the jaguarundi in Venezuela. Again birds were very important (in 54 per cent of stomachs) and included passerines, bobwhite quail and domestic fowl. Mammals (in 46 per cent of stomachs) included cottontail rabbits and rodents (*Sigmodon*, *Sigmomys*, *Rattus*, *Oryzomys*). Reptiles (in 46 per cent of stomachs) included iguanas and teiid lizards.

PUMA

The puma is one of the most widely ranging cats, being found from the north of North America to the southernmost tip of South America. Not surprisingly, its diet varies considerably depending on the prey species available.

Yanez *et al.* (1986) looked at puma scats from two study sites in Chile. In the Torres del Paine National Park, the introduced European hare (*Lepus europaeus*) was the commonest prey, followed by ungulates (mainly guanaco (*Lama guanicoe*), but also guemul (*Hippocamelus bisculus*) and domestic sheep), birds, rodents and carnivores (Table 5.17).

Table 5.17 An analysis of the contents of puma scats from the Torres del Paine National Park, Chile and adjacent ranches

Prey	Frequency (%)			
	National Park		Ranches	
	Winter	Spring/summer	Unknown	Unknown
Cattle	0	0	0	1.2
Guemul	0	2.4	0.4	1.7
Guanaco	5.5	11.9	13.4	2.1
Sheep	2.1	4.8	8.7	26.7
Carnivores	4.9	16.6	7.2	13.6
Hare	73.0	38.1	50.8	44.0
Horse	1.4	0	0.4	0.8
Rodents	4.1	12.0	4.4	1.7
Birds	6.2	9.4	11.9	8.2
No. of scats	99	28	164	414

Source: Yanez *et al.* (1986)

By collecting scats at different times of the year, it was possible to see the seasonal variation in diet. In winter the main prey animal was the hare, but in the spring and summer other prey animals were better represented (Table 5.17). The presence of sheep remains showed that the park pumas hunted on adjacent ranches.

On the ranches adjacent to the National Park, the composition of the diet was quite different. Although hares were still the major component, many more sheep and other domestic livestock were eaten (Table 5.17). However, some of the domestic livestock may well have been scavenged from animals that had died from disease or parasites.

In the Big Bend National Park in Texas, the main prey animals are desert mule deer, collared peccary and to a lesser extent lagomorphs, porcupines and rodents. In a year of low deer population the

Table 5.18 Seasonal variation in the percentage occurrence of prey animals in the diet of pumas in the Big Bend National Park in Texas. Also shown is the effect of a low deer population (1980–1) on the composition of the diet

Prey	1972–4 (%)	1980–1 (%)	Prey	1972–4 (%)	1980–1 (%)
Mule deer			Lagomorphs		
Spring	73	44	Spring	3	16
Summer	58	27	Summer	8	20
Late summer	84	47	Late summer	0	10
Winter	85	36	Winter	2	9
Peccary			Porcupine		
Spring	20	41	Spring	6	0
Summer	25	38	Summer	13	0
Late summer	5	35	Late summer	5	0
Winter	9	36	Winter	2	2
Rodents			Birds, reptiles, arthropods		
Spring	10	1	Spring	0	3
Summer	17	6	Summer	4	3
Late summer	5	4	Late summer	11	2
Winter	6	7	Winter	0	2

Source: Leopold and Krausman (1986)

proportion of peccaries and lagomorphs increased in the diet (Table 5.18) (Leopold and Krausman, 1986).

In the north of its range in the Idaho Primitive Area, the puma feeds mainly on mule deer and wapiti (Hornocker, 1970). About 70 per cent of the winter diet consists of deer and 6 per cent snowshoe hares. Fewer deer are consumed in the summer because they are less vulnerable to predation and other food is easily available such as rodents and lagomorphs. Most of the mule deer (62 per cent) and wapiti (75 per cent) that are killed belong to either the very young or very old age categories, reflecting their greater vulnerability. Adult male mule deer are also predated upon more heavily than adult females, because they are found alone at higher elevations where the snow is deeper, thus making it difficult to notice and escape from a predator. Also male deer are often weakened and injured during the autumn rut, which increases their vulnerability.

Mature male wapiti are, however, probably too strong and big to be preyed upon. However, pumas do not apparently select diseased or weak deer to prey upon, since these are killed in the same proportion as they occur in the population.

In the Peruvian rainforest, a small sample of puma scats was analysed by Emmons (1987). The main prey items were cricetid rodents (*Mesomys*), pacas, agoutis and a bat (*Micronycteris*) mostly in the region of one to ten kilogrammes. There is clearly still much to learn about rainforest pumas.

CHEETAH

The cheetah preys on relatively small prey considering its body size. In East Africa it chases after Thomson's gazelle, but this species is restricted to East Africa, so what do cheetahs eat elsewhere in Africa? The results of a number of studies of cheetah diet are summarised in Table 5.19 (Wrogemann, 1975).

From Table 5.19, it is clear that cheetahs feed mainly on Thomson's

Table 5.19 The percentage frequency of different prey species killed by the cheetah in different parts of Africa

Prey	Area			Prey	Area		
	1	2	3		1	2	3
Thomson's gazelle	91.2	–	–	Zebra Yearling	–	3	1.3
Grant's gazelle	2.3	–	–	Oribi	–	3	–
Reedbuck	0.8	12	5.9	Warthog	–	6	0.7
Wildebeest	1.9	3	0.7	Puku	–	45.5	–
Juvenile	–	3	–	Waterbuck (young)	–	–	11.2
Topi	0.8	–	–	Kudu (young)	–	3	4.6
Kongoni	0.4	9.1	–	Duiker	–	3	2.6
				Bushbuck	–	3	–
Dik-dik	0.4	–	–	Others	–	–	4.6
Impala	1.1	6	68.4				
Hare	1.1	–	–				

Notes
Area 1—Serengeti (Schaller, 1972) (n=261 kills)
Area 2—Kafue National Park (Mitchell, Shenton and Uys, 1965) (n=33 kills)
Area 3—Kruger National Park (de Pienaar, 1969) (n=50 kills)

Source: Wrogemann (1975) (modified)

gazelle in the Serengeti, puku in the Kafue National Park and impala in the Kruger National Park. Schaller (1972) found that most of the prey caught by cheetah in the Serengeti are in good condition.

LYNXES

Lynxes are thought to have evolved to prey on rabbits and hares (Kurten, 1968). The predominant prey of most lynxes is indeed lagomorphs, with only two exceptions: the very large Eurasian lynx preys on reindeer and roe deer, which is a recent historical development (Haglund, 1966), and the bobcat feeds on small rodents as well as lagomorphs (Maehr and Brady, 1986).

BOBCAT

The bobcat is widely distributed throughout the USA and into Mexico. Many studies have been carried out on this species, because of its importance in the fur trade and because it is viewed as vermin in many areas.

Litvaitis *et al.* (1986) studied the winter diets of bobcats in Maine and in particular looked at variations in diet due to age and sex. The main prey was the snowshoe hare for all bobcats. However, white-tailed deer was the most important component of the diets of both adult and yearling males, but was insignificant in the diet of females and juveniles, who ate correspondingly more rodents (Table 5.20). Thus there is a partitioning of prey resources which depends on body size. Female bobcats specialise by feeding mainly on lagomorphs and rodents, whereas males utilise deer, thus reducing intersexual competition.

Litvaitis *et al.* (1986) found that while the body condition of most bobcats declined during the winter due to a lack of food, male bobcats did not appear to suffer, because of their ability to kill deer.

Table 5.20 The percentage frequency of occurrence of the main prey items in the guts of male (m) and female (f) bobcats from Maine

Prey	Juveniles		Yearlings		Adults	
	m	f	m	f	m	f
Snowshoe hare	40	50	33	67	50	54
Deer	5	8	28	0	24	9
Small mammals	20	19	17	8	6	10
Porcupines	10	15	6	0	0	14
Birds	20	15	0	33	15	9
n =	20	26	18	12	34	57

Source: Litvaitis *et al.* (1986)

Female bobcats do not appear to enter winter with the large fat deposits that adult males have, owing to the cost of lactation earlier in the year. Males have better fat deposits, are able to prey on deer and can thus survive the vicissitudes of most winters. Their larger body size means that they can defend carcasses from conspecifics and most other predators.

In Arkansas, the main prey consists of cottontail rabbits, squirrels, rats and mice, opossums and white-tailed deer (Table 5.21) (Fritts and Sealander, 1978a). There were, however, some regional differences. In the Gulf Coastal Plain, where fewer squirrels were available due to a lack of suitable habitat, the bobcats preyed upon more rabbits and rodents than the bobcats of the Interior Highlands. Seasonal variations show that rabbits were more important in autumn than in winter or spring. Deer consumption was greater in autumn and winter, squirrels in spring and winter, and rats and mice in spring and autumn. Similar intersexual differences to those in Maine were also found: male bobcats consume more deer than females, who consume more lagomorphs and rodents.

Table 5.21 Percentage occurrence and percentage volume of prey animals in 150 bobcat stomachs from Arkansas

Prey	% Occurrence	% Volume	Prey	% Occurrence	% Volume
Rabbits	39	41	Deer	7	16
Fox squirrel	9	3	Birds	7	2
Grey squirrel	7	3	Raccoon	5	4
Eastern chipmunk	5	3	Skunk	4	2
Southern flying squirrel	1	1	Domestic mammals	3	4
(total squirrels	22	10)	Other mammals	3	1
Rats and mice	21	10	Snakes	2	2
Opossum	9	8			

Source: Fritts and Sealander (1978a)

In central Arizona, the most important prey are rodents (67 per cent of scats) and lagomorphs (38 per cent) (Jones and Smith, 1979). Rodents include mostly wood rats (*Neotoma*) (32 per cent), heteromyids (16 per cent), gophers (5 per cent), ground squirrels (6 per cent) and deer mice (3 per cent). Lagomorphs include jack rabbits and cottontail rabbits. Mule and white-tailed deer, peccaries and skunks are also found in the scats. Despite great variations in the abundance of the rodent prey, bobcats maintain similar proportions of rodents in their prey throughout the year.

In the Big Bend National Park in Texas, bobcats feed mainly on rodents and lagomorphs, although peccaries and deer also make up a part of the diet (Table 5.22) (Leopold and Krausman, 1986). As the deer population declined in 1980–81, the bobcats switched their prey preference to lagomorphs.

Table 5.22 Relative frequency by season of the prey animals of the bobcat in the Big Bend National Park in Texas. The effect of a low deer population (1980–1) on the composition of the diet is also shown

Prey	1972–4 (%)	1980–1 (%)	Prey	1972–4 (%)	1980–1 (%)
Mule deer			Lagomorphs		
Spring	35	5	Spring	44	78
Summer	15	2	Summer	55	78
Late summer	24	3	Late summer	57	72
Winter	22	0	Winter	47	84
Peccary			Arthropods		
Spring	4	3	Spring	5	21
Summer	15	0	Summer	30	24
Late summer	0	3	Late summer	10	3
Winter	6	0	Winter	9	22
Rodents					
Spring	33	26			
Summer	25	31			
Late summer	33	25			
Winter	34	30			

Source: Leopold and Krausman (1986)

Maehr and Brady (1986) studied the stomach contents of 413 bobcats from Florida. They found that mammals were the most important prey (72 per cent frequency), followed by birds (16 per cent) and reptiles (<1 per cent). The two most common prey species were cottontail rabbits (*Sylvilagus* spp.) and cotton rats (*Sigmodon hispidus*) (Table 5.23) (Maehr and Brady, 1986).

Table 5.23 An analysis of the stomach contents of 413 bobcats from Florida

Prey	% Volume of stomach	Frequency (%)	Prey	% Volume of stomach	Frequency (%)
Rabbits	29.1	25	White-tailed deer	2.8	2
Cotton rat	31.8	26	Birds	10.4	16
Rodents	8.7	10	Snakes	0.1	<1

Source: Maehr and Brady (1986)

Maehr and Brady (1986) also reviewed variations in diet throughout the bobcat's range (Table 5.24). There is surprisingly little difference despite the wide variety of habitats that these studies cover.

Table 5.24 Variation in bobcat diet throughout its USA range

Locality	Frequency (%)			
	Lagomorphs	Small mammals	Deer	Birds
Coastal plain	36	31	3	11
Southern Appalachians	25	40	11	5
Northeastern USA	32	22	26	5
Western USA	28	34	7	9

Source: Maehr and Brady (1986)

EURASIAN LYNX

In Sweden the main component of the Eurasian lynx's diet is not lagomorphs, but deer (Haglund, 1966). In the north of Sweden, lynx feed mainly on reindeer during the winter months, because they are abundant and vulnerable to predation, finding it difficult to move through deep snow. In the south of Sweden, the main winter prey is roe deer. During the summer fewer deer are caught and the diet comprises more rodents. From October to November rodents, hares and birds predominate, in December and January deer and small prey are equally important, but in February six times as many deer are killed as small prey. The Eurasian lynx is the largest lynx and has only recently switched to feeding on deer, due to a scarcity of its normal prey, the mountain hare (Haglund, 1966). The superabundance of cervid food means that Swedish lynx actually get fatter through the winter rather than thinner. Similar results were obtained for lynx in Norway (Table 5.25) (Birkeland and Myrberget, 1980). From May to November deer formed 51 per cent of the diet, but from December to April it reached 73 per cent. Again the vulnerability of deer to predation in winter, due to disease and their lack of mobility in the snow, in addition to a lack of alternative prey, led to this dietary switch. In fact most of this change was due to females changing from a diet low in deer (39 per cent) in the summer to one high in deer (67 per cent) during the winter; the male diet contained the same proportion of deer throughout the year. Although Eurasian lynx tend to get fatter during the winter due to a superabundance of deer to feed on, their Canadian cousins are not so lucky. Canada lynxes feed almost exclusively on snowshoe hares, but become thinner during the winter due to a lack of hares and suitable weather for hunting (Brand and Keith, 1979).

Table 5.25 The percentage frequency of occurrence of prey in the scats of Eurasian lynx from Norway

Reindeer	31	Rodents	8
Roe deer	17	Carnivores	4
Moose	0.5	Birds	10
Mountain hare	19		

Source: Birkeland and Myrberget (1980)

In southeastern Finland there are no deer for lynx to prey on, so that mountain hare and to a lesser extent brown hare are the main food sources, forming 80 per cent of the diet by frequency and 86 per cent by volume (Pulliainen, 1981).

In the Upper Volga region of the Soviet Union, lynx feed predominantly on mountain hare (68 per cent frequency) during the winter (Zheltukhin, 1986). During the summer hares make up 42 per cent of the diet, the shortfall being made up of birds and rodents.

Delibes (1980) examined over 1,500 Spanish lynx scats and found that rabbits were the most important component of the diet (Table 5.26). There were seasonal variations in diet. Rabbits were the commonest prey from July to October, ducks from March to June and deer from November to February.

Table 5.26 An analysis of the contents of 1,537 scats from the Spanish lynx

Prey	% Frequency	% Volume
Rabbit	79	85
Duck	9	7
Deer	3	5

Source: Delibes (1980)

CANADA LYNX

The Canada lynx feeds almost exclusively on one prey species throughout its enormous North American range. More than 50 per cent of its diet consists of the snowshoe hare, and in some areas the percentage is much higher (Table 5.27) (Saunders, 1963a; Nellis and Keith, 1968; Nellis *et al.*, 1972; Parker *et al.*, 1983).

Table 5.27 The percentage frequency of occurrence of snowshoe hares in the diets of Canada lynx from Newfoundland and central Alberta, Canada

	Newfoundland				Alberta
	Spring	Summer	Autumn	Winter	
Snowshoe hare	72	65	56	85	76
Vole	14	30	9	5	0.1
Muskrat	2	2	5	1	–
Beaver	2	2	–	+	–
Moose (carrion)	18	3	42	13	–
Reindeer	7	4	–	5	–
Livestock	2	–	8	2	–
Birds	28	47	22	3	11
Insects	+	3	3	–	–
Squirrel	–	–	–	–	3
Flying squirrel	–	–	–	–	0.3
Carrion	–	–	–	–	10

Source: Saunders (1963a); Nellis *et al* (1972)

In the autumn and winter, Canada lynx are able to take advantage of deer killed or wounded by human hunters. Many deer also die due to disease or poor body condition. In Newfoundland, Canada lynx kill caribou calves in years when snowshoe hares are not abundant (Bergerud, 1983).

On Cape Breton Island near Newfoundland, Parker *et al.* (1983) found a very high incidence of snowshoe hare in 75 stomachs (97 per cent) and almost 500 scats (93 per cent winter; 70 per cent summer). White-tailed deer, voles, squirrels, jumping mice, ruffed grouse and other birds made up the rest of the diet.

CLOUDED LEOPARD

There has been no detailed study of the diet of the clouded leopard, but in Borneo it is known to prey on sambar, muntjac, chevrotain, bearded pig, palm civet, leaf monkey, porcupines and fish

(Rabinowitz *et al.*, 1987). This demonstrates that it may hunt both in the trees and on the ground. Clouded leopards also feed on domestic cattle and chickens.

SNOW LEOPARD

Schaller (1977) gives details of the diets of snow leopards from three different areas of the Himalayas (Table 5.28).

In areas where the natural prey is rare, snow leopards kill domestic sheep and goats, bringing them into conflict with local people (Fox and Chundawat, 1988).

Table 5.28 The diets of snow leopards in three different areas of the Himalayas from an analysis of their scats

Species		Locality	
	Shey	Lapche	Chitral
Bharal	50	73	–
Markhor	–	–	40
Livestock	13	9	45
Marmot	31	9	5
n=	16	11	20

Source: Schaller (1977)

JAGUAR

Although the jaguar ranges widely through Central and South America, remarkably little is known about its diet in the wild. The most intensive study to date, in Cockscomb Basin in Belize, has revealed that the most common item of prey in jaguar scats is the nine-banded armadillo (*Dasypus novemcinctus*) followed by paca (*Agouti paca*), tamandua (*Tamandua mexicana*) and brocket deer (*Mazama americana*) (Table 5.29) (Rabinowitz and Nottingham, 1986). The occurrence of prey animals in the diet mirrored their occurrence in the wild as determined by 27 one-kilometre transects through the forest (Table 5.29). Other food items included opossums, turtles, rodents, skunks, kinkajou, snakes and chickens.

In the Pantanal, Mato Grosso, Brazil, prey items included cattle, dog, capybara (*Hydrochoeris hydrochaeris*), tapir (*Tapirus terrestris*), marsh deer (*Blastocerus dichotomus*), white-lipped peccary (*Tayassu pecari*), collared peccary (*Pecari tajacu*), La Plata otter (*Lutra platensis*), douroucouli (*Aotus* sp.) and tortoise (*Geochelone* sp.) (Schaller

Table 5.29 An analysis of 228 jaguar scats from the Cockscomb Basin, Belize, compared with the natural occurrence of prey animals in the wild

Prey	Frequency in scats (%)	Natural occurrence (%)	Prey	Frequency in scats (%)	Natural occurrence (%)
Armadillo	54	53	Brocket	6.5	11.0
Paca	9.3	30	Peccary	5.4	0.5
Tamandua	9.3	0.5	Agouti	4.3	4.0

Source: Rabinowitz and Nottingham (1986)

and Vasconcelos, 1978). The most important prey item was the capybara.

In the Manu National Park in Peru, 85 per cent of the jaguar's prey is greater than 10 kg in weight (Emmons, 1987). In a sample of 25 scats, Emmons (1987) found opossums, squirrels, pacas, agoutis, capybara, brocket deer, tamandua, peccary, black spider monkey and olingo.

LEOPARD

Being a very widespread species, the leopard has a very variable diet. Leopards also appear to be more opportunistic than other cats and prey on a wider range of different species than any other felid.

In the Himalayas, Schaller (1977) found that leopards fed mainly on wild goats, but livestock and porcupines were also important components (Table 5.30).

In the Royal Chitwan National Park, prey includes wild pig, sambar, axis deer, hog deer, muntjac and domestic cattle (Seidensticker,

Table 5.30 The frequency of occurrence of prey species in the scats of the leopard from Karchat in the Himalayas

Prey	Frequency %	Prey	Frequency %
Wild goat	70	Hare	1
Urial	1	Porcupine	9
Chinkara	1	n=	70 scats
Livestock	14		

Source: Schaller (1977)

1976). In the Wilpattu National Park in Sri Lanka, 60 per cent of the 51 observed prey were axis deer (Muckenhirn and Eisenberg, 1973). Most of the rest of the diet comprised wild pig, but other prey included buffalo calves, sambar, muntjac, langur, porcupine, squirrel, hare and birds.

In the Kalahari Desert, the diet is very varied and no obvious preferences are observed owing to the small sample sizes (Table 5.31) (Bothma and Le Riche, 1986).

Table 5.31 The percentage of kills of different prey species in the diet of the leopard in the Kalahari Gemsbok National Park

Prey	Male	Female and cubs	Prey	Male	Female and cubs
Bat-eared fox	14	20	Steenbok	7	–
Jackal	–	10	Duiker	14	20
Genet	7	–	Gemsbok calf	21	–
Aardwolf	–	20	Springbok lamb	–	10
Springhare	–	10	Eland calf	–	10
Porcupine	29	–	n=	14	10
Aardvark	7	–			

Source: Bothma and Le Riche (1986)

In the richer hunting grounds of the Serengeti, a very varied diet has been recorded, but the main prey item is male Thomson's gazelles (Table 5.32) (Kruuk and Turner, 1967; Schaller, 1972).

Hoppe-Dominik (1984) has recently analysed 215 scats from leopards in Le Parc National de Tai in the Ivory Coast. More than 30 different prey species were found over a 15-month period and included duikers, monkeys, rodents and other carnivores (Table 5.33) (Hoppe-Dominik, 1984).

It is clear from Table 5.33 that over 99 per cent of the prey of a rainforest leopard consists of mammals. The great diversity of species that are consumed is evidence of just how opportunistic leopards can be.

TIGER

In the past, before it was so heavily persecuted, the tiger was very widespread. Today it is confined to a few relict populations. We only know of the diets of the Siberian tiger and Bengal tiger in any detail (Ewer, 1973; Sunquist, 1981).

Rod Williams/Howletts and Port Lympne Zoo Parks Kent

1 A young female ocelot – a typical South American cat

Jane Burton/Bruce Coleman Ltd

2 The European wildcat from Scotland is threatened by hybridisation with domestic cats

Kilverstone Wildlife Park

3 The pampas cat from South America resembles the unrelated European wildcat

Rod Williams/Howletts and Port Lympne Zoo Parks, Kent

4 The jungle cat ranges from North Africa to India

Gordon Clayton/Howletts and Port Lympne Zoo Parks, Kent

5 The Indian desert cat is probably an ancestor of Asian breeds of the domestic cat

Rod Williams/Howletts and Port Lympne Zoo Parks, Kent

6 The black-footed cat from southern Africa is also known as the ant-hill tiger

Rod Williams/Howletts and Port Lympne Zoo Parks, Kent

7 The pupils of the eyes of Pallas's cat close up to a circular opening rather than a vertical slit as in many other small cats

Shigeki Yasuma

8 Leopard cats are very widespread on the mainland and islands of southern Asia – this one is from Borneo

Shigeki Yasuma

9 Although only discovered in the 1960s, less than one hundred Iriomote cats survive today

Rod Williams/Howletts and Port Lympne Zoo Parks, Kent

10 The rusty-spotted cat is one of the smallest cats in the world

Gordon Clayton/Howletts and Port Lympne Zoo Parks, Kent

11 The fishing cat catches fish in its mouth

Rod Williams/Howletts and Port Lympne Zoo Parks, Kent

12 Fishing cat kittens playing in the water

Shigeki Yasuma

13 A flat-headed cat from Borneo – another fishing cat

Rod Williams/Howletts and Port Lympne Zoo Parks, Kent

14 One of the many colour varieties of the African golden cat

Evan Robinson/Howletts and Port Lympne Zoo Parks, Kent

15 The serval's large ear flaps help it to pinpoint prey in tall grasses

Rod Williams/Howletts and Port Lympne Zoo Parks, Kent

16 A male Temminck's golden cat from southern Asia

Gordon Clayton/Howletts and Port Lympne Zoo Parks, Kent

17 The caracal uses its black, tufted ears to communicate with other caracals

Gordon Clayton/Howletts and Port Lympne Zoo Parks, Kent

18 The clouded leopard uses its long tail for balancing when climbing

Rod Williams/Howletts and Port Lympne Zoo Parks, Kent

19 Tufted ears have evolved independently in the lynxes and the caracal – this lynx is from Siberia

Rod Williams/Howletts and Port Lympne Zoo Parks, Kent

20 The small marbled cat is really a big cat

Table 5.32 The percentage of kills of different prey species in the diet of the leopard in the Serengeti National Park

Prey	Kruuk and Turner, 1967 Frequency (%)	Schaller, 1972 Frequency (%)
Thomson's gazelle	27	63
Grant's gazelle	4	6
Reedbuck	11	12
Wildebeest	9	7
Topi	2	2
Hartebeest	–	1
Zebra	7	1
Waterbuck	–	1
Impala	16	–
Bushbuck	2	–
Warthog	–	1
Baboon	4	1
Golden jackal	1	1
Black-backed jackal	2	1
Bat-eared fox	–	1
Serval	–	1
Cheetah	2	–
Rock hyrax	2	–
Springhare	2	–
Python	2	–
Secretary bird	2	–
White stork	4	2
Helmeted guineafowl	2	–
Vulture	2	–
n=	55 kills	164 kills

Sources: Kruuk and Turner (1967); Schaller (1972)

The main prey of tigers in India is deer. Although calves are occasionally killed, the rhino and gaur are virtually immune to predation by the tiger. There did not appear to be any selection for deer of particular ages (Table 5.34) (Seidensticker, 1976; Sunquist, 1981).

The diet of the Siberian tiger mainly comprises wild pig and Manchurian wapiti (Table 5.35) (Abramov, 1962 in Schaller, 1967).

Table 5.33 The percentage frequency of occurrence of prey species in 215 leopard scats from Le Parc National de Tai, Ivory Coast

Species	% Occurrence	Species	% Occurrence
Ungulates:		Rodents:	
Maxwell's duiker	14.4	Brush-tailed porcupine	7.0
Black duiker	2.3	Porcupines indeterminate	0.5
Black/zebra duiker	10.7	Giant pouched rat	1.9
Ogilby's/bay duiker	4.7	Murids indeterminate	2.3
Duikers indeterminate	6.0	Rodents indeterminate	0.5
Royal antelope	2.3	Giant squirrel	1.4
Bongo	1.4	Ebien squirrel	0.9
Water chevrotain	1.9	Red-legged sun squirrel	0.9
Giant forest hog	0.5	Western ground squirrel	0.5
Bushpig	0.9	Squirrels indeterminate	3.3
		Dormouse indeterminate	0.5
Primates:		Carnivores:	
Diana monkey	7.9	Leopard	2.8
Mangabey	4.2	Pardine genet	0.5
Red colobus	3.7	Tigrine genet	0.5
Black and white colobus	2.3	Cusimanse	0.5
White-nosed monkey	2.3	Genet indeterminate	1.9
Campbell's monkey	1.9	Viverrid indeterminate	1.9
Monkeys indeterminate	1.4		
Potto	0.5		
		Others:	
		Tree hyrax	1.4
		Pangolins	4.2
		Mammals indeterminate	12.1
		Birds	0.9

Source: Hoppe-Dominik (1984)

Table 5.34 The frequency of occurrence of mammalian prey in the diet of the tiger

Prey	Kanha	Chitwan	Chitwan
Gaur	9	–	–
Swamp deer	10	absent	absent
Sambar	12	29	20
Axis deer	58	33	–
Hog deer	absent	15	62
Muntjac	–	4	–
Wild pig	1	11	4
Porcupine	3	1	–
Hare	–	1	–
Langur	7	6	4
Cattle/Buffalo	9	–	2
Civet	–	–	2
n=	300	123	55

Sources: Sunquist (1981); Schaller (1967)

Table 5.35 The percentage frequency of occurrence of prey in the diet of the Siberian tiger

Prey	Sikhota-Alin	Primorsk	Prey	Sikhota-Alin	Primorsk
Wild pig	36	30	Roe deer	3	2.5
Wapiti	22	50	Lynx	2	–
Moose	10	2.5	Sika deer	–	5
Brown bear	8	5	Badger	–	2.5
Musk deer	14	2.5	n=	59	40
Hazel hen	5	–			

Source: Abramov (1962) in Schaller (1967)

LION

There have been many studies of the diets of lions in the wild in Africa. Table 5.36 summarises some of these:

Table 5.36 The percentage frequency of the commoner prey killed by lions in different parts of Africa

Prey	Manyara 1967–9	Serengeti Plains 1967–9	Kafue 1960–3	Kruger 1954–66
Wildebeest	2	57	6	24
Zebra	16	29	7	16
Thommie	–	7.5	–	–
Impala	11	–	2	20
Waterbuck	1	–	6	10
Eland	–	3	3	0.5
Hartebeest	–	0.4	16	–
Warthog	–	–	10	2
Giraffe	2	–	–	4
Buffalo	62	–	30	9
Bushbuck	–	–	0.2	0.3
Bushpig	–	–	2	+
Duiker	–	–	0.2	0.1
Hippo	–	–	1	+
Kudu	–	–	1	11
Lechwe	–	–	0.5	–
Puku	–	–	1	–
Reedbuck	–	–	2	0.3
Roan Antelope	–	–	6	0.3
Sable Antelope	–	–	5	2
Tsessebe/topi	–	1.4	–	0.4
Grant's gazelle	–	1.1	–	–
Baboon	6	–	–	+
Lion	–	0.4	–	–
Hyaena	–	0.4	–	–
Ostrich	–	–	–	0.1
n=	100	280	410	12,313
No. prey species	7	9	19	38

Source: Schaller (1972)

In the Kalahari Gemsbok National Park, the number of large prey animals available is limited and lions feed on smaller animals (Eloff, 1973). Schaller (1972) found that in general lions kill prey in good or only slightly poor condition in the Serengeti. A much longer discussion of the variations in diet and preferences shown by lions can be found in Schaller (1972) and Bertram (1978).

6

LIVING APART . . . LIVING TOGETHER

The social life of wild cats is usually dismissed as being solitary, with animals only meeting each other at infrequent intervals either for mating or for a disagreement about territorial boundaries. This would clearly be a very dangerous social system and would probably result in damaging aggressive conflicts between neighbouring animals, whenever they should meet. It may come, therefore, as a surprise to learn that even so-called solitary cats are polygynous or even promiscuous. In other words, male cats may mate with more than one female, and females may mate with more than one male. However, polygyny is not necessarily maintained by direct competition between males as it is in many ungulates and pinnipeds. Instead, it is achieved indirectly by defending an area where males attempt to maintain exclusive mating rights.

All cats have highly developed means of communication using more than one sensory channel, which help them not only to convey their moods and intentions to other cats, but also to maintain their individual living space. The land tenure system of wild cats is usually very stable and its maintenance is not necessarily reliant on frequent direct social interactions.

At close range, cats use visual and vocal signals, and contact patterns, to convey aggressive, defensive and amicable information to a conspecific. At a longer range, olfactory signals are a much more important means of communication to maintain the integrity of home ranges and to convey information about reproductive and dominance status, and sexual and individual identities. Our understanding of the dispersed social system of wild cats has come a long way from the earliest studies on lynxes whose footprints were followed through the winter snow (e.g. Haglund, 1966). The development and miniaturisation of radiotelemetry technology has meant that it is now possible to study the movements and social behaviour of all wild cats even in dense tropical forests (e.g. Emmons, 1988; Izawa and Ono, 1989). In future years, we can only make ever more exciting discoveries about the social lives of all wild cats.

Although most wild cats from the rusty-spotted cat to the tiger have adopted a dispersed social system, three species of cat have chosen to live together for their own good reasons. The lion, feral domestic cat and male cheetah all share home ranges and interact positively with conspecifics. Why are they the only cats to do so? This chapter will explore the basis for cat sociality and discuss the current theories that attempt to explain it. Firstly though, we explore the different means of communication open to cats.

VISUAL COMMUNICATION

As we may know from watching domestic cats, they have a wide range of expressive visual behaviours, which are unambiguous in their meaning even to us. The relaxed look of a domestic cat coupled with its persistent purring and rubbing against its owner or nearby objects cannot be confused with the flattened ears, snarling face and lashing tail of a furious, spitting beast. Paul Leyhausen has studied this visual semaphore to construct a chart of both the facial and body expressions of domestic cats with increasing levels of aggression and fright, and varying combinations of both (Figure 6.1) (Leyhausen, 1979). Some of these signals are very difficult to distinguish from others, but a cat may not always want to be totally understood and may want to hedge its bets depending on the reaction of another individual. It may not be advantageous to tell another cat that you are ready to retreat, if by giving a confusing or intermediate signal you may be able to stand your ground successfully and call your opponent's bluff. In all, domestic cats have some 25 different visual signals which are used in 16 different combinations. Visual signals are often used in conjunction with contact patterns or vocal signals to convey a cat's moods and intentions (Eisenberg, 1973). Lions have a similar diversity of visual signals (17), which are of vital importance to this social cat (Schaller, 1972). Leyhausen (1979) has also observed similar visual signals in a wide variety of different wild cats in captivity (Figure 6.2).

Body markings often emphasise specific visual signals. The ears of cats are particularly important social signallers. Many cat species have white or pale spots on the backs of their dark ears. In aggressive and defensive encounters, flattened ears display their contrasting spots to great effect to reinforce the message. The ear tufts of lynxes and caracals seem to have taken over from their tails in social signalling. Kingdon (1977) discusses in great detail how the caracal uses its elegant, black-tufted ears in head-flagging displays.

a

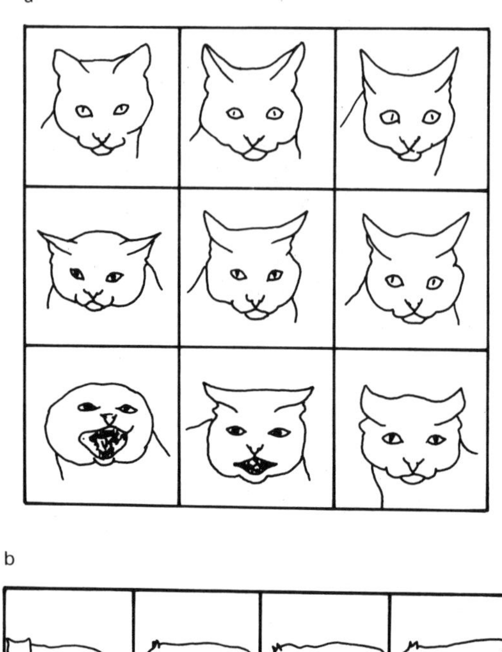

b

Figure 6.1 The variations in the facial (**a**) and postural (**b**) signals of the domestic cat with increasing levels of aggression across the page and increasing levels of fear down the page (Leyhausen, 1979) (A.K.)

The jungle cat has a shortish tail and slight ear tufts. Perhaps this species represents an intermediate stage in the shift in communication function from the tail to the ears?

Tails are also used to convey aggressive or defensive intentions depending on whether they are wafted about or held low (Kiley-Worthington, 1976). When domestic cats lash their tails they are usually excited or aggressive. They hold them high when greeting

Figure 6.2 A medium intensity threat by a Fishing cat (After Nowak and Paradiso, 1983) (S.A.)

another individual or investigating their surroundings, and they hold them low when acting defensively or during prey capture (Kiley-Worthington, 1976). Many cats have a white-tipped tail (e.g. cheetah) or a tail with a pale or white underside whose tip is just held clear of the ground (e.g. Temminck's golden cat) (Figure 6.3). Leyhausen (1979) has suggested that this may help kittens to follow their mothers in the dark or through tall grass. Female aardvarks (*Orycteropus afer*) are said to have a white-tipped tail for the same reason.

Figure 6.3 The white-tipped tail of the cheetah (**a**) and the white underside of the tail of the Temminck's golden cat (**b**) may help cubs and kittens follow them through tall grass and in dark forests (after Leyhausen, 1979) (S.A.)

VOCAL COMMUNICATION

There has been very little investigation into the vocal communication of cats. Again we may be familiar with the purring of a contented cat and the yowling and spitting of an angry cat, but angry cats may purr, too! In all, domestic cats have eight different vocalisations and some intermediates (Eisenberg, 1973). Wemmer and Scow (1977) studied the vocal communication of a variety of wild cats in the Brookfield Zoo, Chicago. These included sand cat, Pallas's cat, leopard cat, Canada lynx, fishing cat, margay, puma, clouded leopard, Temminck's golden cat and jaguarundi. They found that there were six basic calls which were characteristic of small wild cats (Table 6.1).

Table 6.1 The characteristics of the vocalisations of small wild cats

Vocalisation	Harmonic structure	Duration	Relative intensity	Frequency modulation
Hiss	–	Variable but usually brief	Moderate	–
Spit	–	'Fixed'	Moderately loud	–
Purr	–	Variable but usually long and repeated	Soft	–
Growl	+	Variable but usually long and repeated	Moderate	+
Miaow	+	Variable	Moderately loud	+
Scream	+	Variable	Loud	+

Source: Wemmer and Scow (1977)

All of these vocal signals vary in their intensity, duration and rate of emission. Wemmer and Scow (1977) found that often one call would change to another via a combined call, and often in a specific sequence. For example, in aggressive encounters, howling nearly always replaced growling, and was followed by screaming and finally spitting. The sequence would be started again with another growl. Most of the observed vocalisations were used in conflict situations rather than amicably, except, of course, purring.

Most vocal signals are used in short-range direct communication, except incidentally. Often female domestic cats in oestrus miaow frequently, which combined with olfactory signals and rubbing may attract several males. Male domestic cats make a chirping call which signals their intentions towards the female. At the Cat Survival Trust, it has been noticed that male jungle cats produce a barking call when the female is in oestrus. It would be interesting to know whether this call is equivalent to the chirping of male domestic cats, or if it is a warning to other males to keep away from the female, or if it may help the female to locate the male in a closed habitat. Schaller (1967) found that jungle cats would gather at 'meeting places' at the boundaries of home ranges during the breeding season in the same way that urban feral cats interact throughout much of the year (Tabor, 1983). This barking call does not appear to have been recorded in the wild.

CONTACT PATTERNS

A typical domestic cat's greeting consists of much enthusiastic head and body rubbing with some delicate sniffing, but there has been very little study of contact patterns in wild cats (Figure 6.4). Wemmer and Scow (1977) studied contact patterns between siblings of three different species of wild cat: Pallas's cat, sand cat and leopard cat. However, they did not discover the social significance of these contact patterns. Four contact patterns were common to all three cats: rubbing, sniffing, biting and patting with the forepaw. However, there were interspecific variations as to which contact

Figure 6.4 The typical feline greeting as shown by two servals (after Geertsema, 1985) (S.A.)

pattern was used most often and to which part of the body contact was directed. For example, rubbing was common in Pallas's cats, but rare in sand or leopard cats. Most rubbing in the Pallas's cat was body to body (79 per cent) (of which 43 per cent of contacts were against the mother's chest), but very rarely against the head and neck.

Sniffing was a rare activity in all cats, especially compared with other carnivores which have a better-developed sense of smell. Meerkats (*Suricata suricatta*), for example, sniff nearly eight times an hour compared with just over twice an hour for leopard and Pallas's cats. All three cat species sniffed the head/neck and tail, but sniffing of other parts of the body showed some interspecific variation (Figures 6.5, 6.6).

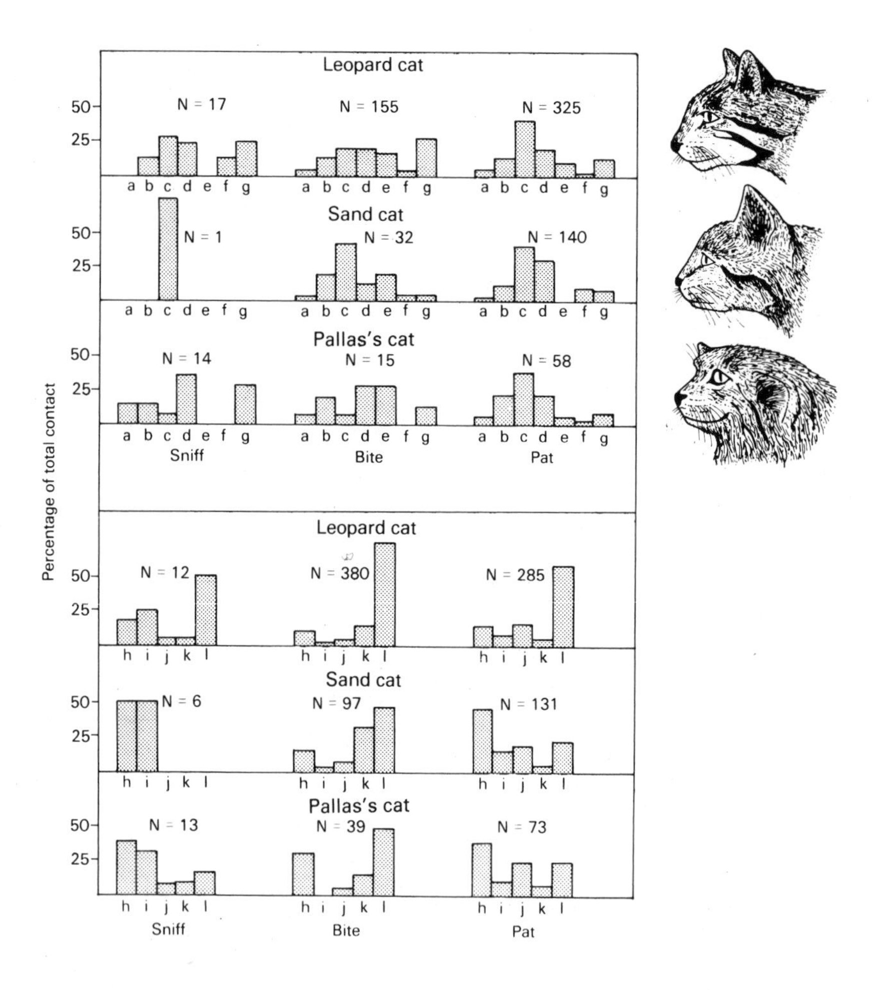

Figure 6.5 The percentage distribution of sniffing, biting and patting to specific regions of the body (upper three graphs) and to specific head regions (lower three graphs): (**a**) chest, (**b**) shoulder, (**c**) back, (**d**) side, (**e**) belly, (**f**) haunch, (**g**) rump, (**h**) ear, (**i**) nose, (**j**) cheek, (**k**) throat, (**l**) neck (after Wemmer and Scow, 1977) (S.A.)

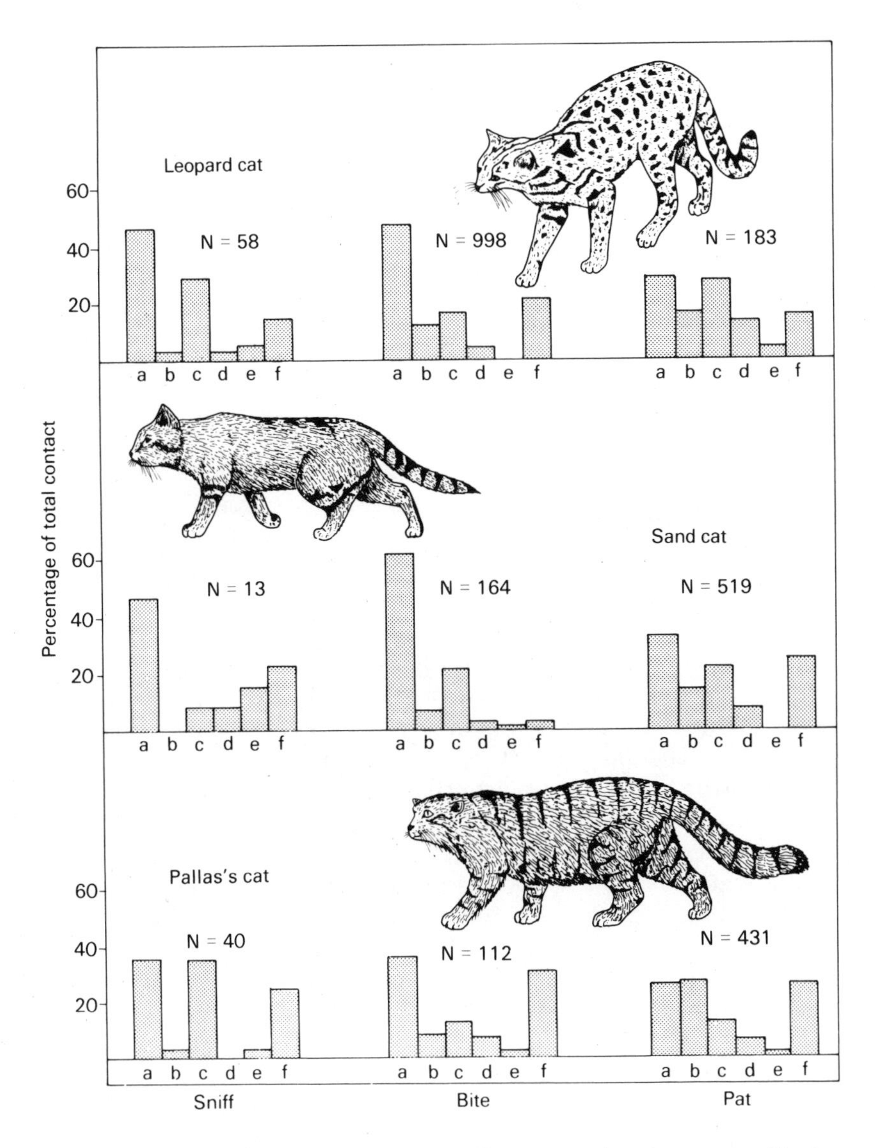

Figure 6.6 The percentage distribution of sniffing, biting and patting to general body areas of siblings in leopard cats, sand cats and Pallas's cats. (**a**) head/neck, (**b**) forelimbs/feet, (**c**) body, (**d**) hindlimbs/feet, (**e**) ano-genital region, (**f**) tail (after Wemmer and Scow, 1977) (S.A.)

Biting was directed against the head and neck, with the body being usually the next most common site.

Contact patterns are a poorly studied area of cat communication, but one which could easily be further elucidated by studies of captive animals. Until such studies are carried out, we can have little idea of their role in the social communication of cats in the wild

OLFACTORY COMMUNICATION

Carnivore bodies are packed with an assortment of skin glands which are used in scent marking. Cats have a relatively poor sense of smell compared with other carnivores, but still rely heavily on scent marking to convey a wealth of information in their absence. This is very important for most wild cats which meet relatively infrequently compared with the social lion and domestic cat, but even in these two species scent marking is very important. Scents can convey information about individual identity, sex, reproductive status and limits of home ranges. Most routine scent marking in home ranges involves those otherwise wasted body products, urine and faeces.

SCENT GLANDS

There are four main sites for scent glands on a cat's body (Macdonald, 1985):

(1) Face. There is glandular tissue (sebaceous glands) around the chin, lips, cheeks, mystacial and genal whiskers. Cats regularly rub their faces and cheeks against objects and other cats, which is the basis for the typical feline greeting, when those parts of the head that are particularly endowed with sebaceous glands are rubbed against other cats and objects (Figure 6.4). Rubbing transfers and mixes body scents between familiar cats, thereby reinforcing and maintaining social bonds.

(2) Anus. There are two sac-like glands which open into the end of the rectum. The secretion from these glands can be mixed with faeces (e.g. McDougal and Smith, 1986; Smith, McDougal and Miquelle, 1988). It is also claimed to be produced in association with urine during the scent marking by, for example, tigers and lions (Schaller, 1967, 1972). However, recently Brahmachary and Dutta (1987) and Smith *et al.* (1988) have failed to find any evidence of this in tigers and doubt its possibility on anatomical grounds! There are also sebaceous glands in the skin surrounding the anus. Albone and Grönneberg (1977) have analysed lion anal sac secretion and compared it with that of the fox. They found that lion anal sac secretion has much more lipid content than that of the fox, and consists of a complex mixture of low molecular weight fatty acid chains which are probably produced as a result of microbial action on sebum produced by sebaceous glands.

(3) Above the tail. Above the base of the tail there is a large sebaceous gland (the supracaudal gland) which is rubbed against objects and familiar cats.

(4) Between the toes. Interdigital glands may produce secretions which are deposited on scratching posts to create a combined olfactory and visual advertisement. Jaguars, tigers and lions may

habitually use certain trees for scratching so that the bark is removed to leave a visually distinctive scratching post (Schaller, 1967, 1972; Schaller and Crawshaw, 1980). When cats use scratching posts, they certainly sniff the area as if sampling the scents they are producing, or those left previously by other cats.

However, for many of their most important olfactory communications (for so-called solitary cats at least), cats reserve cheap and otherwise wasted smelly substances, which may be elaborated by other secretions and substances to communicate different meanings. Urine and faeces are the common languages in the felid world, which they use to maintain a stable living space where they survive and breed.

HOME RANGES

Wild cats occupy an area which is usually described either as a territory or home range. This area consists of a series of trails which link together hunting areas, drinking places, resting areas, lookout positions and denning sites where kittens or cubs can be safely reared. Although cats may drive out most intruding cats from these areas, some are tolerated. However, these areas may vary in time, position and space and may overlap considerably with those of conspecifics. Therefore the looser term of home range is usually ascribed to the living area of most cats, rather than the more strictly defined 'territory'. Within the home range there may be an area which does not vary and is fiercely defended. This core area may be more properly regarded as a territory. For example, in South Carolina it was found that female bobcats occupied a definite core area, but males moved in a linear fashion over a much wider area and did not maintain a core range (Figure 6.7) (Fendley and Buie, 1986). Female domestic cats on a farm in Cornwall ranged over areas of 0.7 to 1.5 hectares, but spent 81 per cent of their time in the tiny core area of 0.2 hectares (Panaman, 1981).

Cats, therefore, maintain home ranges that provide for all their needs during the course of a year. However, male and female home ranges have quite different functions. Female home ranges are usually smaller and are used to provide sufficient prey and denning sites for rearing kittens every year, including years of low prey density. Female cats do not usually tolerate other females in their home ranges, but they do tolerate males.

Male home ranges overlap with two, three or more female ranges. Clearly this larger area will provide sufficient food for a male, even if he feeds on larger prey which may occur at a lower population density (see Chapter 4). This pattern of home range use has been recorded in tigers, servals, ocelots, wildcats and feral domestic cats (Figure 6.8) (Corbett, 1979; Sunquist, 1981; Liberg, 1984a; Geertsema, 1985; Emmons, 1988).

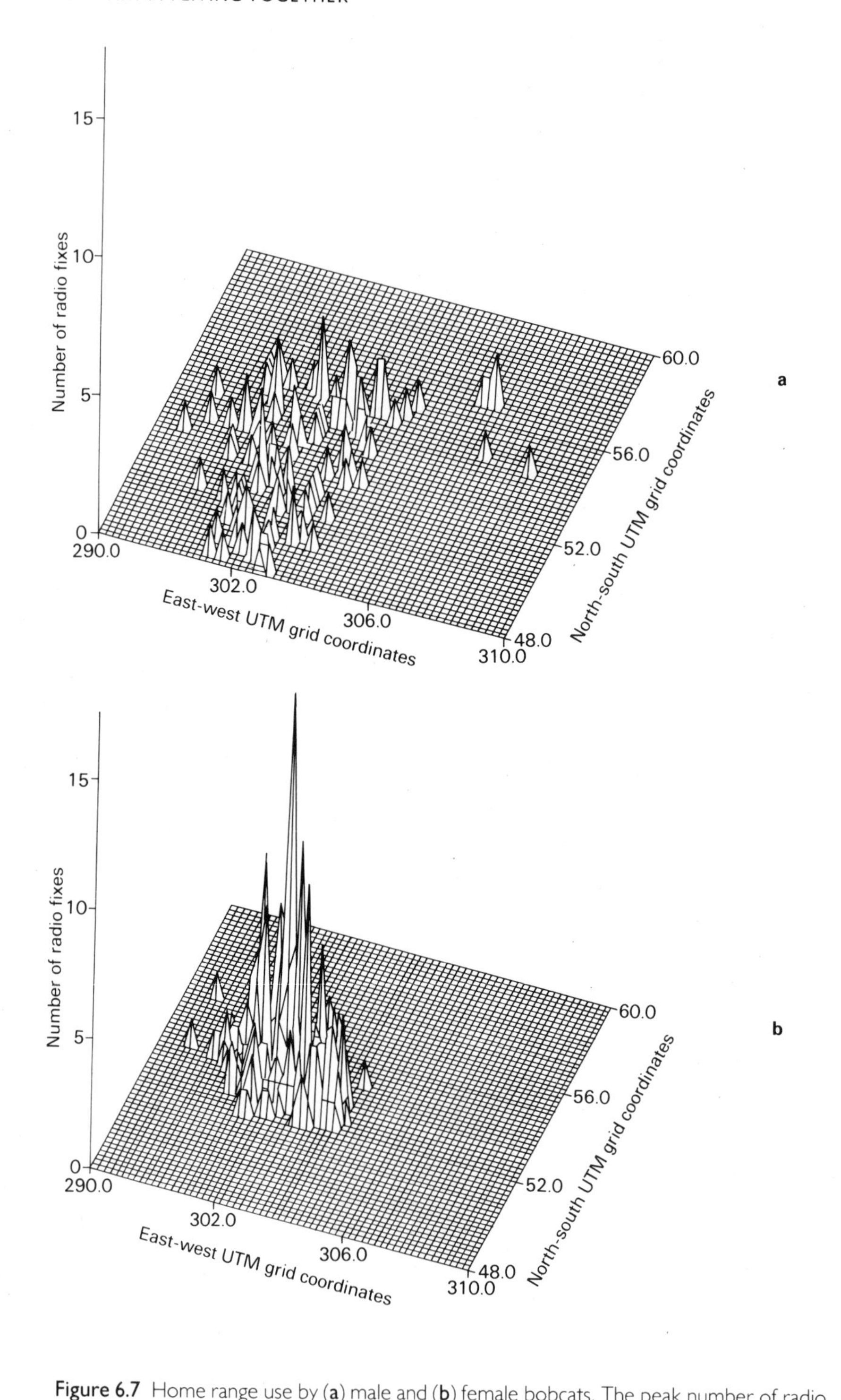

Figure 6.7 Home range use by (**a**) male and (**b**) female bobcats. The peak number of radio fixes for the female bobcat coincides with the den where she reared her kittens (Lancia *et al.*, 1986) (S.A.)

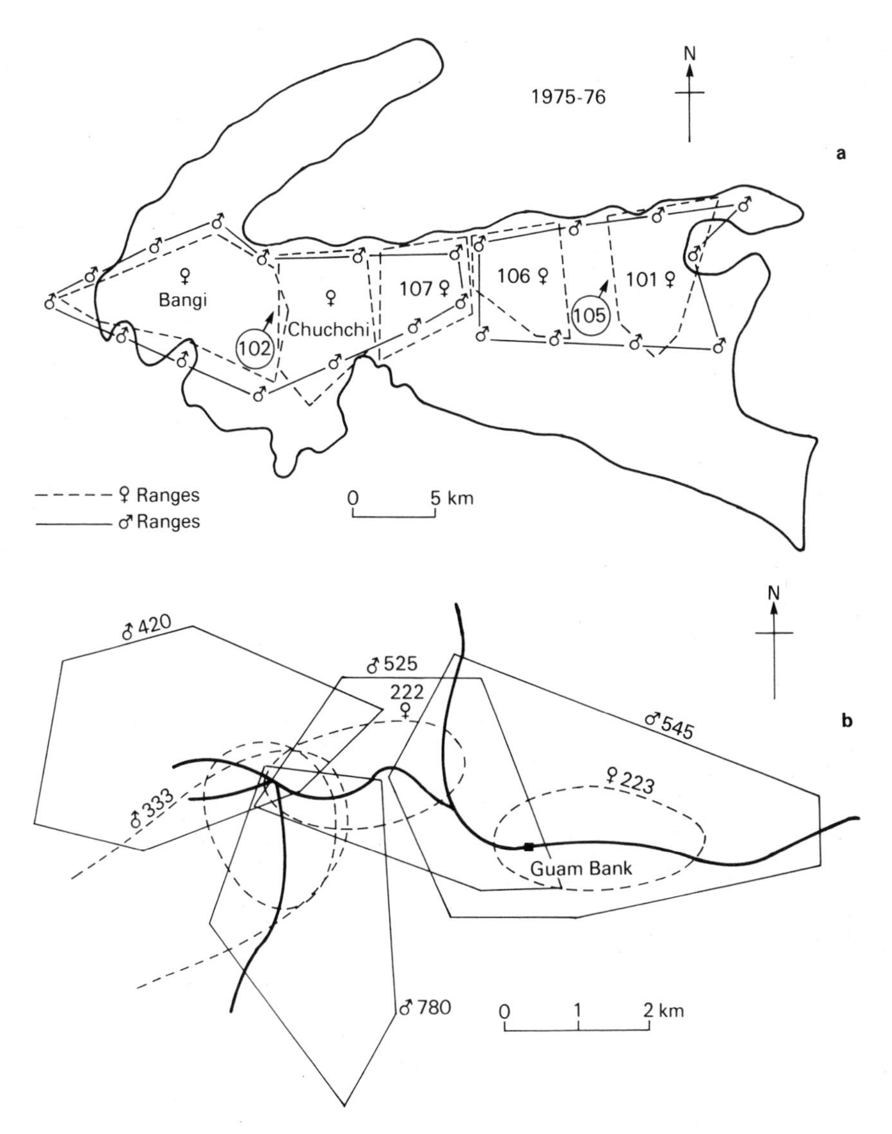

Figure 6.8 The layout of home ranges of wild cats: (**a**) tiger, (**b**) jaguar (Sunquist, 1981; Rabinowitz and Nottingham, 1986) (S.A.)

The function of a male's home range is to provide an area where he can mate with as many females as possible without interference from surrounding males. Male home ranges, therefore, do not usually overlap. Thus, far from having a chaotic, random system of home ranges driven by the need for 'solitary' cats to avoid each other at all costs, most wild cats maintain a predictable system of land tenure, which promotes social stability and maximises the reproductive success of both males and females. Males mate with more than one female to maximise the number of progeny they

Table 6.2 Home ranges and population densities of wild cats. Means of ranges shown in brackets

Species	Locality	Male home range (km²)	Female home range (km²)	Population density (100km⁻²)	Reference
Ocelot	Texas	3.7	2.8		1
	Texas	3.5	2.1		1a
	Costa Rica	–	–	14–25	1a
	Peru	1.2–8.1	1.1–2.5		27
Wildcat	Scotland	1.76*	1.75*		28
	France			3–70	49
	(spring)	0.3–0.5	–		9
	(summer)	1.2–3.3	–		
	(autumn)	0.25–1.35	–		
	(winter)	1.8	0.5–0.9		
	Czechoslovakia			1–70	49
	Germany			33.3	49
African wildcat	Kenya	4.3			12
Feral cat	Portsmouth	0.0008–0.24(0.08)	0.0003–0.04(0.008)		3
	Dassen I.	0.32–0.63(0.44)	0.11–0.32(0.19)	165–228	6
	Marion I.			9.8	34
	S Sweden			250–330	35
	Devon	0.6	0.02–0.07		8
Pallas's cat	–	4.0			13
Amur leopard cat	Siberia	9.0			13
Iriomote cat	Iriomote	2.96+/–1.8	1.75+/–0.8		28
Serval	Ngorongoro	8.1	4.2		29
Jaguarundi	Costa Rica			25–330	1a
Puma	Idaho	453	173–373	1.4	10
	Utah	826	685(396–1454)	0.3–0.5	54
	California	82–301	62–71	1.2–2.3	36
	Colorado			2.1–2.3	38
	Arizona	189 (142–236)*	67+*	1.7	39
	Nevada	211 (181–259)		1.0	40
	California			1.2	41
	California			2.3–2.7	37
	California	152(109–238)	66(57–74)	1.5–3.3	55
	California	178(78–277)	94(54–119)	3.5–4.4	56
Bobcat	Idaho	6.5–107.9(42.1)	9.1–45.3(19.3)	5.4	4
	S Carolina	62(52–76)	14(8–22)		43
	S Carolina	2–4	3–5		19
	S Carolina	25.6	11.4		5
	California	0.3–5.6	0.2–3.4	115–153	11
	California	73(39–95)	43(26–59)	5	52
	N Carolina	37.7	22.1		30
	Arizona			24–28	45
	Arizona	9.1	4.8	26	53
	Florida			500	46
	Louisiana	4.9	1.0		49
	Tennessee	30.9	14		50
	Minnesota	62(13–201)	38(15–92)	4–5	51

Table 6.2 Cont.

Species	Locality	Male home range (km^2)	Female home range (km^2)	Population density ($100km^{-2}$)	Reference
Canada lynx	Newfoundland	18–21	16		7
	Newfoundland			3.9–7.7	47
	Alaska	14–25	10–26	0.9	20
	Minnesota	145–243	51–122		24
	Alaska	783	70(51–89)	0.9	25
	Cape Breton Is.				
	(summer)	25.6	32.3	20	32
	(winter)	12.3	18.6		32
	Alberta	28(11.1–49.5)		3.8–10	48
Eurasian lynx	N Alps	275–450	96–135	1.2	42
	C Alps	–	46		42
	USSR			5	32
Snow leopard	–	138			13
Leopard	Sri Lanka	9–10	8–10		17
	Serengeti		40–60		18
Jaguar	Brazil	50–76	25–38	4–8	2
	Belize	33.4(28–40)	10–11		34
Tiger	USSR	800–1000	100–400	1.0	16
	Nepal	60–72	16–20	2.8	14
	India	78	65		15
Lion	Serengeti	–	30–400		18
	Kalahari	–	119–275		21
	Nairobi NP	–	19–31		22
Cheetah	Serengeti	–	60–65		18
	Serengeti	12–36	800		23

* Seasonal range

References:

(1) Tewes and Everett, 1986
(1a) Tewes and Schmidly, 1987
(2) Schaller & Crawshaw, 1980
(3) Dards, 1978
(4) Bailey, 1974
(5) Fendley and Buie, 1986
(6) Apps, 1983
(7) Saunders, 1963b
(8) Macdonald and Apps, 1978
(9) Artois, 1985
(10) Seidensticker *et al.*, 1973
(11) Lembeck, 1986
(12) Fuller *et al.*, 1988
(13) Goszczynski, 1986
(14) Sunquist, 1981
(15) Schaller, 1967
(16) Matjushkin *et al.*, 1977
(17) Eisenberg and Lockhart, 1972
(18) Schaller, 1972
(19) Provost *et al.*, 1973
(20) Berrie, 1973
(21) Eloff, 1973
(22) Rudnai, 1973
(23) Frame, 1984
(24) Mech, 1980
(25) Bailey *et al.*, 1986
(26) Izawa and Ono, 1989
(27) Emmons, 1988
(28) Corbett, 1979
(29) Geertsema, 1985
(30) Lancia *et al.*, 1986
(32) Parker *et al.*, 1983
(33) Rabinowitz and Nottingham, 1986
(34) van Aarde, 1979
(35) Liberg, 1980
(36) Hopkins *et al.*, 1986
(37) Sitton and Wallen, 1976
(38) Currier, 1976
(39) Shaw, 1973
(40) Ashman, 1975
(41) Koford, 1977, 1978
(42) Haller and Breitenmoser, 1986
(43) Griffith and Fendley, 1986a
(44) Brand and Keith, 1979
(45) Jones and Smith, 1979
(46) Beasom and Moore, 1977
(47) Bergerud, 1971
(48) Brand *et al.*, 1976
(49) Hall and Newsom, 1978
(50) Kitchings and Story, 1978
(51) Berg, 1979
(52) Zezulak and Schwab, 1979
(53) Lawhead, 1978
(54) Hemker *et al.*, 1984
(55) Kutilek *et al.*, 1980
(56) Sitton *et al.*, 1976

sire, and females are defended from intruding males who might kill their young. Far from being strangers, neighbouring cats probably know each other very well from their own distinctive smells.

Although cats do not usually tolerate members of their own sex in their home ranges, there are exceptions to this rule. In Idaho and California, male bobcats may have home ranges which overlap slightly (Bailey, 1974) or up to 89 per cent (Lembeck, 1986). It is suggested that the very large home ranges in Idaho make it impossible to defend them against intruders, but this is not the case for the smaller ranges in California. Male jaguar home ranges may also overlap by up to 80 per cent with those of neighbouring males in the Cockscomb Basin, Belize (Figure 6.8) (Rabinowitz and Nottingham, 1986). However, prey density is so high that male jaguars commonly occupy areas of as little as 2.5 km^2 out of their annual home range of 30 km^2 for periods of up to two weeks. Presumably by not reinforcing the scent marks throughout the home range all year round, other males are apt to wander in to 'occupied' areas. Recent studies of the Iriomote cat have shown that although the home ranges of females do not overlap with those of other females, male ranges overlap extensively (Izawa and Ono, 1989).

The home ranges of male pumas in the Mount Hamilton area of the Diablo Range of California also overlap, whereas females have exclusive home ranges (Hopkins, Kutilek and Shreve, 1986). In contrast, female pumas in the Idaho Primitive Area may share home ranges, whereas male ranges are exclusive. Seidensticker *et al.* (1973) thought that this might be due to the high degree of mobility of the puma's favourite prey (deer) and the limited availability of hunting sites within the habitat. If female pumas occupied smaller discrete home ranges, they might go hungry waiting for dinner to arrive. The young may also settle (for a while) in the home range of their mother until they are able to fend for themselves or are eventually driven away. Occasionally a young female may be able to settle permanently in part of her mother's range.

Home ranges vary considerably both between and within species. In general, larger species of cats occupy larger home ranges than smaller species. Tigers may occupy home ranges of 1000 km^2, whereas the Iriomote cat ranges over a mere 2.9 km^2 (Sunquist, 1981; Izawa and Ono, 1989). Since optimum prey size increases with body size and population density of larger prey falls with body size, we would expect larger cats to have to move further between meals and so range over a larger area (see Chapter 4). However, even within a species the home range size can vary a great deal. For example, Canada lynx have been recorded occupying home ranges which vary from 30 km^2 to more than 700 km^2 (Table 6.2) (Saunders, 1963b; Berrie, 1973; Brand and Keith, 1979; Bailey *et al.*, 1986). This variation may be mediated by availability of prey, in this case snowshoe hare, and suitable cover and conditions for huning (see Chapter 4). The sizes of home ranges and population densities of various species of wild cat are shown in Table 6.2

Home range size may vary with season. In the Idaho Primitive Area, pumas occupy different summer and winter home ranges. In summer, the pumas and their deer prey are found at high elevations, but in winter bad weather forces both deer and puma to lower elevations (Figure 6.9) (Hornocker, 1969). Deep snow restricts movement, so that winter home ranges are smaller than summer ones (Figure 6.9) (Seidensticker *et al.*, 1973). Litvaitis, Major and Sherburne (1987) looked at how bobcat home range size varied with time of year in Maine. Females occupied larger areas during gestation and training their kittens and minimum areas when nursing and weaning. During the winter, snow limited bobcat movements to an intermediate home range size. Strangely, male bobcats showed an identical pattern of annual home range use (Figure 6.10). Males should not be restricted in their movements while females are nursing their kittens. However, the main prey species of the bobcat, the snowshoe hare, is also producing young at this time, so that there is a glut of available food. In other words, male bobcats do not have to range over large areas to find sufficient prey, and females are timing the births of their kittens to when prey is most abundant, energetic costs in lactation are high and movements away from the kittens need to be most limited.

Home range size was also found to be limited for a female serval with kittens in the Ngorongoro Crater, Tanzania (Geertsema, 1985). The need to return to the kittens frequently to feed them, and also to defend the area near the den from other cats and predators, probably restricted the female's movements. Once the kittens were sufficiently grown and were safe from infanticide or predation, they followed their mother on hunting trips and her home range increased in size once more (Figure 6.11) (Geertsema, 1985).

Although a home range is usually expressed as an area covered by a cat over a particular period of time (usually one year), the cat does not use all parts of its range equally. Heller and Fendley (1986) found that bobcats in South Carolina primarily used bottomland hardwoods even though they represented only 17 per cent of the total home range. In contrast, 5–14-year-old or approximately 15-year-old pine plantations were avoided, although they occupied 64 per cent of the area. The reason for this selection is not hard to find, since the bobcats' main prey species were mostly found only in the bottomland hardwoods. A similar pattern of habitat use was found for bobcats on the Quabbin Reservation, Massachusetts (McCord, 1974). Bobcats spent most of their time hunting deer in hemlock hardwood where prey density was highest. They rested and hunted snowshoe hares in the spruce plantations, bred in the

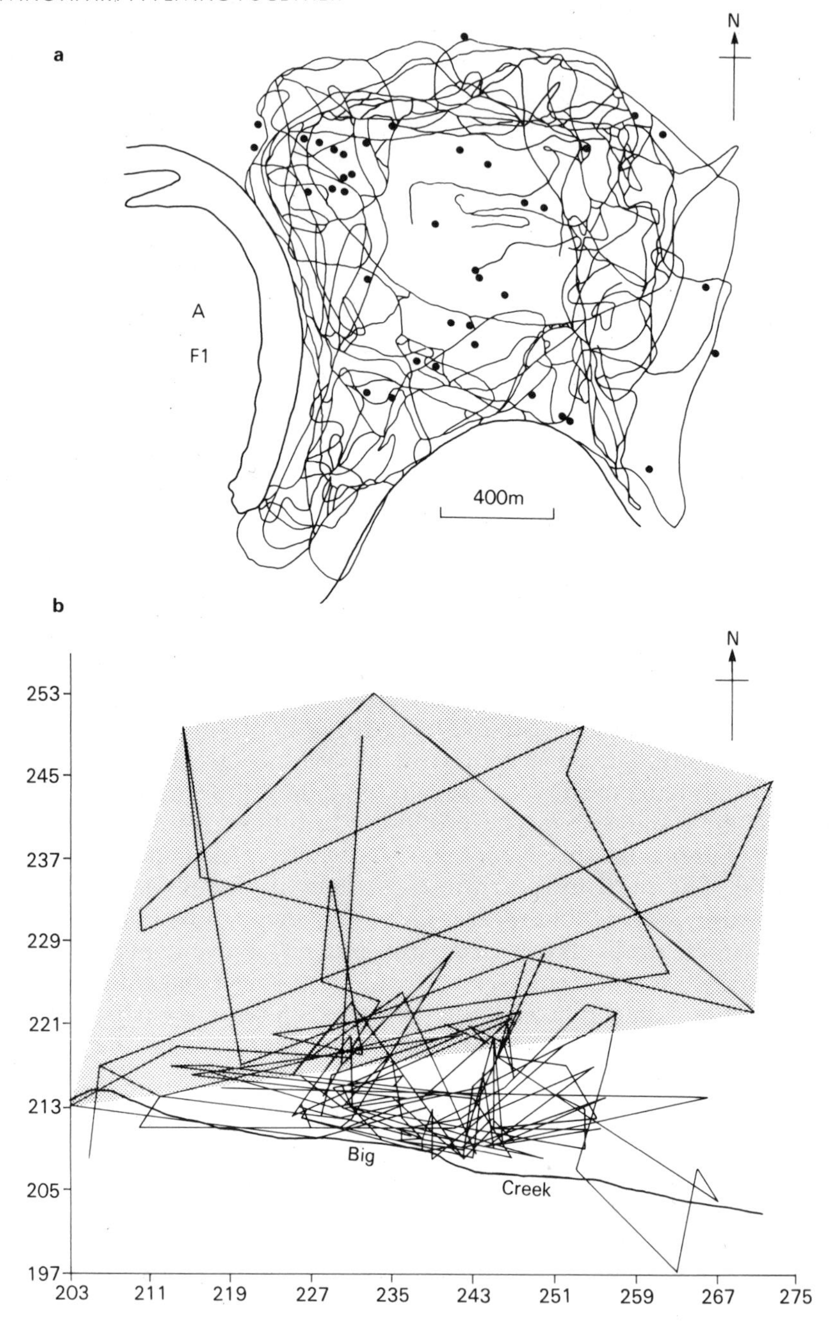

Figure 6.9 (**a**) Home range use by a female ocelot showing that most activity is confined to the boundaries (Emmons, 1988). (**b**) Home range use by a male puma in the Idaho Primitive Area, showing separate summer (stippled) and winter ranges (Seidensticker *et al.*, 1973) (S.A.)

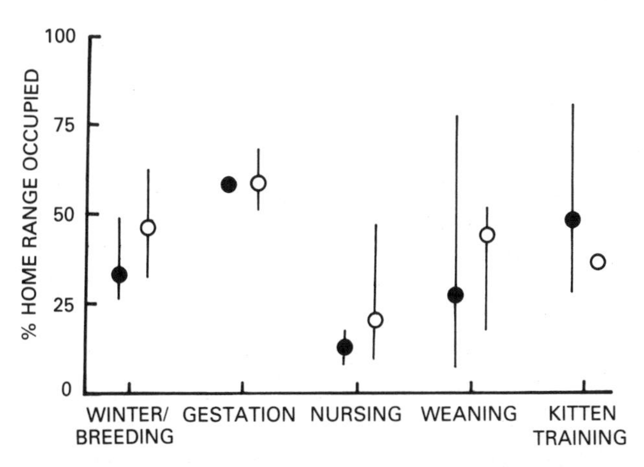

Figure 6.10 Percentage of home range used by female ○ and male ● bobcats during the course of a year (Litvaitis *et al.*, 1986) (A.K.)

cliffs and used man-made roads in winter to conserve energy rather than moving through snow.

Home range size and use may also be affected by dominance where cats share home ranges. In a study of domestic cats on North Uist, dominant cats had larger home ranges than subordinate cats (Corbett, 1979). Dominant cats were thus able to have priority use of larger areas containing the best hunting sites and use the optimal mobile hunting strategy to kill more rabbit prey than subordinate cats.

When young cats gain independence from their mother, they must go through a transient stage, continuing to move through the home ranges of other cats until they are able to find a vacant area to establish as their own. This is a critical stage in life for all cats as their longevity and breeding success is severely limited if they remain as transients. In fact it is unlikely that transient cats breed successfully, if at all. Geertsema (1985) found that young servals dispersed up to eight kilometres from their natal range, moving erratically through the home ranges of residents. It was never discovered where or if they became settled during five years of study. Only in populations of wild cats which are subject to heavy human exploitation are transients going to find it easy to establish a new home range, because the stability of the land tenure system and potential long life of most species of wild cat probably provide few openings except in marginal habitats.

SCENT MARKING

Although cats will actively defend their home range from other cats, if they should ever stumble across them, home ranges are usually too large to be continually defended. Some tropical cats

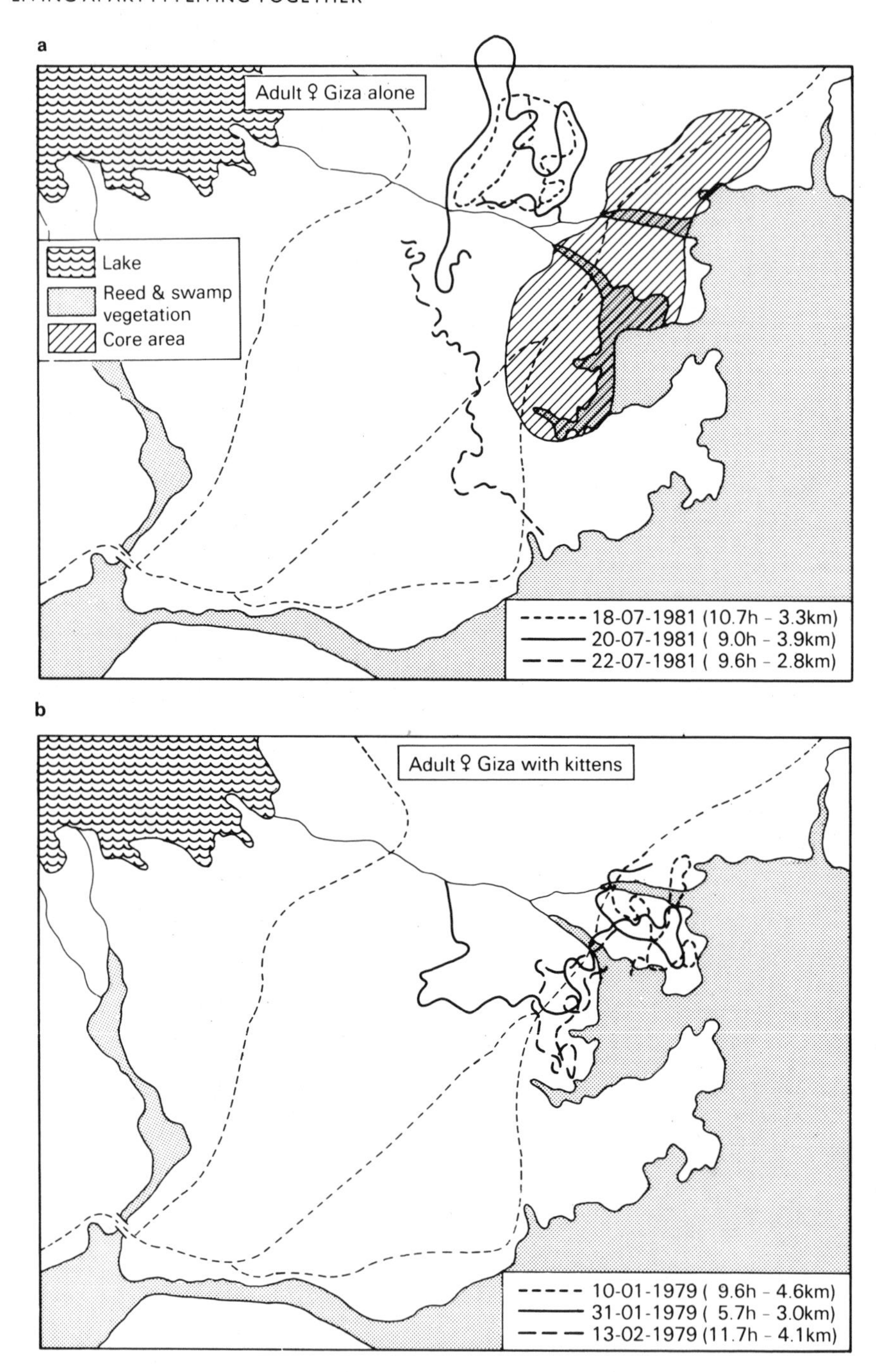

Figure 6.11 (**a**) The home range use by a female serval, showing the core area and movements outside this area. (**b**) Home range use by the same female with kittens, showing most activity confined to core area (Geertsema, 1985) (S.A.)

such as the ocelot have small stable home ranges which they can completely patrol every two days (Figure 6.9) (Emmons, 1988), but many temperate species cannot do this and may use different areas of their total home range throughout the year. How then do cats tell intruders that they are at home and deter them from trying to settle in their home range? The answer seems to be a sophisticated system of scent marking which is continually updated to remind potential intruders who is at home.

As a cat proceeds along the trails in its home range, it will stop periodically and spray urine onto the surrounding vegetation to make a visually conspicuous mark as well as a smelly one (Figure 6.12). Male servals may spray up 41.2 times per kilometre or 46 times per hour compared with only about 15 times per kilometre and 20 times per hour for females. Bobcats spray urine from 1.9 to 7.5 times per kilometre (Bailey, 1974) and Canada lynx from 10.6 to 12.5 times per kilometre (Saunders, 1963b). Tigers spray mark up to 11 times per 30 minutes (Schaller, 1967). Male cats have specially retractable penises which allow them to direct the urine at a height convenient for other curious noses to sniff. I remember very well an occasion at Marwell Zoo when two visitors were asking each other why the Siberian tiger was showing its rear end to them. The ensuing shower of urine answered all their questions. I am glad to say that there are now warning signs on that enclosure!

Figure 6.12 The typical urine spraying behaviour of wild cats as shown here by a cheetah (after Eaton, 1968) (S.A.)

Cats may also urinate onto the ground and scrape their hind feet on the damp earth to leave an obvious advertisement for other cats to see (Figure 6.13). In the Idaho Primitive Area, pumas scrape together a pile of dead leaves on to which to urinate (Seidensticker *et al.*, 1973). However, spraying seems to be the preferred method of scent marking in tigers, scraping being restricted to open areas where no trees are available for spraying (Smith *et al.*, 1988).

Faeces may also be deposited with scrapes. Jaguars in Belize leave faeces with 50 per cent of their scrapes and never deposit any urine (Rabinowitz and Nottingham, 1986). Tigers mark their scrapes mostly with urine (54 per cent) and rarely with faeces (17 per cent) (Smith *et al.*, 1988). Scraping with urine or faeces was only rarely recorded for servals (Geertsema, 1985).

Unlike domestic cats, wildcats also tend not to bury their faeces, but leave it on exposed sites such as on logs or tussocks of grass next to trails where they will be noticed by intruding cats (Corbett, 1979). African wildcats, ocelots and bobcats may even concentrate their faeces in one site to produce large smelly latrines which may be used by more than one cat (Bailey, 1974; Stuart, 1977; Emmons, 1988). Sand cats leave small piles of faeces on heaps of sand, which stand out well in the otherwise featureless desert (Leyhausen, 1979). Robinson and Delibes (1988) studied the pattern of faecal deposition for the Spanish lynx in the Coto Donaña National Park. Most of the faeces (82 per cent) were deposited singly, but the rest consisted of double faeces. It was found that faeces were more likely to be deposited at the intersections between trails in the home ranges even after taking into account the fact that these intersections would be used more frequently than the rest of the trails. Using a computer model of a lynx randomly moving through a home range, it was found that this pattern of scent marking maximised the probability of an intruder locating the scent marks. Female cats may bury their faeces and may not otherwise scent mark in the core area of their home ranges, particularly if they have kittens. By avoiding any olfactory advertisements, kittens and cubs are less exposed to the dangers of infanticide and predation.

Scent marks must be continually updated and rival's marks are always overmarked in a sort of smelly territorial war. Corbett (1979) documents the aftermath of just such a skirmish between a resident and an intruding Scottish wildcat. The resident increased his rate of spraying tenfold while the intruder was in his home range and overmarked every one of the intruder's scent marks.

An important part of scent marking is to transfer some of the scent to the cat that has made its mark. Many cats rub their heads against urine marks, or scrape urine and faeces with their hind legs (Figure 6.13). They may roll their bodies on the ground, thereby flattening the vegetation and leaving their body scent. Scratching posts are also another form of smelly advertisement. Cats may rub

Figure 6.13 Scent marking patterns of the snow leopard: (**a**) urine spraying, (**b**) scraping with the hind feet, (**c**) head and neck rubbing, (**d**) sniffing scent mark (after Wemmer and Scow, 1977) (S.A.)

their bodies against strongly smelling objects found within the home range, thereby transferring the scent to their own bodies, providing yet another olfactory link between a cat and its home range (Rieger, 1979; Gosling, 1982).

But how to these smelly conversations work? After all, scent marks do not deter other cats from investigating neighbouring home ranges at intervals. It is quite common for male pumas to travel far into a rival's home range to investigate the possibility of shifting their home range or in pursuit of a female in oestrus (Seidensticker *et al.*, 1973).

When a cat moves into a strange area, it can tell whether the area is occupied and probably how long ago it was marked by the freshness of the scent. Under laboratory conditions, it has been found that cats can distinguish between new and old scent marks, but those of intermediate age are less distinguishable (de Boer, 1977). As it proceeds into the area, it can weigh up whether to try and take advantage of the resources offered by the new area, such as hunting opportunities or females in oestrus, against the potential costs of meeting the resident cat and risking injury in any such encounter. By sniffing scent marks, the intruder will gain an olfactory profile of the resident cat, so that should the intruder meet another cat, it will know immediately whether that cat is the resident by matching the scent marks to the owner's smell (Gosling, 1982; Smith *et al.*, 1988). The intruder can then decide whether to fight or run.

SCENT MARKS AND HOME RANGES —A CASE STUDY

Although there have been several studies of the home ranges of wild cats or their patterns of scent marking, there has only been one study that has combined home range layout with scent marking patterns to see how a wild cat maintains its home range. Smith *et al.* (1988) looked at scent marking and home range layout of tigers in the Royal Chitwan National Park in Nepal. The social system among tigers is typical for a wild cat: two or more non-overlapping female resource territories overlapping with a larger and exclusive male home range (Figure 6.8). Moreover, this system of land tenure had been maintained for several years, allowing a long-term study of scent marking patterns with respect to home range layout.

Tigers scent marked their home ranges in a number of different ways. Of the two commonest forms of scent marking, urine spraying was more prevalent (78.2 per cent). Scraping (21.3 per cent) was much less frequent—54 per cent of scrapes were marked with urine, 17 per cent with faeces and 19 per cent with no apparent scent mark. All other forms of scent marking were very uncommon.

Anal gland secretion was often mixed with faeces deposited in scrapes and not with urine as suggested by Schaller (1967). Other forms of scent marking included flattening of the vegetation by rolling the body beside a trail, clawing tree trunks and cheek rubbing, which comprised only 0.5 per cent of the observed scent marks.

Urine spraying was often directed to the underside of leaning trees in an attempt to protect the scent mark from the rain. Tigers even selected certain trees to mark; *Bauhinia malabaricum* comprised 60 per cent of the tree population but was only sprayed on three occasions (0.3 per cent), whereas 80 per cent of the *Adinia cordifolia* trees were sprayed. The rate of urine spraying was slightly higher for male tigers (2.4/km) than females (1.8/km). Scrapes were only made where there were no suitable spraying posts, so that spraying is obviously the preferred form of scent marking. Scrapes not only take longer to produce, but their scent lasts for only a few days compared with up to three weeks for a more persistent spray. Scent marks were always a combined visual and chemical signal which were placed along trails where vegetation or topography would direct any intruding tigers. Perhaps the limited sense of smell of wild cats makes the visual component essential, if the scent mark is to be noticed.

Apart from studying the scent marks themselves, Smith *et al.* (1988) were able to see scent marking in action. For example, in the case of two adjoining female ranges, 80 scrapes were recorded over a five-week period along a road 520 metres long which ran through the boundary area (Figure 6.14(a)). Only six weeks later, over 100 scrapes were found in this same area. A similar pattern of scent marking was found along a 1.3-kilometre road separating the home ranges of two recently dispersed males. Almost 50 scrapes were found in the contact zone between their home ranges and none along other roads that abutted onto their home ranges (Figure 6.14(b)).

In another case, a dispersing male settled in November in an area adjacent to, and partly in the home range of, a resident male who had settled during the monsoon in July and August. By December onwards they had evidently sorted out any territorial disputes so that there was now no overlap in their territories (Figure 6.15). As the boundary between their home ranges shifted westwards, the diffuse pattern of scent marking also shifted.

Although tigers scent mark throughout their home ranges, the rate of scent marking is greater at the boundaries than in the centre. For example, a female tiger marked her home range 11.1 (eastern section) and 9.4 (western section) times per 100 metres in the boundary areas, compared with only 1.6 times per 100 metres in the central area. The tigress did not visit these boundaries more often, but she did mark at a higher rate when she was there: 4.8 (eastern)

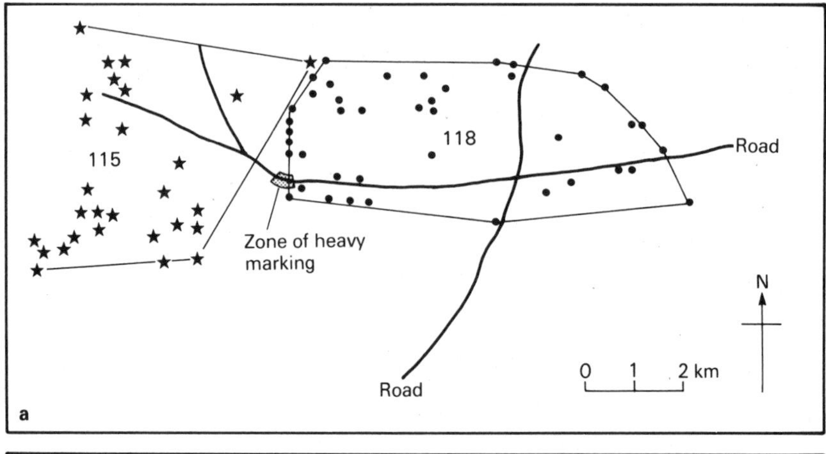

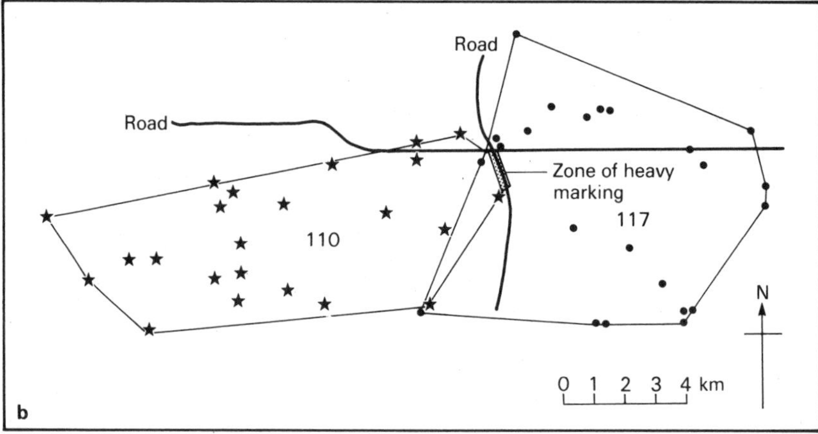

Figure 6.14 (**a**) Concentrated scent marking at the home range boundary of two female tigers (115 and 118). Dots and stars record radio locations of the two tigers (Smith *et al.*, 1988). (**b**) Concentrated scent marking at the home range boundary between two male tigers (110 and 117) (S.A.)

and 4.4 (western) times per kilometre at the edge, compared with 0.85 times per kilometre in the centre. There are, therefore, three patterns of territorial scent marking. Firstly, concentrated marking in a narrow boundary zone. Secondly, a shift in the pattern of marking as boundaries also shift. Thirdly, a five-fold increase in the rate of scent marking in boundary regions compared with the core area of the home range.

When a female tiger began to visit the boundary area of another female, the resident doubled her rate of urine spraying from 19 trees per month to 37 trees per month to let the visitor know she was there. Increased rates of scent marking also occur when home ranges are established. Three-year-old females were observed to scent mark at a high rate as they established a home range within their natal range.

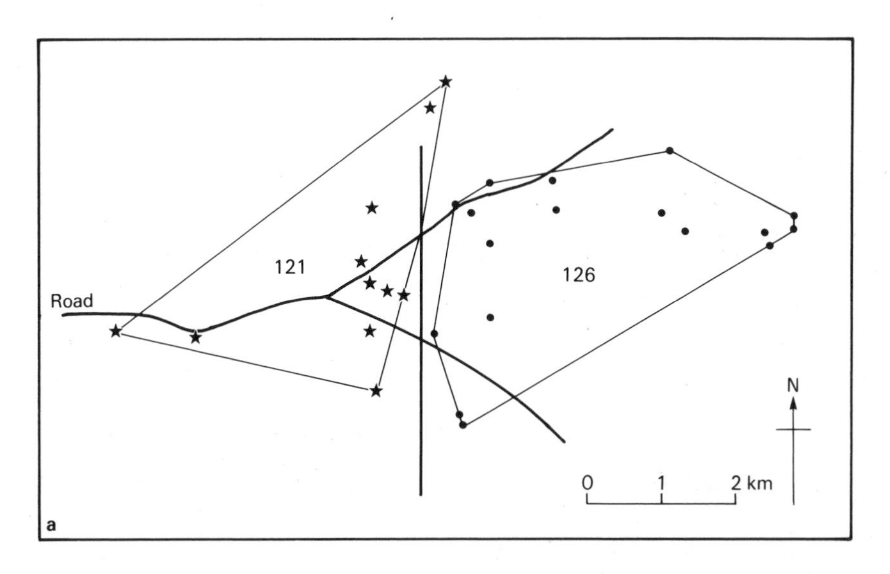

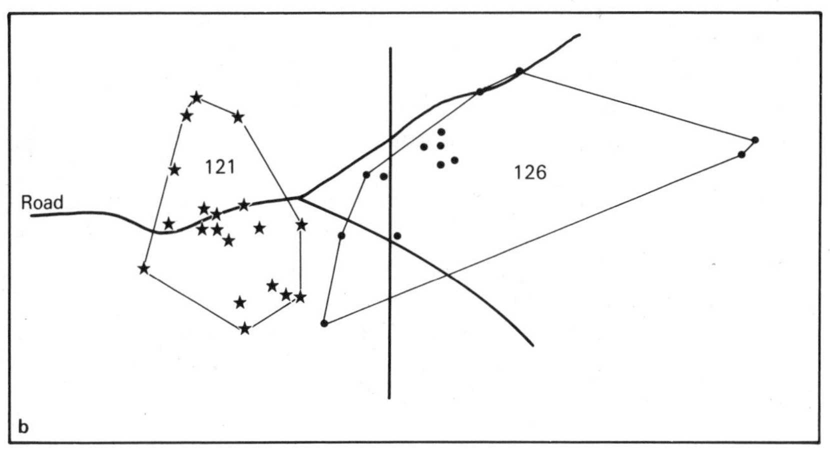

Figure 6.15 Changing territorial boundaries. Male 126 settled on the edge of the home range of male 121. There is a gradual shift in the boundaries from February (**a**) to April (**b**) as this border dispute is eventually resolved (Smith *et al.*, 1988). Dots and stars indicate radio locations (S.A.)

OTHER FUNCTIONS OF SCENT MARKS

Although scent marks are important in proclaiming home ranges to other cats, it is by no means their sole function. A female cat in oestrus either begins to scent mark or scent marks more frequently to tell the resident male that she is ready to mate. A tigress in the Royal Chitwan National Park doubled her rate of urine spraying a few days before oestrus, but did not scent mark during oestrus (Smith *et al.*, 1988). In contrast, a male increased his rate of scent marking fourfold when the female was in oestrus and visited the area she was in twice as often. The smell of the female's urine and other glandular secretions changes; this distinctive smell tells the male that she is in oestrus (see Chapter 7). A male domestic cat can tell whether a female is in oestrus just by the smell of the secretions from her cheek glands (Verbene and de Boer, 1976).

The frequency of scent marking can also be used to determine a cat's dominance and it is probable that cats can identify other individuals from their own unique smells. Therefore, scent marking is not only used in maintaining home ranges, but can convey much other information including sex, age, reproductive status and probably individual identity.

TIME SHARES

Not all cats occupy exclusive home ranges. Feral cats and many domestic cats share home ranges, although they do not necessarily occupy the same area at the same time. Leyhausen and Wolff (1959) suggested that 'solitary' cats could share home ranges by using the same area at different time of the day. By treating old scent marks as green traffic lights and new ones as red lights, more than one cat could share an area without coming into conflict. It is probable that such a system would be based on dominance. A dominant cat would use an area at will, hunting and resting wherever it chose, but subordinate cats would have to fit in their activities around the dominant cat, using scent marks and sightings as a way of avoiding any direct contact and possible attacks (Corbett, 1979).

Female cheetahs occupy huge home ranges in the Serengeti, which overlap extensively with those of other females. Most male cheetahs are territorial, but the non-territorial males also wander freely over large areas. Female cheetahs avoid each other by sight or by urine spraying in order to prevent any conflict. Female cheetahs follow the movements and migrations of their main prey, the Thomson's gazelle, in order to provide their growing cubs with sufficient food (Durant *et al.*, 1988). They commonly spray prominent objects with their urine and keep a look out for other cats, which includes not only other cheetahs, but also leopards and lions which are potential predators of cheetahs.

LIVING TOGETHER

Three species of cat are known to be gregarious and have developed a relatively complex social behaviour compared with the rest of the Felidae—the cheetah (males), the domestic or feral cat and the lion. All the other cats are found together at some time in their lives, but the associations are usually transient and do not necessarily involve any positive interactions apart from mating and the rearing of young. Lynxes commonly hunt together, but this usually occurs when there is a female with kittens who are learning to hunt, or if it is a consorting pair (Haglund, 1966; Barash, 1971; Parker *et al.*, 1983). Servals have been observed hunting together, but not so as to increase their hunting efficiency (Geertsema, 1985). Recently, a group of tigers has been observed socialising in Ranthambhore National Park in India, where open habitat and large prey are probably the important factors in this unique example of social behaviour in the otherwise solitary tiger (Thapar, 1986).

CHEETAHS

Although female cheetahs are solitary and are not territorial, male cheetahs are often found in groups of two, three or more rarely four in small territories (Frame, 1984; Caro and Collins, 1987a, b). These groups are called coalitions and the territories are defended fiercely from other males, intruders risking death if they do not retreat (Frame, 1984). Territories are occupied for much of the year and are only abandoned when no prey is available or when they go to seek out water in the dry season (Figure 6.16). Just as a coalition of male lions remains in association with a pride of lionesses for longer if there are more males in the group, so cheetah males retain their territories longer if they are in a larger group. A single male may occupy a territory for four months, two males for seven and a half months and three males for about 22 months. Caro and Collins (1987a, b) found that even though males lived together in territories, they were not more likely to meet and mate with females than non-territorial cheetahs. So why do they live together? By looking at the health of cheetahs living in groups and the ecological characteristics of the territories, it was discovered that cheetahs living in groups were about 10 kg heavier (i.e. healthier) and were also older than single non-territorial male cheetahs. Therefore, it seems that by living in groups male cheetahs remain healthier and live longer, and thus are better able to achieve a higher lifetime reproductive success compared with a solitary or nomadic male. In general, territories are centred on areas with a fairly good supply of food, but if no food were available or during a drought the cheetahs may abandon their territories.

LIONS

Lions are always regarded as the only truly social cat, but not all lions are sociable and domestic/feral cats may show the same spectrum of social behaviours as their larger cousins. In East Africa (particularly the Serengeti), lionesses live in groups of two to twelve with their cubs. These groups are called prides. Living in association

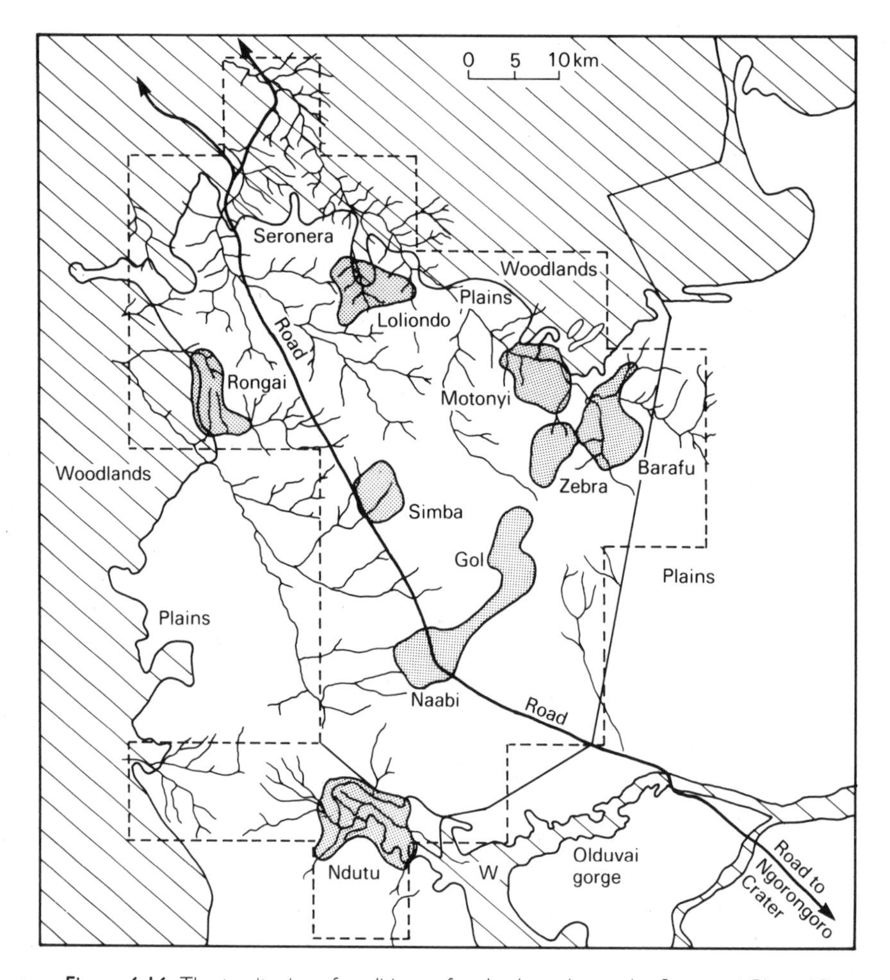

Figure 6.16 The territories of coalitions of male cheetahs on the Serengeti Plains (Caro and Collins, 1987a) (S.A)

with a pride is a group of males called a coalition (Schaller, 1972; Bertram, 1975a, b, 1978). Some large coalitions of males may live in association with more than one pride. The difference between the social lion and the asocial smaller cats is tht the home range is shared by all the pride members. The pride does not necessarily stick together all the time, but may split and coalesce at any time within the common home range. Unlike the tiger, lions do not use their faeces to mark their home range. Scent marking involves primarily urine sprays or scrapes. Presumably living in a group means that a long-term signal such as faeces is not as effective a means of communication as urine, which has to be updated at much more regular intervals. The coalitions of males are important in defending the prides from other coalitions which threaten the survival of cubs, and as such they retain exclusive although shared mating rights with the pride members. It has been found that the larger coalitions remain with prides longer than smaller coalitions or single males. If a coalition takes over a pride, all the cubs under six months of age are killed. Therefore, larger coalitions are likely to achieve a greater reproductive success than smaller ones. Also an individual male within a large coalition is more likely to achieve a greater reproductive success despite having to share matings (Bygott, Bertram and Hanby, 1979). The lionesses within a pride do not appear to have a dominance hierarchy. This is especially obvious when feeding at a kill. This event turns into a ferocious free-for-all. Male lions do, however, steal a large part of their food from kills made by females.

A number of altruistic social behaviours have been observed in lion prides. Females suckle each other's cubs, share other nursing duties, groom each other and hunt together for food, which is shared, all be it on a 'first come, first served' basis. Bertram (1976) has suggested that an important basis (although not necessarily the only one (Bertram, 1983)) for these altruistic behaviours is kin selection, i.e. pride members help each other and cooperate because they share a high proportion of their genes or in other words because they are related. Therefore, any benefit that accrues to a fellow pride member is likely to benefit an individual lionesses's own genes. Usually the females within a pride are related to one another. They may be sisters, mother and daughter, half sisters, cousins, etc. On average the lionesses in a pride share one-seventh of their genes with other pride members. It is, therefore, an advantage for lionesses to cooperate in the rearing of their young and the sharing of food to promote the survival of their genes.

But how has this social behaviour developed? After all, the tiger is similar in size to the lion, but remains the archetypal solitary cat (Sunquist, 1981), although recently incipient socialisation has been observed in Ranthambhore National Park (Thapar, 1986). Why could lions not live a solitary life in small, discrete home ranges? The answer seems to be related in some way to the availability of food and the openness of the habitat in which they live. Not all lions are as sociable as those in East Africa. Variations in lion social behaviour are shown in Table 6.3.

In the Kalahari Gemsbok National Park, lions live in much smaller prides or are solitary (Eloff, 1973). Prey animals are scarce and usually small in size, making it difficult to support large prides of lions. It is claimed that the now-extinct Barbary lion from North Africa was also solitary. Again the lack of large prey items may be part of the explanation.

Over the years many different explanations have been put forward to explain why lions bother to live and hunt together. The advantage of cooperative hunting in exploiting very large prey is the commonest explanation for lion social behaviour. However, as we have seen, Caraco and Wolf (1975) and Elliott *et al.* (1977)

showed that pride members were in fact disadvantaged by sharing food from a carcase (see Chapter 4). A dynamic model based on the body reserves a lion needs to survive for the next 30 days and which optimises the utilisation of different-sized carcases may revitalise this theory (Houston *et al.*, 1988).

Kruuk (1986) pointed out that some lion prides do not contain related lionesses, so that kin-selection cannot be the primary mover in association. He suggested that in the open savanna, there would be limited hunting sites with cover for stalking prey. The only way that lions could survive on the otherwise abundant supply of food was to hunt together and share food resources. From this basis, other social behaviours could have developed.

Table 6.3 Lion social structure in different habitats

Habitat	Group size	Structure
Desert with dispersed waterholes (N and SW Africa)	1 male 1 female + cubs	Permanent
Semidesert (SW Africa)	1 male 2–3 females + cubs	?
Semidesert with one regular rainy season (Kalahari (Botswana))	1–3 males 2–6 females + cubs variable, nomadic	Form prides during rains. Dry season
Savanna (Zaire, Uganda)	2–3 males 3–5 females + cubs	Form prides; pride males stay 6 years
Savanna (E Africa from Kenya to Transvaal)	2–5 males 5–20 females + cubs nomads >50% of population	Form prides; pride males stay 2–3 years; variable groups
Bush with scattered grasslands and forest (Gir Forest Reserve, India)	2 males 3–4 females + cubs	Permanent?

Source: Leyhausen (1988)

Recently, Packer (1986) has suggested a slightly different theory to explain lion sociality. In the open savanna, prey animals that have been killed or have died of disease are easily spotted by other predators and scavengers (including lions), who are only too happy to get some easy food. In the Serengeti, lions steal about 12 per cent of their food from other predators (Schaller, 1972). In the Ruwenzori National Park, male lions need to steal 60–75 per cent of their food from lionesses in order to survive (van Orsdol, 1986). Obviously the abundance of food and the higher efficiency of cooperative hunting mean that females can sustain these losses. Packer (1986) believes that the primary basis for lion social behaviour is that it is better for a lioness or lion to share its kill with a close relative than with a complete stranger, if it is going to lose the kill or part of the kill anyway due to conspecific or other scavenging. By living in groups with close relatives, it would be possible to defend carcasses from other lions and predators.

Clearly we must continue to research just why lions live together by looking at differences in lion sociality throughout its wide range, but the study of domestic and feral cat social groups may be able to offer us some clues.

THE SOCIAL BEHAVIOUR OF DOMESTIC CATS

Domestic cats have travelled the world with humans and colonised virtually every place at which they have arrived, to survive as feral animals. Domestic cats seem to be able to survive anywhere from subantarctic islands to the inner city. But with this variation in habitat comes a huge variation in social behaviour (Table 6.4). At

Table 6.4 Characteristics of domestic cat social groups

Location	Food	Density (cats/km^2)	Range size (ha)	Social organisation
Japan	Fish dumps at fishing village	2350	M—0.31–1.7 F—0.06–1.8	Stable membership, kinship unknown. Several adult males in each group
Portsmouth	Rubbish and provisions by cat lovers	200	M—0.08–24.0 F—0.03–4.24	Female prides. Sometimes one adult male attached.
Devon	Handouts (milk), mice, birds	6	M—60 F—2–7	Experimental group of three related females
Sweden	Rural households with regular food. Some rabbits, etc.	3–7	M—84–990 F—50 (mean)	Female prides. Males usually leave when 1–2 years old.
Dassen Island	Rabbits, seabirds	30	M—32–63 F—11–32	Solitary, overlapping ranges.
Monach Islands	Mosaic habitat, rabbit prey, birds	4	M—no data F—24–60	Solitary, non-overlapping ranges.

Sources: Corbett (1979); Izawa and Ono (1986); Apps (1983); Liberg and Sandell (1988)

one end of the spectrum, there are the feral cats on islands such as Heisker, one of the Monach Islands just to the west of the Hebrides (Corbett, 1979). These cats live mainly on rabbits in a mosaic of habitats. Food is very hard to come by, and the cats live in large non-overlapping home ranges marked with faeces, just like any wild species of solitary cat (Table 6.4) (Corbett, 1979). The food resources are so poor that Corbett (1979) did not record any successful breeding in a three-year study.

At the other end of the extreme are the feral cats in an urban environment such as Portsmouth Dockyard (Figure 6.17). The cats here may have been isolated within the walls of the dockyard for 200 years (Dards, 1978, 1981). Within these walls, the cats do not hunt much of their prey except for the odd bird. They do not eat rodents, which are controlled by poison. Instead, the feral cats survive on the generous handouts of dockworkers and the rubbish that builds up in skips around the dockyard. The female cats live in prides of from two to seven animals within shared home ranges in the dockyard. The males are solitary, but range over a wider area (about ten times greater) than females, so as to increase their

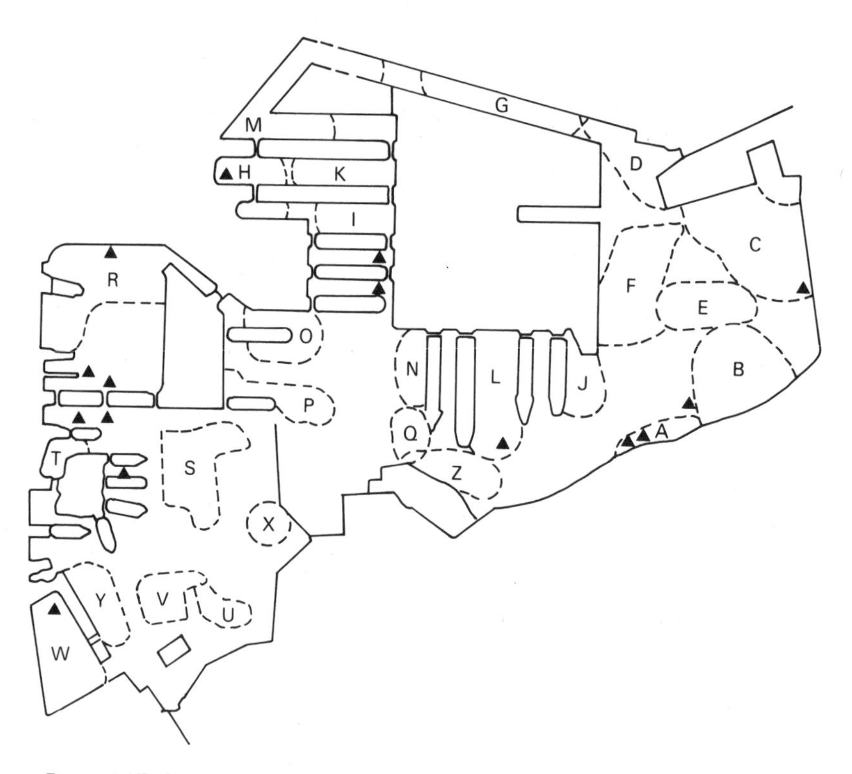

Figure 6.17 Group home ranges of feral domestic cats in Portsmouth dockyard. There is no overlap between group home ranges. Triangles mark the sites of feeding skips (after Dards, 1981) (S.A.)

opportunities for mating. Macdonald and Apps (1978) studied some semi-independent farm cats in Dorset. These cats occupied larger home ranges than the Portsmouth cats, but much smaller ranges than the Heisker cats. The female cats shared their home ranges and the males wandered over a much wider area. These cats survived partly on milk from the milking sheds, but mostly on animal prey caught in the fields. Two female cats shared the responsibility for rearing their kittens just like the members of a lion pride. They suckled each other's kittens, 'babysat' while the other was away hunting, and even brought food back to the 'babysitter'. Therefore, domestic cats can be either totally solitary like most other wild cats, or live together like lions. Many other studies of feral cats behaviour have revealed variations in the degree of sociality between these two extremes. These studies have recently been reviewed by Corbett (1979) and Liberg and Sandell (1988) and some of them are listed in Table 6.4. Much other information concerning the social behaviour of domestic and feral cats is contained in Turner and Bateson (1988). As with lions, the key to social behaviour seems to be access to an abundant and clumped supply of food, whether it be herds of large ungulates or skips full of discarded food.

7

A CAT'S NINE LIVES

Although most wild cats are solitary hunters, when it comes to reproduction, they are polygynous. As we have seen in Chapter 5, the social organisation and system of land tenure of most wild cats ensures that males hold larger home ranges that overlap with those of several females, so that they can mate with as many females as possible. After mating, the male probably contributes little to the upbringing of his progeny except by preventing other males invading the female's home range and killing the kittens or cubs. Indeed, it is now thought that infanticide by male cats is a reproductive strategy whereby males bring about an increase in their reproductive fitness at the expense of that of rivals.

The biggest problem for solitary hunters is finding each other at the right time for mating and risking an intimate association without being injured while being armed with fearsome teeth and claws. Social odours provide the main means of communication at long range, but vocal and visual signals are obviously also used, at closer range, to ensure that intentions are understood on both sides. Most wild cats are thought to be induced ovulators, so that even though the female may come into oestrus, no ovulation occurs unless the vagina and cervix of the female are stimulated repeatedly during mating. As a consequence of oestrus lasting several days and ovulation being induced, the chances of a successful fertilisation can be maximised.

Female cats rear their kittens or cubs in a secure den for the first few weeks. The young are born in a semi-altricial state, but start to develop their senses and motor actions immediately. The kittens play and learn, developing skills useful for socialisation and predation later in life. These skills are essential if the young cat is to stand any chance of survival to adulthood. As with all mammals, cats have their own fair share of diseases and parasites and the other hazards of life. We mostly know about the mortality mediated by humans and in some cases this can be very severe. But let us start at the beginning with finding a mate.

FINDING A MATE

A considerable problem for solitary hunters is finding each other when the female is in oestrus. Before female cats come into oestrus, they either begin to or increase the rate at which they spray urine (Corbett, 1979; Smith *et al*, 1988). The addition of sex hormonal products to the urine and body secretions brings about a change in their odour to tell the male that a female is ready to mate. By rubbing her body against objects within her home range, the female advertises her presence to the male. A vaginal secretion has been observed in domestic cats which may be yet another odour source to help the male to locate the female (Tabor, 1983).

Male cats may also increase their rate of urine spraying at this time, and increase their home range to its maximum extent in search of females during the breeding season (Litvaitis *et al.*, 1987; Smith *et al.*, 1988). In this way, they advertise their presence and ownership of their home range in an attempt to deter other males from interfering, and possibly to tell the female that they are nearby. This does not always work, and even in wild species of cat more than one male has been seen to pursue a female in oestrus, e.g. tiger (Schaller, 1967), puma (Seidensticker *et al.*, 1973).

Cats test the urine marks of other cats by drawing the odour over the vomeronasal organ in the roof of the mouth. This olfactory organ senses the presence of sex hormones in the urine (Estes 1972). The action of using the vomeronasal organ in this way produces a characteristic grimace on the face of the cat which is called 'flehmen'.

OESTRUS

All cats seem to be polyoestrous, but there are temporal differences with latitude. Tropical cats tend to be aseasonal and can have kittens or cubs at any time of the year, although there may be a tendency towards a main birth season. The cats of the temperate regions of the world tend to be seasonally polyoestrous. In domestic cats, day length seems to be the most important factor in controlling the oestrous cycle. Increasing day length or long days bring about oestrus in domestic cats, or decrease the length of the anoestrous period (Scott and Lloyd-Jacobs, 1959). This is probably also the case for other felids in temperate habitats.

Some cats appear to be seasonally polyoestrous, but they usually mate successfully at the beginning of the breeding season and rear their kittens throughout the rest of the summer months. Scottish wildcats were said to be able to have two or even three litters in a year. However, Corbett (1979) found that they invariably had only one litter per year. If a litter is lost, the female may come into oestrus again, mate and have a second litter, but she could not successfully rear two litters in a year. The appearance of litters at different times during the summer is probably due to this latter effect and the fact that younger females breed later than older ones. A similar scheme has been suggested to explain apparent multiple births in the bobcat (Figure 7.1) (Crowe, 1975).

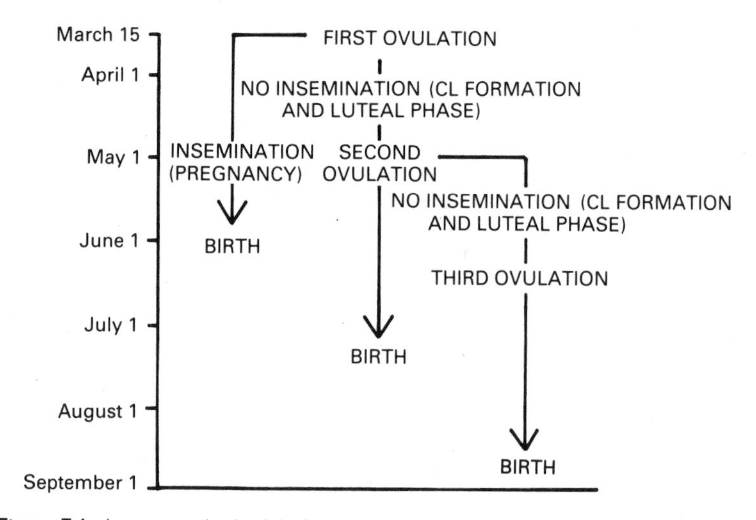

Figure 7.1 A proposed schedule for the ovulation and parturition of bobcats in Wyoming (Crowe, 1975). This schedule may also apply to many of the other smaller temperate cats (A.K.)

Male cats from colder climates may also be seasonal breeders, only producing sperm during the breeding season, e.g. bobcats (Crowe, 1975). Again, day length is probably the main factor in controlling spermatogenesis.

CATNIP

As a matter of interest it has been known for a long time that catnip or catmint (*Nepeta cataria*) is strangely attractive to cats. The active ingredient in catnip is a monoterpene, *cis*, trans-nepetalactone, which is active at concentrations of one part in 10^9 to 10^{11} (Figure 7.2) (Albone and Shirley, 1984). However, not all cats respond universally, the ability of a cat to detect catmint being governed by a dominant autosomal gene (Todd, 1962).

Catmint has a profound effect on the behaviour of cats. Palen

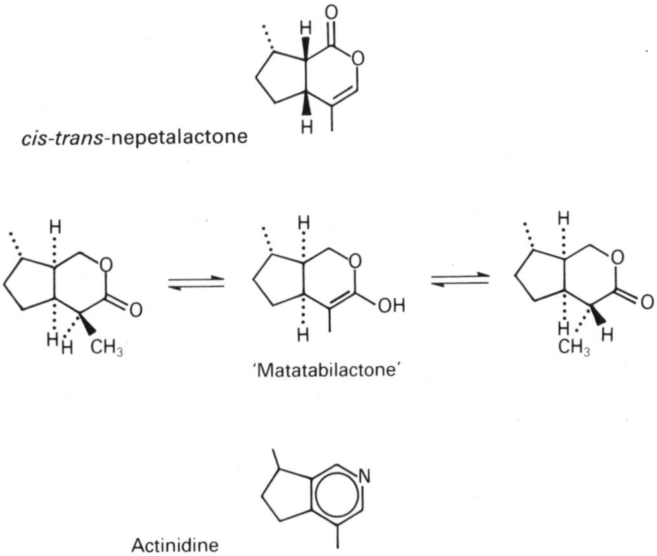

cis-trans-nepetalactone

'Matatabilactone'

Actinidine

Figure 7.2 The chemical compounds which elicit the catnip response in cats (Albone and Shirley, 1984) (S.A.)

and Goddard (1966) found that the rolling and rubbing of the body, and shaking of the head are very similar to the motor actions of a female cat in oestrus. Mysteriously, even male cats act in this way.

Recently Cherfas (1988) has reviewed the literature on the effect of catnip on cats. It is now known that several wild species also respond, including lions, snow leopards and jaguars (Todd in Ewer, 1973; Cherfas, 1988). Tigers, pumas and bobcats were found not to respond during trials in captivity but are said by hunters to be very attracted to catnip (Cherfas, 1988). Leopards are said to be weakly attracted to catnip. Perhaps the differences can be accounted for by the presence or absence of the inheritable gene responsible for the catnip response, or the genetics of the catnip response in other cats may not be as simple as it is in the domestic cat.

Other substances produce the catnip response, including actinidine from the plant *Actinidia polygama* and matatabilactone (Figure 7.2) (Albone and Shirley, 1984).

MATING

When the male and female cat locate each other, there is usually a prolonged courtship. At first, the female resists the male's advances, despite acting provocatively towards him (Leyhausen, 1979). There is a considerable amount of visual and vocal communication, which is essential in reassuring these dangerous carnivores that their intentions towards each other are mutual. An example of the courtship sequence by male and female cheetahs is shown in Table 7.1 (Wrogemann, 1975). The male and female may spend several days together even hunting side by side. For example, male and female serval have been observed hunting together (but not cooperatively) in the Ngorongoro crater in Tanzania (Geertsema, 1985). Lynx courting couples may, however, hunt cooperatively (e.g. Haglund, 1966). By testing the male's persistence, the female may be trying to ensure that it is the home range owner who is her mate, since any intruder is likely to be driven away by the resident male. If a strange male did mate successfully with her, the resident male might kill the young, thereby wasting her investment in her young (see section on development of the kittens, below). However, female wild cats may mate with more than one male, which could make the males uncertain as to the paternity of the young. This, in

Table 7.1 The courtship sequence of male and female cheetahs[1]

Cycle	Female	Male
	Interaction minimal	
	Start of courtship	
Pro-oestrus	Vaginal discharge (not always visible)	Smelling of the vagina
(1st phase)	Stutter call	Stutter call
	Rolling?	Erections and spraying Mound building?
	Non-receptive and swats at approaching males	Charging at the female
		Inter-male agression
Oestrus[2]		Male aggression reaches peak
(2nd phase)	Receptive	
	Lateral tail displacement	Partial mounting
	Copulation (infrequent)	
Met-oestrus	Hormonal change	Aggression minimal
(3rd phase)	Interest minimal	
	Courtship ceases[3]	

Notes

1 This chart is derived from observations of certain captive animals.

2 Most carnivores are induced ovulators, which means that the females will ovulate when actual mating takes place. But it is not clear if cheetah ovulate freely during the second phase of the cycle, or if liberation of the ova occurs only when copulation takes place.

3 If fertilisation does not occur, the female will, within about ten days, start another cycle.

Source: Wrogemann (1975)

particular, applies to lions and domestic cats, and will be discussed later.

As oestrus progresses, the female becomes very provocative and more receptive to the male's approaches. However, even this may be too much for some males. I remember watching a female leopard present her rear end to a male at London Zoo, wrapping her tail sensuously around his head. Despite this flagrant invitation, the male was totally unimpressed. The mating behaviours of domestic cats and some species of wild cats in captivity are described in some detail by Leyhausen (1979).

Most wild cats are probably induced ovulators. In other words, mating is necessary to bring about ovulation for the successful fertilisation of the eggs. Although most carnivores are also probably induced ovulators, they tend to have highly-developed penis bones or bacula for stimulating the walls of the vagina and the cervix, which is necessary for inducing ovulation. Cats have poorly-developed bacula (Figure 7.3), but make up for this by having a penis covered in keratinous spines. It has been claimed that these spines cause the female to cry out in pain after mating as the penis is withdrawn (Tabor, 1983), but it seems very unlikely to me that mating would cause such pain. Leyhausen (1979) found that in captivity a female would not scream out or attack the male after mating if she knew him well. It seems that this behaviour is a defensive one brought on by the anxiety of mating with a strange male who might confuse her with his next meal. The fact that cats may copulate many times and that the female resumes her provocative behaviour after copulation seems to confirm that females do not experience any pain. If they did, surely this would be maladaptive by tending to reduce copulation frequency and thereby reduce the chances of induced ovulation. In general, the more frequently mating occurs, the greater the chances of inducing ovulation and successfully fertilising the eggs. If unsuccessful, the female may have a false pregnancy presumably induced by the mating (Ewer 1973, Leyhausen, 1979).

Not all cats are induced ovulators. Recently it has been discovered that not only domestic cats, but also bobcats and Canada lynx may ovulate spontaneously (Scott and Lloyd-Jacobs, 1959; Fritts and Sealander, 1978b; Quinn and Parker, 1987). Perhaps in some species from northern latitudes the breeding season is so restricted that there is no point in ovulating at different times. Alternatively, bobcats and lynx may be induced ovulators in years of low prey density, when there is presumably less chance of meeting a mate, and spontaneous ovulators in years of high prey density, when there are good chances of meeting a male and of survival of kittens. Type of ovulation might also vary with age, younger cats being more likely to be induced ovulators. More research is needed on the reproductive behaviour of wild cats to see if our assumptions about

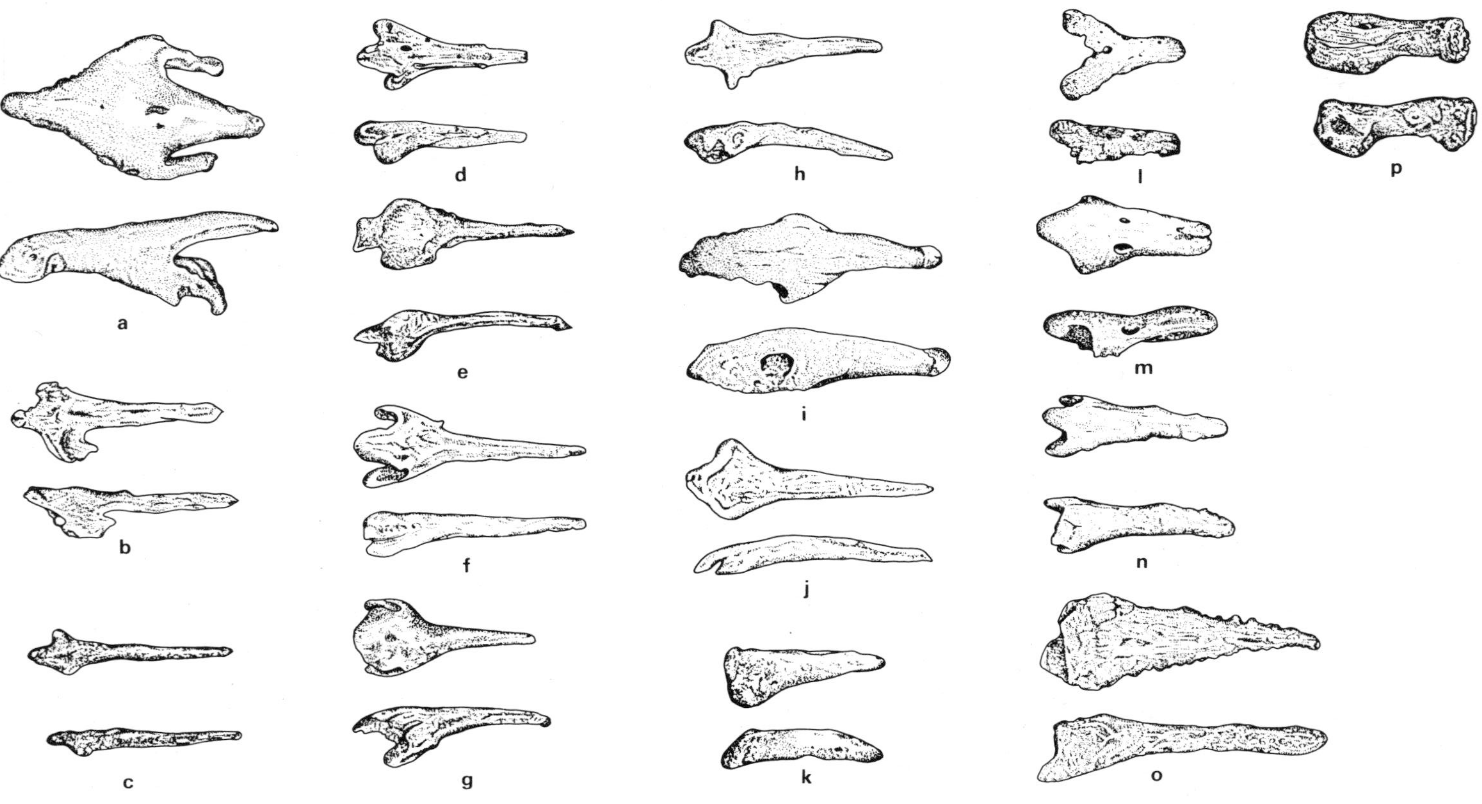

Figure 7.3 The bacula of wild cats, distal end facing right, showing dorsal and lateral aspects (Didier, 1949): (**a**) ocelot (7 mm long), (**b**) wildcat (5.5 mm), (**c**) African wildcat (*caffra*) (5 mm), (**d**) domestic cat (4 mm), (**e**) African wildcat (*rubida*) (5 mm), (**f**) leopard cat (*chinensis*) (6 mm), (**g**) leopard cat (4.5 mm), (**h**) fishing cat (5 mm), (**i**) serval (6 mm), (**j**) caracal (6 mm), (**k**) puma (4 mm), (**l**) clouded leopard (4 mm), (**m**) leopard (6.5–7.5 mm), (**n**) jaguar (8 mm), (**o**) tiger (11.5–12.5 mm), (**p**) lion (6–9.5 mm) (S.A.)

induced ovulation are consistent for all species and under all conditions.

Eventually the female accepts the male for copulation by adopting her mating position (lordosis) with her belly on the ground, her rump slightly raised and her tail turned aside (Leyhausen, 1979). The male mounts her by placing his front paws on the ground astride her and his hind feet on her rump. By making treading movements with his hind feet, he stimulates the female to adopt her mating position, if she has not already done so (Ewer, 1973). The male may fear that the female could turn on him, and often grips the skin of the back of her neck to make her quiescent—just as mother cats hold their kittens (Figure 7.4). (This carrying reflex makes sure that the kittens do not struggle and injure themselves.) For her part, the female's mating position ensures that the male cannot place his front paws on her and confuse her with a prey item

Figure 7.4 (**a**) The typical mating position of the domestic cat. Note that the male is holding the scruff of the female's neck. (**b**) A female domestic cat carrying its kitten (After Leyhausen, 1979) (S.A.)

(Ewer, 1973). As I have said already, after copulation, the female may turn on the male to drive him away just in case she is mistaken for a tasty morsel.

Copulation frequency varies considerably in the Felidae (Table 7.2) (Eaton, 1976a). Lions may copulate hundreds of times during the female's oestrus, which lasts for several days, but the black-footed cat dare not risk mating more than a dozen times during a one and a half day oestrus. This difference is assumed to be due to the vulnerability of small cats to predation by other carnivores. Cheetahs are particularly vulnerable to attacks from leopards and lions, which may explain their brief courtship compared with other cats of similar size.

Table 7.2 Copulation frequency, average length of oestrus and probability of conception of felids

Species	Copulation frequency per day	Average length of oestrus (days)	Probability of conception per oestrus
Ocelot	5–10	5	0.6+
Black-footed cat	low	1.5	?
Puma	50–70	7–8	0.67
Cheetah	3–5	5–14	?
Bobcat	5+	?	?
Snow leopard	5–15	3–8	0.5+
Jaguar	100	7	?
Leopard	70–100	7	0.65
Tiger	50+	7	?
Lion	100	6–7	0.2–0.25

Source: Eaton (1976a)

On average, lions may copulate as much as once every fifteen minutes throughout day and night. Schaller (1972) recorded that one lion copulated 157 times during a 55-hour period with two different females. Despite this, mating success is low and it is believed that high copulation frequency coupled with low mating success promotes social bonding in the pride (Eaton, 1976b; Packer, 1986). In Nairobi National Park, only four out of fourteen lionesses produced cubs after mating (Rudnai, 1973).

The breeding success of the Canada lynx is inextricably linked to the population densities of snowshoe hare. For example, after a 1962/3 peak in showshoe hare density in Alberta, Canada, there was a peak in lynx reproduction during the spring of 1963 to produce a maximum number of kittens during the winter of 1963/4 (O'Connor, 1986). The lynx population did not peak until the winter of 1965/6, some two years after the hare population peak (Figure 4.12). In years of hare abundance, 99 per cent of young females (ten months old) were reproducing (i.e. corpora lutea were present), but in years of low hare density only 10 per cent of young females were breeding. Although older females bred every year, they had smaller litters in years of low hare density.

GESTATION

The female cat is probably already aware of all the potential denning sites in her home range. In fact, it is essential that a cat obtains a home range with suitable denning sites, if she is to breed successfully. Gestation length varies from 58 days in the smaller cats to more than 90 days in the big cats (Table 7.3). Not surprisingly, bigger cats take longer to grow.

Female cats may alter their behaviour, increasing their food intake to build up a fat store for energetically expensive lactation. Oftedal and Gittelman (1989) assessed the energetic investment in

Table 7.3 Gestation period, birth weight, daily weight gain, ages at eye opening, first walk and female sexual maturity, and nipple number in felids

Species	Gestation period (days)	Weight at birth (g)	Daily weight gain (g)	Eyes open (days)	First walking (days)	Female sexual maturity (months)	Nipple number (pairs)	Litter size
Ocelot	75 (70–80)	250	—	—	—	—	—	1–2
Tiger cat	75 (74–76)	—	—	17	—	—	—	1–2
Geoffroy's cat	75 (74–76)	65	—	—	14–21	—	—	—
European wildcat	68 (63–69)	100 (38–165)	11 (8–13)	11 (9–12)	18 (16–20)	11 (10–12)	— —	4/5 (1–8)
African wildcat	58 (56–60)	—	—	10	—	—	4	3 (1–5)
Indian desert cat	62	—	—	—	—	—		
Domestic cat	63 (55–70)	90 (80–120)	13	9 (7–20)	9–15	10 (7–12)	4	4 (1–8)
Sand cat	61 (59–63)	39	12	14 (12–16)	21 (20–22)	—		
Black-footed cat	67 (63–68)	60–84	8	7 (6–8)	—	21	3	2 (1–2)
Jungle cat	66	136	22	—	—	—		

Table 7.3 Cont.

Species	Gestation period (days)	Weight at birth (g)	Daily weight gain (g)	Eyes open (days)	First walking (days)	Female sexual maturity (months)	Nipple number (pairs)	Litter size
Leopard cat	66 (63–70)	80 (75–95)	11	10 (5–15)	—	32 (29–34)	—	3
Amur leopard cat	58 (56–60)	—	—	—	—	—	—	4 max.
Fishing cat	63	92	11	—	—	—	—	2(2–3)
Flat-headed cat	—	—	—	—	—	—	4	—
Serval	73 (66–77)	250 (230–265)	—	9	—	—	2	3 (1–4)
Caracal	71 (68–78)	—	21	7 (1–10)	9–23	—	—	1–6
Temminck's golden cat	—	250	—	9 (6–12)	—	—	2	1
African golden cat	—	—	27	—	—	—	—	—
Cheetah	92 (90–95)	270 (250–300)	50 (40–50)	7 (4–11)	12 (12–13)	22 (21–22)	6	3/4 (1–5)
Puma	93 (89–98)	400 (230–500)	32	10 (7–14)	—	29	—	3 (1–5)
Jaguarundi	67 (63–70)	—	—	—	—	—		
Eurasian lynx	69 (63–75)	260 (69–306)	30 (13–54)	13 (10–17)	26 (25–28)	27 (21–34)	—	1–4
Canada lynx	61	—	—	—	—	—		
Bobcat	63 (50–70)	280–340	25	6 (3–9)	—	—	—	4 (1–6)
Clouded leopard	88 (85–92)	145 (140–150)	23	2–11	19–20	—	—	2 (1–4)
Snow leopard	99 (93–110)	470 (320–708)	48 (29–75)	8 (7–9)	—	—	—	1–4
Jaguar	101 (91–111)	800 (680–990)	48 (34–69)	8 (1–13)	18	37	—	2 (1–4)
Leopard	96 (90–112)	300 (400–700)	33 (20–42)	7 (0–9)	13	33 (30–36)	2	2/3 (1–4)
Siberian tiger	103 (93–112)	1359 (785–1760)	86 (50–106)	11 (4–20)	17 (12–22)	48 (37–60)		
Sumatran tiger	—	750	75 (66–83)	—	—	—		
Bengal tiger	95–109	1100–1800	—	0–17	—	—	—	2/3(1–4)
Lion	110 (100–114)	1400 (1150–1785)	106	6 (0–11)	13 (10–15)	42 (33–50)	2 (3)	3/4 (1–6)

Source: Ewer (1973); Hemmer (1976b)

the young of different carnivores by looking at the calorific value of the young, litter size, gestation and lactation lengths. They found that cats invest a relatively low amount of energy in their young before birth. Ungulates, for example, invest much more energy in their young because they are born in a more precocious state. Presumably a female cat should not be too encumbered during gestation, so that she can still hunt efficiently. Female cats also bury their faeces and cover their urine in the vicinity of the denning site, in order not to attract the attention of other cats, particularly males, and other predators who may threaten their young (e.g. Corbett, 1979).

PARTURITION AND LACTATION

Kittens or cubs are born into a den in a cave, tree root or even a thicket—basically anywhere that is safe, dry and secluded. The den site may change after birth, if it becomes fouled with food or faeces, or if it is disturbed by, for example, a strange male. Lembeck (1986) records the instance when a female bobcat with kittens changed her den site two or three weeks after birth, but the reasons for his move are not known. Before giving birth, the female licks her nipples and genitalia clean. After giving birth, she cleans away the birth membranes, severs the umbilical cord, lies with the young to keep them warm, suckles them and calls, licks and nuzzles them for reassurance. The parturition behaviour in domestic cats is described in great detail by Schneirla, Rosenblatt and Tobach (1966) and Leyhausen (1979).

The number of young varies considerably. Some cats have only one or two young (e.g. ocelots and margays), but domestic cats, cheetahs and lions may have litters of up to seven or more. In general, cats seem to have twice as many nipples compared with their modal litter size (Table 7.3). Most of the rest of this account applies to the domestic cat.

Kittens are born in a semi-altricial state (Martin and Bateson, 1988). Their eyes are closed, and they have poor motor coordination and thermoregulatory control. Therefore, kittens rely on their mother for warmth, as well as food. She may spend virtually all her time with them for the first two days, the kittens suckling for up to eight hours per day (Figure 7.5). The first milk, as in all mammals, is colostrum, which provides the young with antibodies for defence against disease in the first few weeks of life. If the young are accidentally removed from the den by, for example, hanging onto a nipple as the mother gets up and leaves, they cry out to attract the mother's attention. However, if they are less than half a metre away, they are able to crawl back using thermal and olfactory cues. If the young require milk, warmth, or their mother to move to avoid squashing them, they will also cry out.

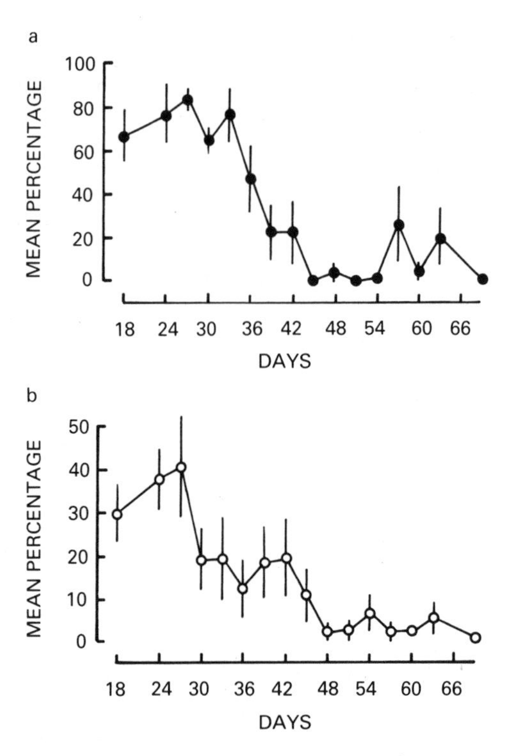

Figure 7.5 (**a**) The average amount of time spent by a female domestic cat in the nest box with her kittens. (**b**) The average amount of time spent by kittens in the suckling or ventral position (Martin, 1982 in Turner and Bateson, 1988) (A.K.)

Within the first few days of birth the kittens develop an attachment to a certain nipple. Ewer (1973) considers this to be adaptive behaviour to ensure the mother does not get ripped to shreds by sharp teeth and claws at feeding time. The kittens also perform a treading movement on the belly of their mother with their front paws while suckling, which promotes milk flow.

The costs of lactation may be quite considerable to any mammalian mother, particularly in its later stages (Figure 7.6). Large litters of domestic cats grow more slowly and are weaned earlier than smaller litters, thereby testifying to the inability of the mother to provide enough milk (Figure 7.7) (Deag, Manning and Lawrence, 1988). A litter of five kittens requires twice as much energy for growth as a litter of only two (Deag, Lawrence and Manning, 1987). Mother cats may lose weight, as body reserves are converted to milk for their voracious young. The energetic requirements of female domestic cats may increase by 1.5 to 1.7 times normal levels during pregnancy (Loveridge, 1986). This reaches a peak of 2.5 to 3 times normal energetic needs during the period of peak lactation for a large litter, i.e. 4–5 weeks after birth. During lactation the big cats have a low energy requirement compared with other carnivores,

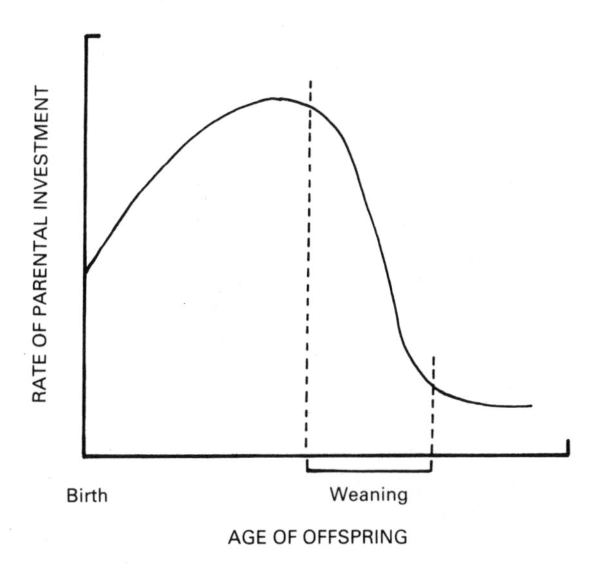

Figure 7.6 A theoretical model to show the rate of parental investment (in terms of provisioning of milk and food, protection against predators, etc.) in current offspring with increasing age (Martin, 1982 in Turner and Bateson, 1988) (A.K.)

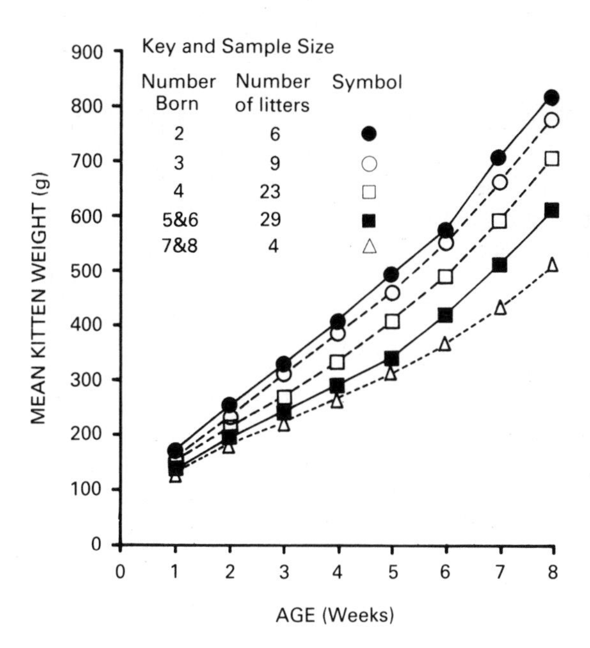

Figure 7.7 The relationship between kitten weight and age for litters of varying size of the domestic cat (Deag *et al.*, 1987) (A.K.)

but small cats have an intermediate energy requirement (Oftedal and Gittelman, 1989). The milk energy content of cats is low compared with other carnivores. However, why these differences exist for felids is not as yet really understood. Composition of some felid milks is given in Ewer (1973).

Another effect of lactation is to increase the hunting effort made by a female cat. Emmons (1988) found that a female ocelot increased her hunting activity to 93 per cent of the day or 113 per cent of normal while lactating. Despite this, her kittens eventually died due to a lack of prey. Geertsema (1985) found that a female serval with kittens hunted over a smaller home range and during daylight hours in order to feed her kittens. She spent only 16 per cent of her time in the 1.3 km^2 core area before she had her kittens, but spent 37.5 per cent of her time there after their birth. By being with them at night, she could not only keep her kittens warm, but also protect them from other carnivores when they were most active (Figure 7.8). Lancia *et al.* (1986) found no difference in home range size or activity pattern for a female bobcat with kittens, but when she was active, she moved faster, indicating a more intensive hunting effort (Table 7.4).

DEVELOPMENT OF THE KITTENS

From birth the kittens develop fast (Figure 7.9) (Martin and Bateson, 1988). After two weeks, their eyes open and they begin to

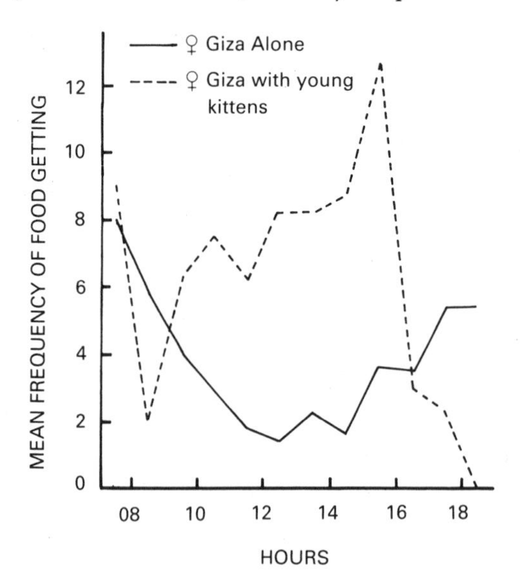

Figure 7.8 The mean frequency of food getting by the female serval, Giza, when she has kittens and when she is alone (Geertsema, 1985). Note that her activity cycle is diurnal when she has kittens (A.K.)

Table 7.4 Variation in home range size of a female bobcat during the breeding season

	Area (km^2)	Travel distance (km)
Before birth	8.8	0.75
After birth den 1	9.2	0.48
After birth den 2	3.9	0.51
After birth + kittens	10.8	0.65

Source: Lancia *et al. (1986)*

develop thermoregulatory control. At about a month old, the female domestic cat brings food back to the nest in order to start the weaning process. Play is also a very important component of kitten behaviour as motor actions and coordination develop. However, the functions of play are poorly understood, although there are plenty of ideas about its purpose (Poole, 1985; Bekoff, 1989). It has been suggested that play is important for physical training, developing competitive skills for fighting or predation, developing cognitive skills or for promoting socialisation. These different proposed functions are not necessarily mutually exclusive and different types of play are apparent at different stages of development. However, Poole (1985) rejects the socialisation hypothesis since even solitary bears indulge in play when young and may play on their own when adult.

However, play is not without its costs and risks, thereby emphasising its importance to young animals. The young risk injury and may increase their demand for milk and energy from their mother. Play may also affect the hunting success of the mother. Caro (1987) found that a mother cheetah failed in 16 per cent of her hunts due to the intervention of her cubs. Like adult activity, play has two peaks in lion cubs, one at dawn and the other at dusk (Schaller, 1972).

Different play actions are apparent at different times during development. Barrett and Bateson (1976) looked at four different play actions and their incidence at 4–7 weeks and 8–12 weeks (Figures 7.10, 7.11). They found that object play, wrestling and stalking increased from the first to the second period, but pawing and arching decreased. Male kittens played with objects significantly more at the 8–12 week stage, but if female kittens were reared with males the difference was less. It has been suggested that this correlates with the earlier independence of male kittens, so that they need to learn prey capture techniques earlier than female kittens. However, more recent research found no difference in prey capture learning between the two sexes (Caro, 1981). Some behaviours, associated with aggression in adulthood, are seen less

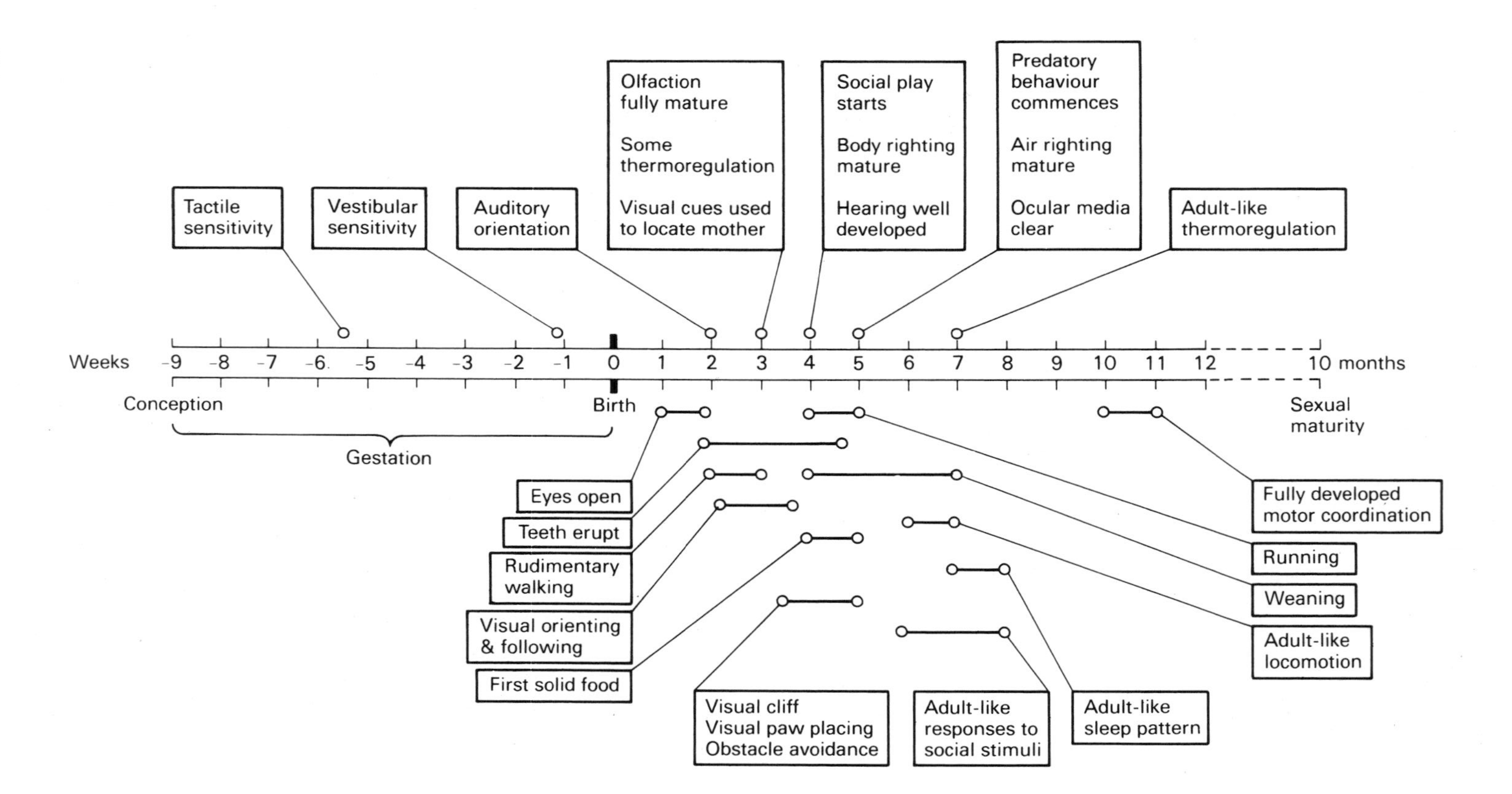

Figure 7.9 A schematic timetable outlining some of the major changes in behaviour and physiology occurring during the development of domestic cat kittens (Martin and Bateson, 1988) (S.A.)

frequently in play as kittens get older, but object oriented play become commoner as a prelude to learning prey capture skills (Figure 7.11) (Caro, 1981). The factors which cause this switch in behaviour are as yet poorly known. Where cats have only one kitten, the mother becomes an important focus for play. Fagen (1981) found that a hand-reared margay kitten's behaviour only began to develop normally when the keepers began to play with it.

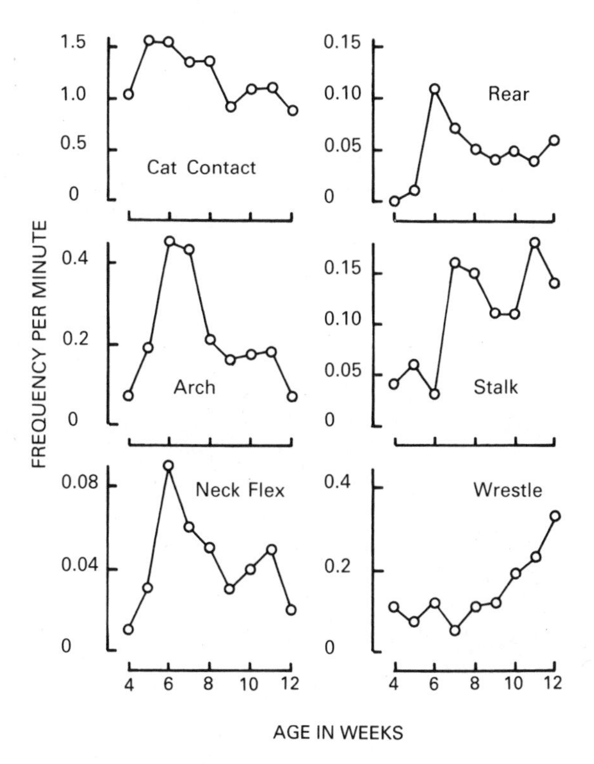

Figure 7.10 Changes in the frequencies of six different measures of play in domestic cat kittens over the first twelve weeks of life (Barrett and Bateson, 1976) (A.K.)

As the young grow, they initiate more of the suckling bouts (Schneirla *et al.*, 1966) (Figure 7.12). However, the mother can prevent suckling by adopting a squatting position. From spending 100 per cent of her time with her young in the first two days after birth, the female spends only 50 per cent with them at 4–5 weeks of age. Weaning begins when the mother starts to bring dead prey back to the nest. Not only does this shift the provisioning of resources for the kittens from milk to meat, but it gives the kittens the opportunity to learn what prey animals look like and what ones are available. With the development of more object play, the kittens benefit from the mother bringing back half-dead prey to play with. They can practice stalking and handling prey items. Although the functions of play are poorly understood, it does seem to play a

Figure 7.11 Some social play patterns become dissociated from motor patterns used in aggressive social encounters after the end of weaning (two months of age for the domestic cat). Simultaneously, social play patterns become increasingly associated with predatory motor patterns such as Approach, Paw and Bite. After weaning, different motor patterns come under separate types of control; some are increasingly controlled by the same factors that control predatory behaviours, others are controlled by factors that control aggressive social behaviours (Caro, 1981) (A.K.)

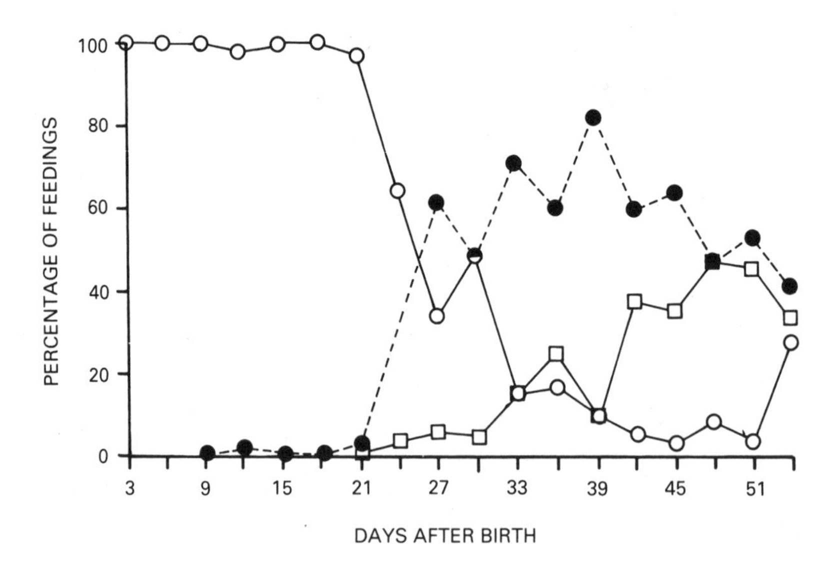

Figure 7.12 The initiation of suckling in domestic cats from birth. Female to kitten (○ – ○), Mutual approach (● ---- ●), kitten to female (□ – □) (Schneirla and *et al.*, 1966) (A.K.)

role in the development of prey capture skills. However, it is not easy to generalise about the way in which kittens or cubs learn these skills. It can take several different routes, but all end up with the same result of an effective hunting carnivore. The different routes may include watching mother or siblings hunt, starvation, sibling motivation and a kitten's own experience (Figure 7.13). There are sensitive periods during the life of a kitten when behaviours are more likely to be learnt and perfected than at other times. However, this does not preclude the development of normal prey capture skills outside this sensitive period. Leyhausen (1979) discusses at great length the development of prey capture skills in the domestic cat.

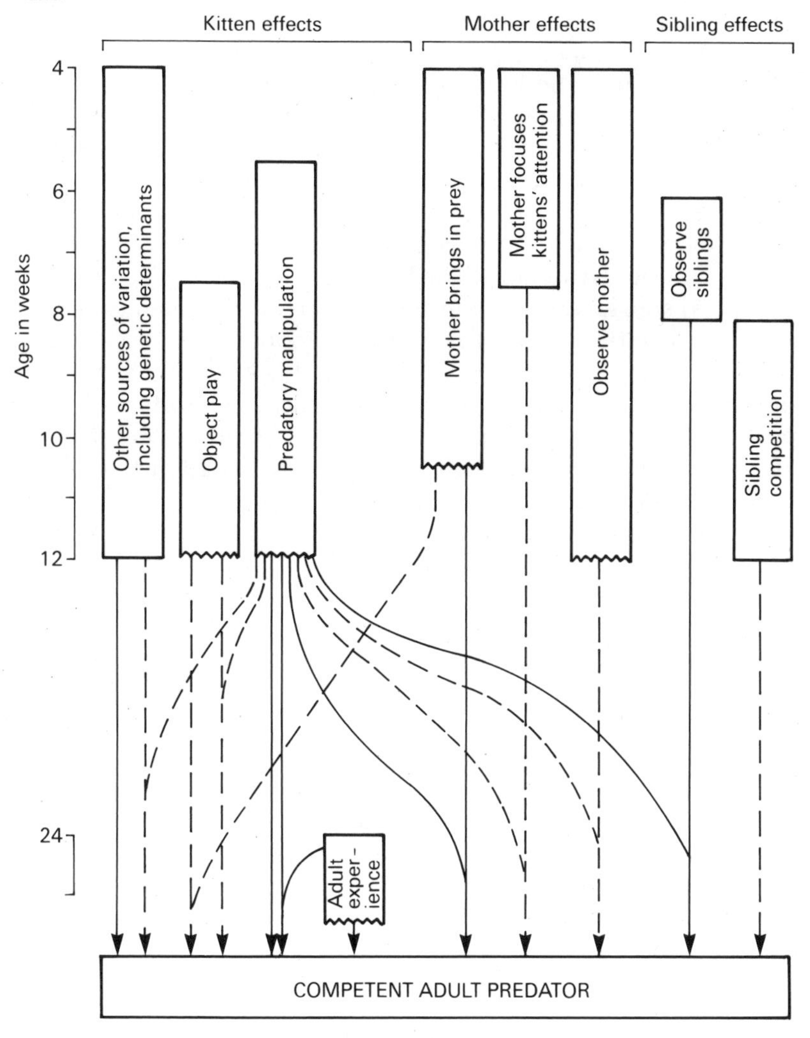

Figure 7.13 The major factors that influence the development of predation skills in domestic cats. Solid lines indicate experimental observations, broken lines indicate hypothetised influences (Martin and Bateson, 1988) (S.A.)

As their motor actions and senses develop, the kittens or cubs begin to explore the area surrounding the den. At six weeks of age, domestic kittens are able to respond to visual and olfactory stimuli like an adult cat. They keep in touch with their mother using a two-phase call—one half of which is ultrasonic (Martin and Bateson, 1988). If the mother senses danger, she growls and the kittens scatter and hide until they hear the all clear. Lions stay longer with their mother than the smaller cats. They begin to watch their mother kill at eleven months of age, but do not begin to kill for themselves until they are at least 15 months old, or when they are even older, just before they disperse, which may be at two or three years old (Schaller, 1972). Lions are able to take advantage of their social situation to learn prey capture skills by direct observation and experience in cooperative hunts.

After weaning, the kittens disperse from their natal range. In a colony of feral cats on Ainoshima Island in Japan, female kittens dispersed much more slowly than males and usually ended up sharing part of their mother's range. At one year of age, male kittens only shared 10–50 per cent of their natal range, but female kittens occupied 50–80 per cent at the same stage (Izawa and Ono, 1986). On the Savannah River Plantation in South Carolina, young bobcats disperse about 8 km from their natal range, moving into gaps between the home ranges of adult bobcats (Fendley and Buie, 1986). These first territories are about 1.5 km^2—about the size of the home range of an adult female bobcat. Therefore, it is quite possible for a young female bobcat to breed in her first year. However, adult male bobcats have a home range of 6 km^2, which therefore probably precludes the breeding of yearling males. Young servals were also observed to disperse about 8 km from their natal range in the confines of the Ngorongoro Crater (Geertsema, 1985). Dispersal cannot occur before the permanent canines have erupted, or else the kittens will be unable to capture their own prey.

INFANTICIDE

There are many records of male lions killing the cubs of a pride which they have just taken over (Schaller, 1972; Bertram, 1978; Packer & Pusey, 1984). This is evident from the much lower survival of lion cubs up to six months of age before or after a pride takeover (Table 7.5). There are also scattered references in the literature to adult male cats killing kittens and cubs including tigers (Schaller, 1967), ocelots (Emmons, 1988), Canada lynxes (Quinn and Parker, 1987) and pumas (Hornocker, 1970; Hemker *et al.*, 1986).

But why do males practice infanticide? Obviously, they are reducing the reproductive fitness of their rivals directly, but they

Table 7.5 The effect of male takeovers on the mortality of lion cubs in the Serengeti

	Mortality of lion cubs	
	Fathers replaced in first 4 months (%)	Fathers remain in pride for first 6 months (%)
All cubs in litter die before 6 months of age	89.5	40.8
At least one cub in litter survives to 6 months of age	10.5	59.2
Number of litters	19	98

Source: Packer and Pusey (1984)

are also potentially increasing their own reproductive success, because the females will come into oestrus sooner than if they had been left to rear their own young. In the Serengeti, lionesses give birth a median of 244 days after losing their cubs in a male takeover, but if they reared the cubs, the male might have to wait a total of 560 days before his cubs were born (Packer and Pusey, 1983a, 1984). Since lions only retain tenure over a pride for an average of two years in the Serengeti, it make sense for lions to kill any cubs that they have not fathered in order to stand any chance of fathering their own and see them reared to an age when their long-term survival is assured.

However, it would be wrong to think that lionesses are powerless in all this. If her cubs are older than six months of age, a lioness will take them away from the pride until they are independent. Also Packer and Pusey (1983a) have observed several lionesses together defending their cubs during a male takeover, but obviously they risk injury in doing so. But the story does not end there. In the Serengeti after a male takeover, there is an increase in sexual activity in the pride, but mating success is much reduced. In other words, the fertility of the lionesses is suppressed for up to 6–9 oestrous cycles before conception. A female which loses her cubs naturally conceives 1–2 oestrous cycles later or a median of 110 days before a female who has lost her cubs in a male takeover (Figure 7.14). Therefore, a female who loses her cubs naturally gives birth again a median of 134 days after the loss, whereas a female who loses her cubs in a male takeover gives birth again a median of 244 days after the loss. What are the advantages of this promiscuous behaviour? Firstly, it brings about paternal uncertainty: no one male can be sure who is father and so any cubs that are born are less threatened. Secondly, it may help to produce a synchrony of births which could be advantageous in communal rearing of young. However, Packer and Pusey (1983a) believe also that it brings about increased competition between coalitions of males until the largest coalition takes over the pride. In other words, it increases the chances of a large stable coalition taking over so that the lionesses stand a chance of rearing their future cubs to independence. Small coalitions or single males retain tenure of a pride for a shorter time than larger coalitions (Bygott, Bertram and Hanby, 1979). It is believed that if the females are promiscuous and have a low fertility, the males are more likely to show fidelity to one pride, because they do not have to search for other females with which to mate (Eaton, 1976b: Packer and Pusey, 1983a). In other words, it helps to reinforce social bonding and stability between the pride and the new coalition soon after a takeover.

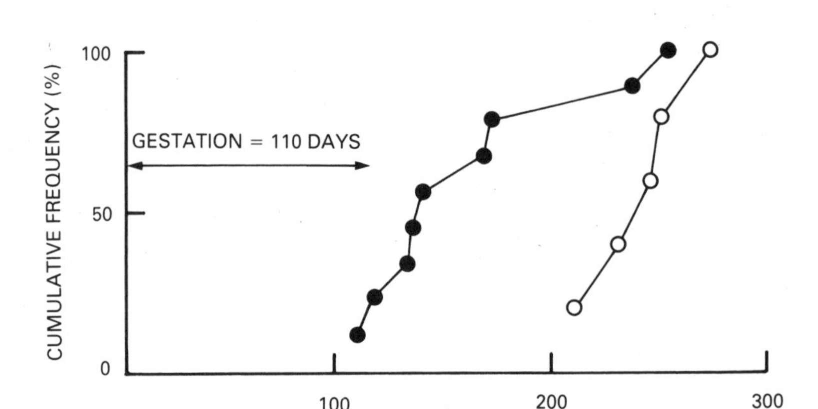

Figure 7.14 The interval from the loss of unweaned cubs by lionesses who mate with lions who fathered the previous cubs (median interval = 134 days), or who mate with incoming males (median interval = 244 days) (Packer and Pusey, 1983a) (A.K.)

In other wild cats, infanticide presumably occurs when a strange male takes over a home range or intrudes into a home range of another male. However, there are no field data to confirm this.

SOCIAL CATS

Reproduction and development are slightly different in the two social cats, the lion and the domestic cat. Firstly, in lion prides oestrus and births tend to be synchronised, but may occur at any time of the year (Schaller, 1972; Bertram, 1975a; 1976). This has the advantage of ensuring that the cubs grow at the same rate: lion cubs in a pride suckle from any female and any nipple, so that some lionesses babysit while other pride members are hunting. Communal suckling and nursing has also been observed in groups of domestic cats (Izawa and Ono, 1986; Macdonald and Apps, 1978; Dards, 1978, 1981). But why do they do this?

One reason put forward by Brian Bertram is kin-selection. The males in a coalition and the females in a pride tend to be related to each other, so that apparently altruistic behaviour can be explained by the fact that group members share a proportion of the same genes. By benefiting a close relative, a group member increases its own inclusive fitness by promoting the survival of its shared genes.

Bertram (1976) calculated that on average lionesses shared 0.18 of their genes with others in the pride. Lions were slightly more closely related (0.22). (However, Packer and Pusey (1982) have recently recalculated this to be 0.17—see also Bertram (1983).) Prides consist usually of mothers and daughters and cousins, but coalitions consist of brothers and cousins. Bygott *et al.* (1979) looked at the length of time a coalition maintained tenure of a pride with the size of the coalition. One or two lions could keep a pride for about two years, a group of three for more than two years, but groups of four to six may keep hold of a pride for nearly four years or more. Moreover, larger coalitions tend to contain more closely related lions and fewer unrelated ones. Indeed, lions in a large coalition may well have a greater lifetime reproductive success than single lions or lions in a small coalition even if they never successfully mate with a female.

It was always thought that there was no competition between males for the females in a pride. However, Packer and Pusey (1982, 1983b) have recorded severe fights between males over a lioness in oestrus. Perhaps the synchrony of oestrus in females helps to reduce intracoalition competition in the pride (Bertram, 1976; Eaton, 1976b). Feral cat colonies do not seem to show the same degree of synchrony of oestrus. For example, in a feral cat colony in Rome only weak synchrony was apparent (Natoli & De Vito, 1988). In this colony, the females might be courted by up to 20 different males, but the males were not organised into coalitions as in lions. A courting male generally kept close to a female in oestrus, but the male's presence was not necessarily decided by rank differences between males. In other words, dominant males did not necessarily retain exclusive mating rights even though they might achieve more matings per year. Dominant breeding males show two alternative mating strategies at different times of the breeding season. At the beginning of the breeding season when breeding females are relatively scarce, they stick close to them throughout oestrus. However, as more females are available to mate with in the peak of the breeding season, a dominant male moves between females to maximise his reproductive success.

Lionesses and feral domestic females may frequently copulate with more than one male. Darie and Boersma (1984) suggested several reasons for this promiscuity. Firstly, it may increase the genetic diversity of the young, which may be advantageous in an unpredictable and changing environment. Secondly, it may test the genetic quality of the males through sperm competition. It may also reduce the chances of infanticide through paternal uncertainty. However, Darie and Boersma (1984) also favour the idea that it increases the degree of relatedness of the young, which is of obvious importance in a social cat where kin selection is thought to be an important factor affecting altruistic behaviour (Figure 7.15).

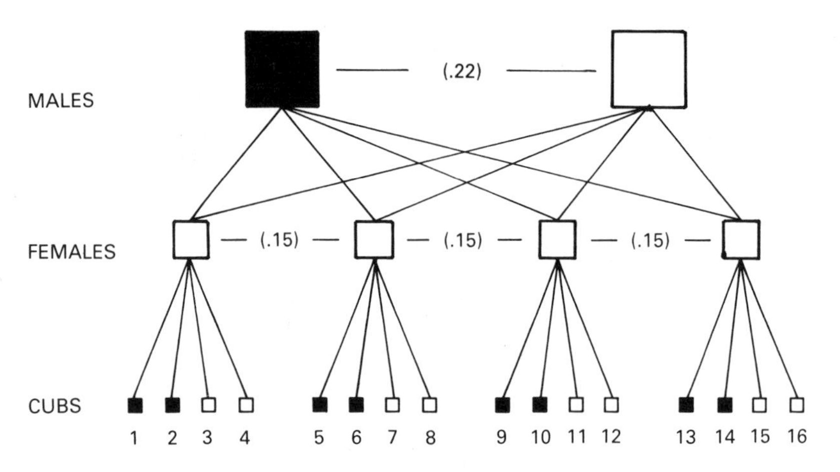

Figure 7.15 A polygamous mating system where females copulate with both males, and both males copulate with all four females (Darie and Boersma, 1984) (A.K.)

Another curious behaviour seen in the Serengeti lions is that although they may steal prey from a female, they may share it with the cubs (Schaller, 1972). Obviously the amount consumed by a cub is far less than that needed by an adult female. However, in the Ruwenzori National Park in Uganda lions never share their food with their cubs (van Orsdol, 1986). The difference between the lions in these two areas seems to be the stability of the prides. In the Serengeti, a male may only have a reproductive life of two years so that each cub represents a large part of his total reproductive success. However, in the Ruwenzori National Park lions may breed for ten years or more, so that each cub is only a tiny part of his total reproductive success. Therefore it is much more advantageous for Serengeti lions to share their meals with their cubs than the Ruwenzori lions. However, more observations are needed to confirm whether this explanation is valid.

In solitary cats both sexes disperse from their natal range although females may take longer to disperse than males. In lions, however, males always disperse and females rarely do so, usually remaining within their pride (Figure 7.16). However, about one-third of females do disperse despite the advantages of staying with the pride. When it comes to breeding, a sedentary female will have a good knowledge of all the cubbing and hunting sites within the pride range, and would gain from the assistance of close relatives (Pusey and Packer, 1987). Some females disperse, but then return to their natal pride. However, they are usually at a disadvantage, because they tend to breed later in life. The other disadvantages of dispersing are considerable. They suffer a higher mortality and are unlikely to breed successfully. In the Serengeti, buffaloes seem to be a very common cause of mortality among nomadic lionesses. Perhaps the wandering females do not have access to refuges from buffaloes where they can avoid being attacked. Even if the dispersing

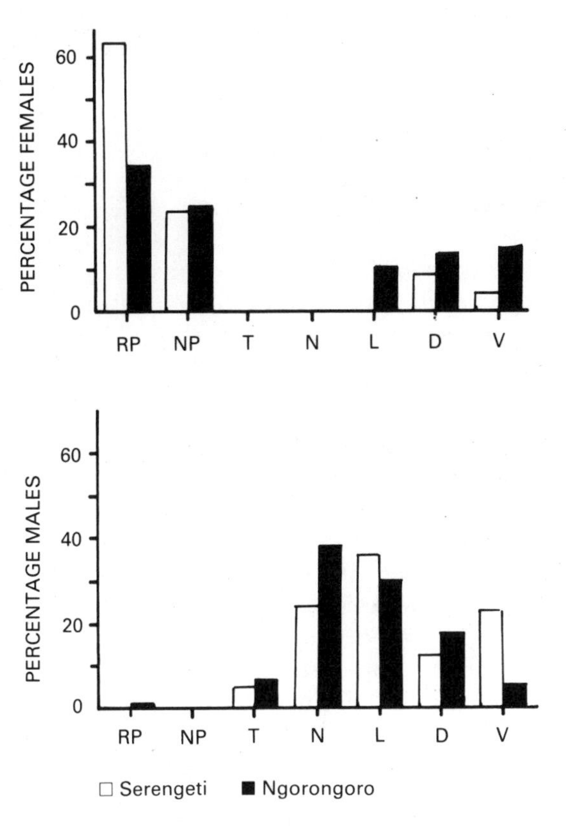

Figure 7.16 The fate of subadult lions and lionesses in the Serengeti and Ngorongoro. Key: RP Remain: still present in natal pride at 48 months. NP New pride: in or adjacent to natal pride range, but do not associate with it. T Transfer: become a member of a pre-existing pride. N Nomad: individuals found well away from their natal range. L Left: last seen outside natal range, but not seen at 48 months of age. D Died: animals in poor health, or that disappear singly or dead. V Disappeared: animals in good health that disappear with another pride member. (Pusey and Packer, 1987) (A.K.)

females eventually cofound a pride, this new pride risks increased injury and mortality from disputes with neighbouring prides until they can establish themselves. Why then do females disperse at all?

Pusey and Packer (1987) found that females may disperse as a result of aggression due to a male takeover or an aggressive lioness with new cubs. Dispersing females would also avoid inbreeding by mating with their father or other close relatives. In prides with more than ten members, the pride reproductive success falls, so that a female might disperse in order to increase her reproductive success elsewhere.

Males always disperse. Although they appear to suffer increased mortality when they do so, they might suffer even more if they stayed within the pride. The older males will simply not tolerate their presence. Therefore, resident males may evict young males, or a new coalition may do this. Some males leave a pride voluntarily, usually two or three closely related males leaving together. This gives them the chance to either establish a new pride or take over another, because larger coalitions are more likely to take over and hold onto a pride than single lions (Bygott *et al.*, 1979).

DISEASES AND PARASITES

All cats suffer their own fair share of diseases and parasites. Wild cats in captivity suffer a variety of ailments including the following (Ashton and Jones, 1980; Seidel and Wisser, 1987).

VIRAL DISEASES

(a) Feline panleucopenia or feline distemper. A highly infectious and fatal disease of cats which produces loss of appetite, vomiting and depression. Diarrhoea may also occur after a few days. Pregnant females may pass this disease to their unborn young, who develop brain damage affecting movement. South American cats are said to show some resistance to this disease.

(b) Feline respiratory diseases or 'cat flu'. There are two common viral diseases of the upper respiratory tract of cats—feline rhinotracheitis virus and a calicivirus. Both produce a nasal discharge, sneezing, conjunctivitis, ulceration of the mouth, raised body temperature, depression and loss of appetite. Severe cases may be fatal.

(c) Feline viral lymphosarcoma. As its name suggests, causes cancer in the lymph glands, etc.

(d) Feline infectious peritonitis.

(e) Cowpox has a variable effect on felids. Cats may either suffer from acute pneumonia leading to death in a few days, or develop ulcerated nodules in the skin and ulcers in the mouth over a prolonged period of time. Animals suffering from the latter symptoms may recover.

(f) Rabies. Recorded at a low incidence in Canada lynxes (Quinn and Parker, 1987) and tigers (Seidel and Wisser, 1087) in the wild.

BACTERIAL DISEASES

(a) Salmonella from contaminated food may result in mild diarrhoea or severe gastroenteritis.

(b) Tuberculosis.

(c) Feline infectious anaemia is caused by *Haemobartonella felis* which is probably transmitted by blood sucking insects. Small cats are especially susceptible to this disease.

ENDOPARASITES

(a) Nematodes or roundworms. Many cats carry the nematode *Toxacaris* which produces very robust eggs capable of surviving for a very long time in the environment. Only in very severe cases can it cause death by obstructing the gut.

(b) Cestodes or tapeworms are less common in cats than nematodes in captivity, because they are usually transmitted by a vector such as fleas. Infection can occur by ingesting cysts in meat.

ECTOPARASITES

(a) Mange mites of sarcoptic and notoedric types cause severe skin irritation and loss of hair.

(b) Ear mites usually cause irritation and discomfort, but may result in ruptured ear drums and a chronic bacterial infection.

In the wild cats suffer from many diseases and parasites which cannot be exhaustively dealt with here, but I will mention a few examples. It is unlikely that any of these ailments is sufficient to kill an otherwise healthy cat. However, in association with starvation, injury, old age or even a combination of diseases, deaths may occur. For example, Schaller (1972) considers that 18 per cent of lions in the Serengeti die due to disease. Other causes of mortality include old age (9 per cent), snaring (4 per cent), hunting injuries (9 per cent), shooting, starvation and abandonment. Schaller (1972) also discusses in great detail the diseases and parasites that afflict lions. These include biting flies, hippoboscid flies, ticks (which may carry tick fever (*Babesia*)), trematodes, nematodes and trypanosomes.

Corbett (1979) found that Scottish wildcats on the Glen Tanar estate in eastern Scotland suffered from cestodes, nematodes, fleas and ticks. Ocelots are afflicted by rabies, feline panleucopenia and pancreatic amyloidosis (resulting in diabetes mellitus) (Tewes and Schmidly, 1987). Jaguarundis suffer from ticks, chewing lice, tapeworms and feline panleucopenia (Tewes and Schmidly, 1987).

This latter disease has also killed bobcats in San Diego County, California (Lembeck, 1986). Canada lynxes may suffer from nematodes, cestodes, hookworms and feline panleucpenia (Quinn and Parker, 1987).

MORTALITY AND SURVIVAL

It probably comes as no surprise that cats need all their nine lives if they are to survive to breed successfully. Not only to they have to put up with diseases, parasites, starvation and injury during prey capture, but humans poison them, snare them, trap them, run them over with their cars and shoot them. Of course data on mortality are biased towards human mediated mortality, but this can be very significant in some areas.

Diamond (1988) has reviewed studies on the rate of mortality among domestic cats that fall from buildings in New York. The cats fell on average 5.5 stories (one storey is equal to 15 feet or 4.5 metres), but the range was 2 to 32 with most cats landing on concrete. An incredible 90 per cent of cats survived. Those that died suffered thoracic injuries and shock. Strangely, cats that fell more than seven stories suffered lower mortality even though they fell a greater distance. Being small, cats reach a fairly low terminal velocity (60 mph compared with 120 mph for humans). Also, falls for a greater distance give a cat time to move its body into the best position for landing on flexed all fours, thereby spreading the force of impact and absorbing the energy from the fall. It has also been suggested that over very long falls, cats relax once they reach terminal velocity and spread their legs out sideways, thereby reducing velocity of the fall and spreading the force of impact over a larger area, so that mortality is minimised.

Corbett (1979) looked at mortality in the Scottish wildcat and domestic cats (Table 7.6). More than 50 per cent of the wildcats and nearly as many domestic cats were snared. Less than 20 per cent were found to have died due to natural causes. In fact, few of the cats were more than two years old, indicating a very high rate of turnover in the population.

Indian desert cats suffer from predation by jackals, wolves and dogs in Pakistan (Sharma, 1979). Canada lynx may be predated upon by wolves, if caught in the open (Quinn and Parker, 1987), and a trapped lynx is recorded as having been killed by a wolverine in Alaska (Berrie, 1973).

Lion cubs may suffer a very high mortality rate. It ranges from only 15 per cent in the Nairobi National Park, where the lack of nomadic males reduces infanticide and there are no marauding hyaenas to predate upon cubs, to 67 per cent in the Serengeti where infanticide, predation, starvation and disease are the main killers

Table 7.6 Probability of mortality in Scottish wildcats and domestic cats on and in the vicinity of the Glen Tanar estate in eastern Scotland

Cause of mortality	Glen Tanar		Glen Tanar and surrounds
	Wildcats	Domestic cats	Wildcats
Snared	0.58	0.48	0.38
Shot	0.08	0.26	0.19
Hit by car	0.08	0.13	0.19
Trapped	0.08	0	0.15
Natural causes	0.17	0.13	0.08
Number of cats	12	23	26

Source: Corbett (1979)

(Figure 7.17) (Schaller, 1972). The dependence of Serengeti lions on migratory wildebeest, zebras and gazelles means that if cubs are born when the migrants are absent, they are quite likely to starve to death either from lack of milk or lack of meat.

The effects of humans on the mortality of Canada lynx can be incredibly high. Mech (1980) found that 50 per cent of the radio-collared Canada lynx in his study were killed by humans in north eastern Minnesota, but these lynxes may have been more vulnerable to mortality because they were moving into a new area. However, on the Kenai Peninsula in Alaska, the situation is far worse. The population of Canada lynx suffers a 90 per cent annual mortality, of which 80 per cent is due to trapping. This includes 88 per cent of

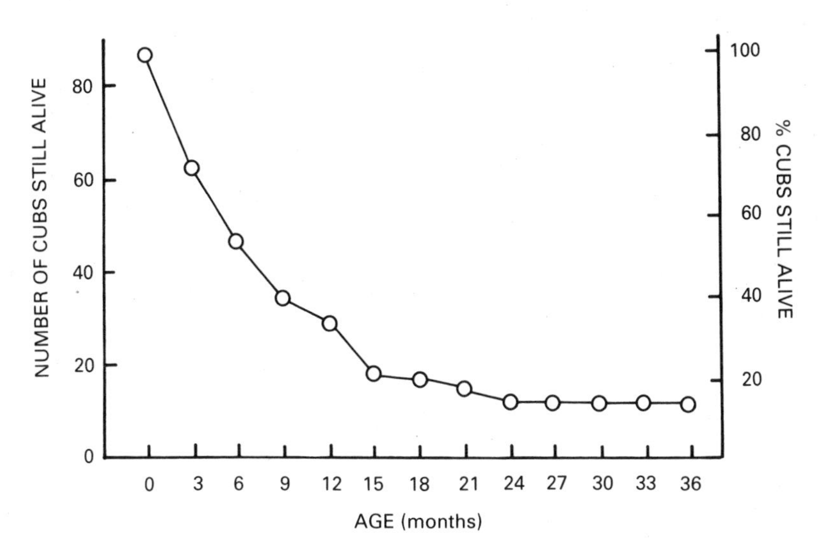

Figure 7.17 The survival of lion cubs in the Serengeti (Bertram, 1975a) (A.K.)

males, 45 per cent of females and 77 per cent of the year's production of juveniles. Given the shorter time of the year that juveniles are available for trapping, they are five times more likely to be trapped than adults. The more intrepid males are twice as likely to be trapped as females. Interestingly, Scottish wildcat males are much more likely to be killed on the roads because of their larger home ranges and less wary nature. On Cape Breton Island, 65 per cent of the Canada lynx population is trapped annually (Parker *et al.*, 1983). In non-hunted populations, the annual mortality is as low as 25 per cent. The survival of juvenile Canada lynx is also affected directly by the density of snowshoe hares (see Chapter 4).

A bobcat population in San Diego County, California, also

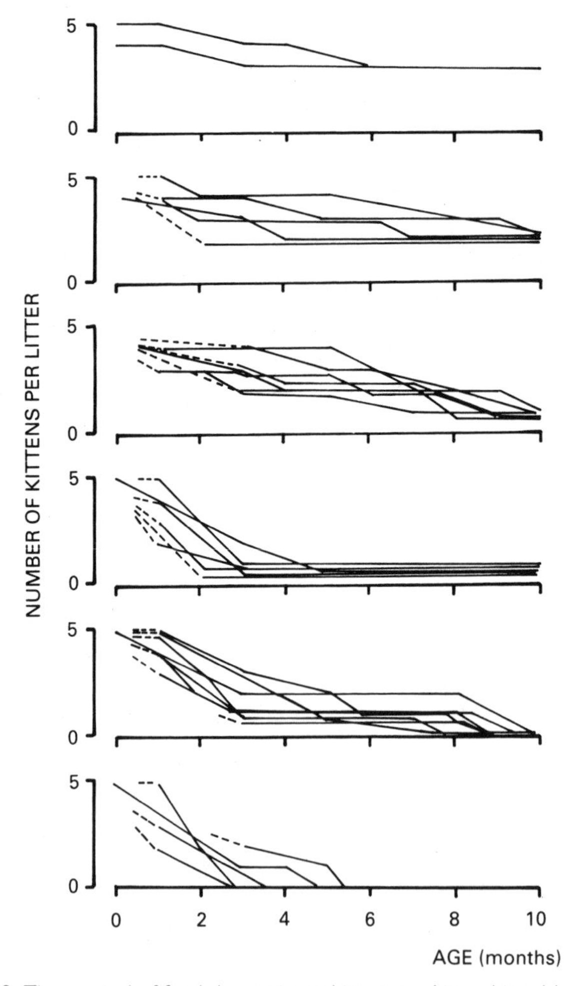

Figure 7.18 The survival of feral domestic cat kittens on Aionoshima Island, Japan. Dramatic mortality is due to feline infectious enteritis spreading through a litter. Gradual mortality is due to other factors including starvation (Izawa and Ono, 1986) (A.K.)

appeared to be mostly regulated by prey availability and other natural causes (Lembeck, 1986). Twenty-nine cases of mortality were recorded, including predation (9), feline panleucopenia (5), starvation (3), trapping (2), shooting (2) and a road casualty.

In a non-hunted population of pumas in southern Utah, ten family groups of pumas were followed by radio tracking and tracking on foot to assess the survival of the cubs (Hemker *et al.*, 1986). The cubs were only mobile at three months of age, so that it was not possible to calculate survival rates and litter sizes from birth. Litter size at three months ranged from one to four with a mean of 2.4. Survival of cubs between three and ten months of age was 72 per cent, but from ten months to dispersal at 16–19 months the survival rate was 92 per cent. Causes of death included accidents, poaching, cannibalism and loss of mother. Again in hunted populations the mortality rate is higher. Hemker *et al.* (1986) discovered that in areas of low prey density, births of puma cubs were delayed, when they would unfortunately coincide with the hunting season for pumas, thereby increasing possible cub mortality due to loss of lactating mothers. In the Idaho Primitive Area, fourteen deaths were recorded by Hornocker (1970). Half of these were due to hunting, although others included an accident while hunting wapiti, being killed by dogs and cannibalism (infanticide).

The survival rate of feral domestic cat kittens on Ainoshima Island, Japan, was only 9.5 per cent at ten months of age. Litters declined either gradually due to starvation or drastically due to diseases such as feline infectious enteritis (Figure 7.18) (Izawa and Ono, 1986). The high population density of these cats undoubtedly contributed to the rapid spread of diseases.

POPULATION REGULATION

Natural population regulation has been studied and compared for the bobcat, puma and leopard (Hornocker and Bailey, 1986). All populations were similar, in that the cats were solitary hunters, there was an adequate supply of food and the populations were stable. The puma population in the Idaho Primitive Area is characterised by males having large non-overlapping home ranges, but females have smaller overlapping ranges. The bobcat population in Idaho has small exclusive female territories and overlapping male ranges. The leopard population in the Kruger National Park showed partially overlapping home ranges. In the puma and bobcat populations, juveniles dispersed from the area, but in the leopard population, the young remained as transients until a home range became available due to adult mortality.

Mortality among the leopards was high. The adults suffered 18 per cent annual mortality, the subadults 32 per cent and the cubs 50 per cent. A third of the leopard cubs were predated upon, the rest starved to death. To a great extent the leopard population was controlled by the availability of food. The bobcat and puma populations remained stable, with virtually no mortality. However, mortality in transients may well have been high if they did not eventually locate a vacant home range outside the study area. In these cases, the system of land tenure and social behaviour maintained the stability of the populations. In the year after this study, the bobcat population crashed following a crash in the population of its main food source, the jackrabbit (*Lepus californicus*). The home range system of the bobcats broke up as they searched for food, and they showed much greater tolerance towards each other as they hunted at the few available hunting sites.

Lions in the Nairobi National Park also seem to retain a stable population. Here the subadults disperse into the Kitengala Conservation Area where they soon succumb to human hunting. Thus the pride system maintains a stable population within the Park and does not seem to respond to changing numbers of ungulate prey (Rudnai, 1973).

8

CATS AND HUMANS

The relationship between the many species of wild cats and humans has been a long and changing one. From the initial and continuing exploitation of cats for their fur and perhaps even competition for the same prey, humans went on to establish a close and mutually beneficial relationship with one species, the wildcat, *Felis silvestris*, which has resulted in the hundred or so breeds of domestic cat known today. Even this relationship has not been constant; from adoration as gods through persecution as harbingers of evil, domestic cats are, today, generally viewed either as pampered pets or pests.

Although the exploitation of the big cats for their fur is now mostly over, the smaller cats are under increasing pressure from the still enormous trade in fur. Hundreds of thousands of wild cats die each year to adorn people's shoulders with fur. But whether big or small, all cats are threatened by habitat destruction.

Humans must now take a much more responsible role in the management of their world, of which cats form a vital part. Conservation and management of wild cat populations is essential if most species are to survive well into the twenty-first century. Most tropical forest cats are highly endangered by habitat loss. Reintroduction programmes are being instigated for a number of species, including the serval, wildcat and Eurasian lynx. Zoos also have their role to play as a possible last refuge or back-up population for small wild populations. Correct genetic management is essential if these captive populations are to remain as possible reservoirs for topping up wild populations from time to time. But let us start first with the origins of the domestic cat.

THE ANCESTOR OF DOMESTIC CATS

The ancestor of the domestic cat is generally believed to be the wildcat from North Africa, *Felis silvestris lybica*, rather than the European wildcat, *F. s. silvestris*. Various other ancestors have been suggested, particularly for the Asian breeds of domestic cats, including the jungle cat, leopard cat and Pallas's cat (Kratochvil and Kratochvil, 1976; Hemmer, 1978b). However, it is now known that only *Felis silvestris* genes are to be found in today's domestic cats. On the basis of cranial characters and the morphology of the baculum, Kratochvil and Kratochvil (1976) have suggested that Asian subspecies of *Felis silvestris* are the ancestors of Asian domestic breeds; namely *F. s. nesterovi* for the Persian breed and *F. s. ornata* for the Siamese.

The process of domestication has brought about some changes in behaviour and morphology, but not to the same extent as in other domestic animals (Clutton-Brock, 1981; Robinson, 1984). In its behaviour, the domestic cat is partly neotenous, retaining juvenile appearance and behavioural patterns into adulthood, so that it is much more docile than its wild ancestor. In part this is thought to be due to relatively smaller adrenal glands which are responsible for producing adrenaline and noradrenaline, the hormones of fright and flight, although Robinson (1984) considers that there has been artificial selection for more placid behaviour as well. Morphological changes include a decrease in body and brain size, and an increase in intestine length to cope with a less carnivorous diet. Hemmer (1978b) suggested that the ancestor of the domestic cat was the subspecies of *Felis silvestris* with the smallest brain size. This just happens to be *F. s. lybica*.

GENETICS OF COAT COLOUR

The most profound effect on wildcat appearance due to domestication is the multiplicity of coat colours known today. These are mostly due to simple mutations at one or more of six different gene loci (Robinson, 1970, 1976a). Other gene loci have mutations which affect the pattern and coloration of the coat less commonly. Each gene locus may be occupied by different alternative genes or alleles. Animals inherit one allele for each gene locus from each of their parents, and these may be different in an individual. Some alleles are dominant in their effects on the appearance of the animal over other alleles. Conventionally the allele at each gene locus is denoted by a letter. Dominant alleles are represented as a capital letter and recessive alleles as a lower case letter with or without superscripts. Recessive genes usually only affect the external appearance of the animal if the same recessive gene is inherited from each parent. In some cases, there is a rank order of dominance for the alleles of a given gene locus. The mutant alleles of a particular gene locus are listed in decreasing order of dominance below:

(1) A Agouti	The typical banded hair colour of many wild mammals, including *Felis silvestris*. Mutations: a—non-agouti (usually black, i.e. melanism)
(2) B Black	Normal expression of black pigment. Mutations: b—chocolate
(3) C Colour	Full manifestation of yellow and black. Mutations: c^{ch}—chinchilla c^{h}—acromelanic albinism c—complete albinism
(4) D Dilute	Normal expression of dark pigment. Mutations: d—blue/pale
(5) E Extension	Normal expression of agouti. Mutations: e—yellow, reddish (erythrism) E^{d}—dominant black
(6) P Dark eye	Normal pigmentation. Mutations: p—red/pink eyes and pale coat

Other gene loci that affect the pattern and colours of cat fur include:

(7) T Tabby (Felid pattern)	Striped tabby pattern (normal coat colour pattern). Mutations: T^{a}—Abyssinian T^{t}—striped tabby t^{b}—blotched tabby
(8) S Piebald	Piebald spotting due to a partly dominant gene, S. Less white spotting in the heterozygote.
(9) Wb Wide band	Width of band on the agouti coat. Wider bands produce lighter coats, possibly including non-extension (e) coat colours.
(10) O Orange	Ginger coloration. Sex-linked to X chromosome in cats. Tortoiseshell cats are always female.

These mutations are occasionally seen in wild felids. The non-agouti mutant gene (a) is very common in many species of wild cats. It is this gene which is responsible for the black panthers, particularly of southeastern Asia, where despite being due to a recessive gene, natural selection has resulted in black coat colour being relatively common (Robinson, 1969b). The famous white tigers of Rewa are not true albinos. Robinson (1969a) considers them to be chinchilla (c^{ch}) mutants which have a limited amount of dark pigmentation and blue eyes. True albinism is rare, even in domestic cats, because it is usually regarded as an indicator of lethal genes causing deafness, blindness and reduced fecundity and survival. The felid pattern (T) is affected by mutations in wild cats. For example, the king cheetah is thought to be a blotched tabby mutant (t^b), as has recently been confirmed by breeding experiments (van Aarde and van Dyk, 1986). The servaline pattern of servals from the drier areas of Africa is also thought to be a mutation at the T locus. Erythrism (e) is also quite common in wild cats, e.g. jaguarundi where the normal grey form and the mutant red form were described as separate species until kittens of both colours were discovered in the same litter. A full list of the commoner alternative coat colours and patterns of wild cats, and the gene mutations thought to be responsible for them are given in Table 8.1 (Robinson, 1976a).

THE PROCESS OF DOMESTICATION

At least two alternative hypotheses have been proposed to explain the domestication process for domestic cats. Baldwin (1975) favours an active domestication process which is divided into the following phases:

(1) Period of competition: before 7,000 BC.

(2) Period of commensality: 'half-wild' cats feeding on rodents attracted to grain stores and scavenging in early human settlements (7,000–4,000 BC).

(3) Period of early domestication: confinement of cats to temples for religious purposes (2,000–1,000 BC).

(4) Period of full domestication: secularisation of the domestic cat in Ancient Egypt and spread through the rest of the world (1,000 BC onwards).

However, Todd (1978) favours a passive domestication process involving the following stages:

Phase 1. Food storage and concentrated waste disposal resulted in the attraction of vermin, particularly rodents, to provide a super-abundance of food to attract wildcats.

Phase 2. Cats begin to rely on scavenging for a greater proportion of their diet. There is 'natural selection' for wildcats that are tolerant of humans.

Table 8.1 Homologous mutant genes in the Felidae

Species	Mutant gene									
	a	b	c^{ch}	c^{h}	c	d	e	p	S	T
Acynonyx jubatus (Cheetah)	.	.	+	.	.	.	.	.	.	(+)
Felis bengalensis (Leopard cat)	+	.	.	.	.	.	.	.	.	.
— *caracal* (Caracal)	+	.	.	.	.	.	?	.	.	.
— *catus* (Domestic cat)	++	++	++	++	+	++	.	+	++	(+)
— *chaus* (Jungle cat)	+	.	.	.	.	.	.	.	+	.
— *colocolo* (Pampas cat)	+	.	.	.	.	.	.	.	.	.
— *concolor* (Puma)	+	.	.	.	+	.	+	.	.	.
— *geoffroyi* (Geoffroy's cat)	+	.	.	.	.	.	.	.	.	.
— *pardalis* (Ocelot)	.	.	.	.	.	.	+	.	.	.
— *serval* (Serval)	+	.	.	.	.	.	.	.	.	(+)
— *silvestris* (European wild cat)	.	.	+	.	.	.	.	.	.	.
— *temminckii* (Temminck's cat)	+	.	.	.	.	.	.	.	.	.
— *yaguarondi* (Jaguarundi)	.	.	+	.	.	.	+	.	.	.
— *tigrina* (Tiger cat)	+	.	.	.	.	.	.	.	.	.
Lynx canadensis (Canadian lynx)	.	.	.	.	.	+	+	.	.	(+)
— *rufus* (Bobcat)	+	.	?	.	+	+	+	.	+	.
Neofelis nebulosa (Clouded leopard)	+	.	.	.	.	.	.	.	.	.
Panthera leo (Lion)	.	.	+	.	.	.	+	.	+	.
— *onca* (Jaguar)	+*	.	.	.	.	.	.	.	.	.
— *pardus* (Leopard)	++	.	+	.	+	.	+	.	.	(+)
— *tigris* (Tiger)	+	.	++	.	+	.	+	.	.	.

Notes
The degree of homology is indicated as follows: ++ – almost certain, + – probably, and ? – suspected. A (+) sign under T indicates a major change in the felid melanoid pattern for this species. Mutant genes are as follows: a – non-agouti, b – brown, c^{ch} – chinchilla, c^{h} – acromelanic albinism, c – albinism, d – blue dilution, e – erythrism, p – pink eye, S – piebald spotting, T – felid pattern. * – melanism is dominant in the jaguar.

Sources: Robinson (1976a) (modified). Additional data from Rabinowitz *et al.* 1987 and pers. obs.

Phase 3. Cats rely almost totally on human handouts and scavenging for food. It is only in this phase that humans can begin to control the breeding and spread of cats.

Therefore, humans created a new niche into which the wildcat moved, so that it evolved almost passively into what we know today as the domestic cat. I think that the combination of two factors was critical at the beginning of the domestication process: the accumulation of rubbish even in moderate-sized human settlements and the abundance of rodent pests in the vicinity of food stores and rubbish tips. This allowed cats to live at very high densities, resulting in passive selection for those cats that were predisposed to domestication.

THE FIRST DOMESTICATION

When did domestication first take place? The Ancient Egyptians in about 2,000 BC are generally held responsible. However, the dog, cow and sheep were domesticated thousands of years before this. It seems very unlikely that the wildcat would have escaped the attentions of the Ancient Egyptians for so long.

Zeuner (1963) has suggested that a single molar found at Jericho and dating from the Prepottery Neolithic of about 6,700 BC may have been from one of the first domesticated cats. Alternatively, it may have been from a wildcat killed for its fur or meat (as were foxes), or a scavenging wildcat. Recently, a cat's mandible was discovered at Khirokitia in Cyprus which dates to about 6,000 BC (le Brun *et al.*, 1987; Davis, 1987). A comparison of the length of the first molar and toothrow length with those of the wildcat and jungle cat reveals that in size it most closely resembles a wildcat (Figure 8.1) (Davis, 1987). However, wildcats are unknown from Cyprus, so that this cat could have been taken there by humans and could, therefore, represent one of the first domesticated, or at least tamed wildcats.

By 1,600 BC, there was pictorial evidence in Ancient Egypt for the domestication of the cat. Those early domestic cats still closely resembled the striped tabby wildcats. At that time, cats were revered as gods in Ancient Egypt: male cats were sacred to the sun god Ra and female cats to the fertility goddess Bast (Robinson, 1984). At first the domestic cats were confined to temples; this was probably important in fully domesticating the wildcat because for the first time they would have been isolated from the rest of the wildcat population. Later, domestic cats spread into the community. It was illegal to kill cats, and if a cat died, all the members of a household would shave off their eyebrows as a mark of respect (Zeuner, 1963; Clutton-Brock, 1981). If cats strayed abroad by, for example, stowing away on a ship, travelling Egyptians brought them back so that they did not spread in any numbers from Egypt at first.

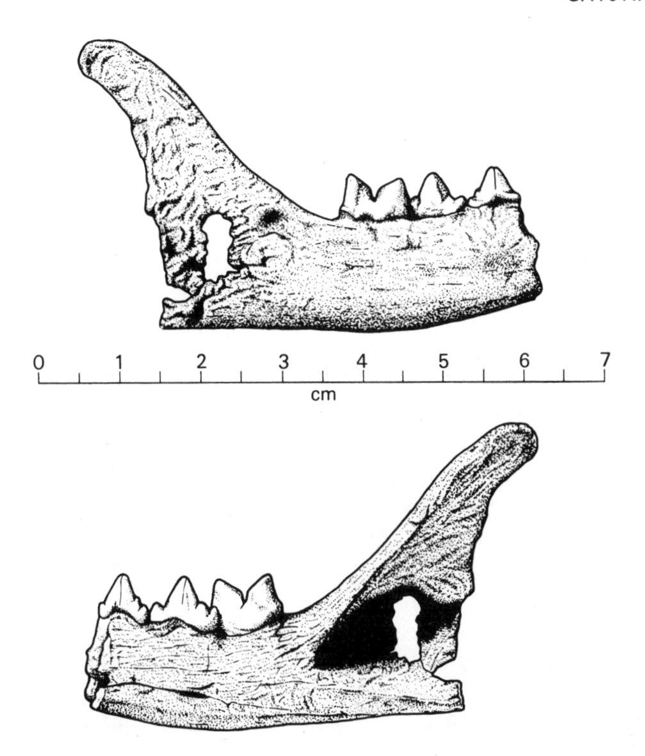

Figure 8.1 The eight-thousand-year-old cat mandible from Khirokitia in Cyprus: the first domestic cat? (Odile le Brun) (S.A.)

Dead cats were mummified and sent to the huge cat cemeteries at Bubastis. So abundant were these remains, that during the nineteenth century mummified cats were used as ballast on ships or ground up to be used as fertiliser. One shipment of 19 tons of mummified cats found its way to Manchester to meet just such a fate. Sadly only one skull survives from this historic cargo. Cats were mummified in two different ways (Armitage and Clutton-Brock, 1981). Some were covered in natron-soaked bandages, while others were first preserved with embalming resin. Despite the cat's sacred status, it was apparent that the mummified cats had been strangled especially for mummification. Cats from two age groups were most commonly used, and only one individual in the sample that was studied was over two years old. Armitage and Clutton-Brock (1981) suggest that cats of one to four months of age were specially killed for mummification. Older cats in the nine to twelve month age group were probably animals that had grown past their optimum size and age, and may have represented excess males that were culled and sold cheaply.

Eventually the domestic cat spread throughout Europe following the rodent pests from Ancient Egypt and preceding the spread of the Roman Empire (Clutton-Brock, pers. comm.). Baldwin (1980) has also chronicled the probable spread of the domestic cat through the islands of the Pacific.

During the Middle Ages, domestic cats were associated with witchcraft and sorcery, with the result that many thousands of cats probably perished in the ensuing persecution. As recently as the end of the seventeenth century, desiccated corpses of domestic cats were built into the walls of new buildings, often with rats or mice, in order to bring good luck or to keep vermin away. Today the cat is largely a reformed character, ranging from a pampered pedigreed pet to the 'streetwise' cat delving into dustbins to survive. Cats are still kept on farms for their skill in controlling rodent populations, although contrary to popular belief, they do not need to be starved to give them an incentive to hunt.

In northern latitudes the striped tabby domestic cats now resemble the forest form of the wildcat, *Felis s. silvestris*. This may be due to hybridisation between domestic cats and local wildcats, or because natural selection has produced a similar phenotype in the domestic population. Todd (1978) suggests that in hybridising populations, gene flow should be from wildcats to domestic cats, because the larger wildcat males will outcompete domestic and hybrid males from breeding with female wildcats. Therefore, the hybrid population should act as a sort of buffer, or even filter, diverting wild genes into the domestic cat population. In the early days of the spread of domestic cats, their population would have been so small that the contribution of wild genes could have been significant. However, today the domestic cat population is so large that this effect is negligible. Robinson (1984), however, suggests that gene flow should be from the domestic to the wild population, because of the smaller wild population. Whatever is happening, hybrids between domestic cats and wildcats occur throughout the entire range of the wildcat. We must discover in which direction gene flow is going in order to implement management procedures that will preserve the wildcat gene pool, if indeed that is necessary.

POPULATION GENETICS

It all started with the Egyptians, but, ever since, domestic cats have travelled the world with humans. Ships in particular have been a very effective means of transport to every corner of the globe.

Todd (1977) found that humans have had very little effect on the proportions of different coat colour mutations seen in local populations of domestic cats. Throughout most of Europe the frequency of each gene mutant responsible for a coat colour was found to be in a stable equilibrium with all the others (the Hardy-Weinberg equilibrium). There was, therefore, no evidence of human selection for a particular coat colour. When Todd mapped

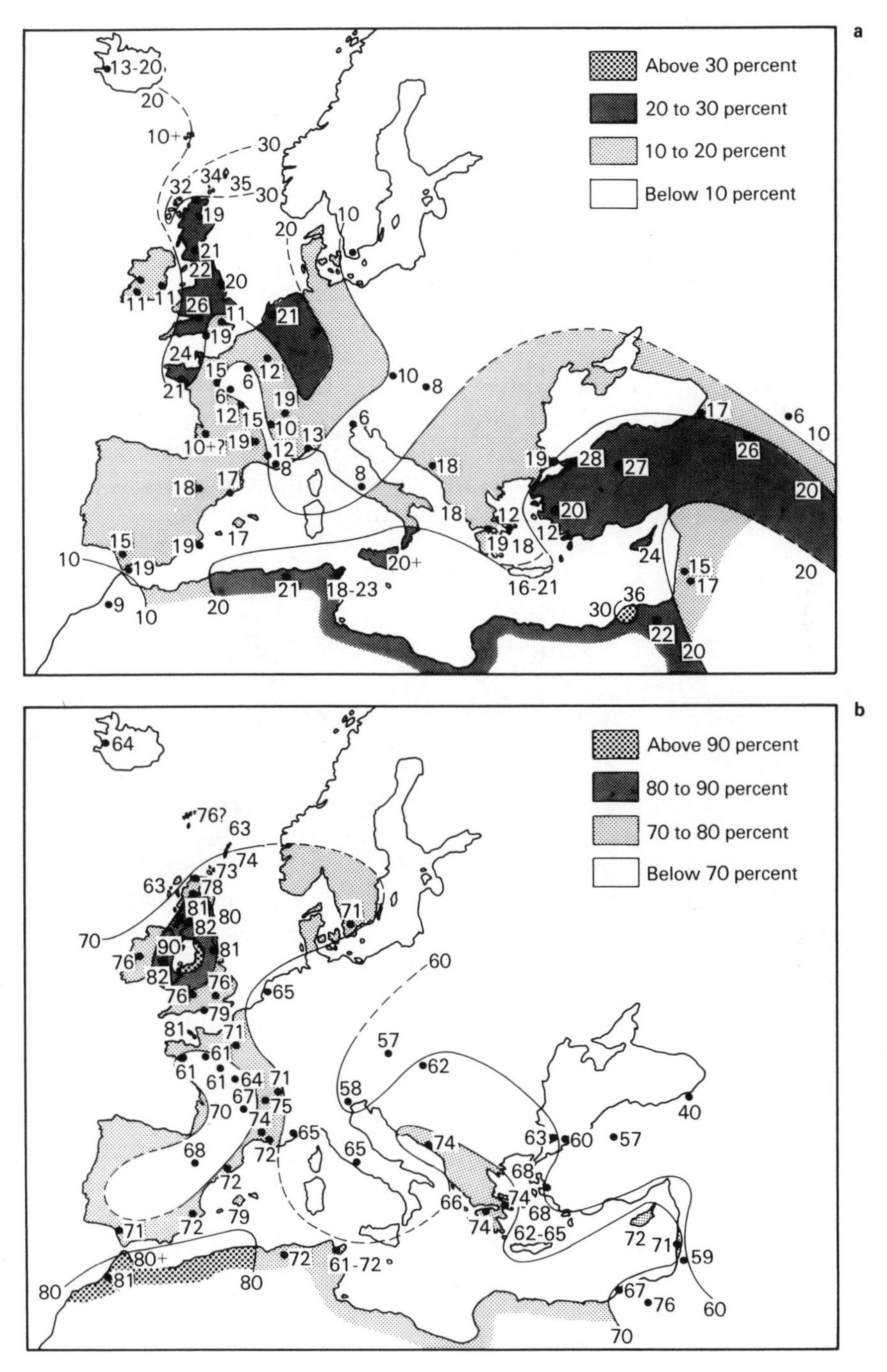

Figure 8.2 (**a**) This cline map represents the frequency distribution of the sex-linked orange allele (O) in Europe. The origin of this mutant gene appears to be in southwestern Asia (Todd, 1977). (**b**) The frequency distribution of the non-agouti allele (a) throughout Europe. This mutant gene probably arose in Phoenecia (the Middle East) or Greece (Todd, 1977) (S.A.)

out the frequency of each gene, he noticed that there were clines in their distribution, which could apparently reveal the possible origin of the mutant gene and its subsequent spread through Europe.

For example, the sex-linked orange gene (O) is found in fairly low frequencies throughout Europe, but a concentration in Asia Minor seems to reveal its origin there (Figure 8.2a). A fairly high frequency along the coast of North Africa shows that it spread by ship throughout the Mediterranean. A higher concentration across France to Britain reveals its route along the valleys of the Seine and the Rhone via the elaborate canal system to be found there (Figure 8.2a). A similar map for non-agouti (a) reveals its possible origin in Phoenecia (the Middle East) and its spread via the same overland route to Britain, where it is very abundant (Figure 8.2b). In fact the occurrence of the non-agouti gene is directly related to the degree of urbanisation of the habitat, suggesting some sort of natural selection for non-agouti in urban environments (Figure 8.3)

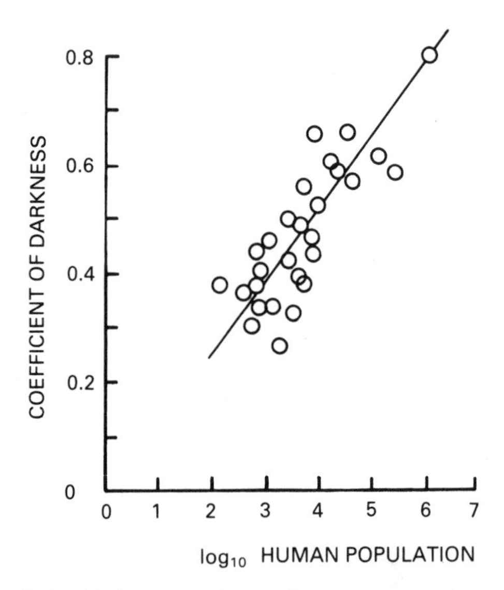

Figure 8.3 The relationship between the coefficient of darkness of cat populations and human population densities in Scotland. More of the cat population is darker in areas where the human population density is greater (Clark, 1976) (A.K.)

(Clark, 1976). Perhaps the high incidence of this gene in Britain reflects the early industrialisation there. The blotched tabby gene's (t^b) highest frequency is centred on Britain, where it almost certainly originated (Figure 8.4). Again the overland route through France shows some spread towards the Mediterranean. Tabor (1983) has suggested that the blotched tabby cat should be called the British Imperial Cat, because of its spread throughout the former British Empire. The frequency of the blotched tabby gene in each former colony reflects the time when it was first colonised.

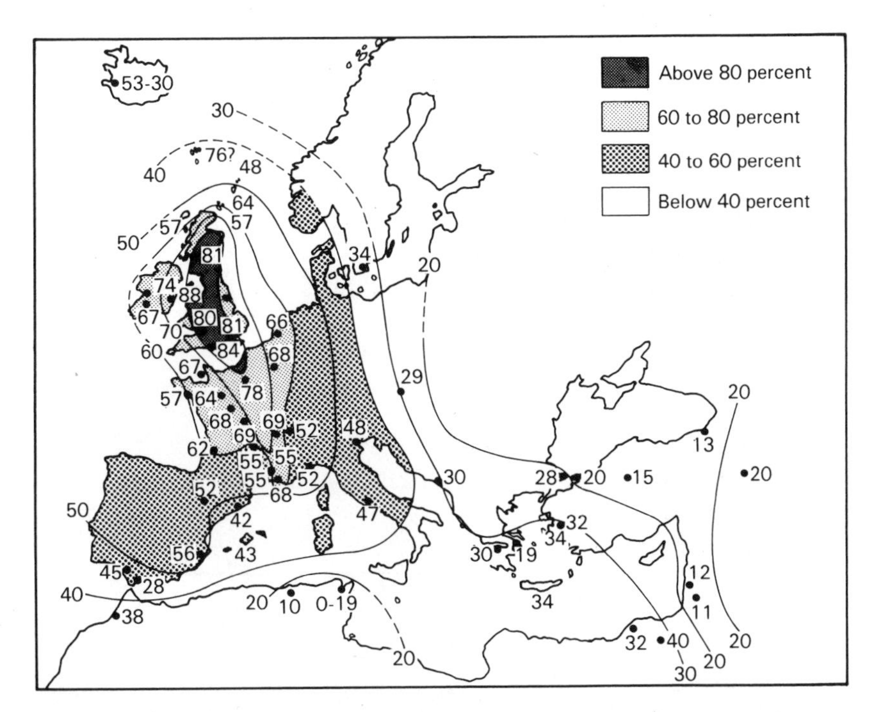

Figure 8.4 The frequency distribution of the blotched tabby allele (t^b) in Europe. Britain seems to be the origin of this mutant gene (Todd, 1977) (S.A.)

Therefore, there are fewer blotched tabby cats in North America than in Australia, reflecting the lower frequency in the gene pool in Britain at the time of the earlier colonisation of North America (Figure 8.5) (Todd (1977)).

EXPLOITATION

Like many other mammals, cats have long been exploited for their fur. Leopard skins are regarded as status symbols in local cultures as well as western civilisations. The Canada lynx has been subject to commercial exploitation since the seventeenth century, involving millions of furs (Elton and Nicholson, 1942; Novak *et al.*, 1987a, b). Exploitation of the spotted and striped big cats did not begin on a large scale until the end of the nineteenth century (McMahan, 1986). By the 1960s, many species of big cat were becoming severely endangered in the wild due to this exploitation and the loss of habitat. Many countries introduced their own legislation during the 1960s and 1970s to protect their big cats. A voluntary ban on the use of spotted/striped big cat furs by the International Fur Traders Association preempted international legislation to stop this trade. Between 1968 and 1970 there was, for example, a five-fold reduction

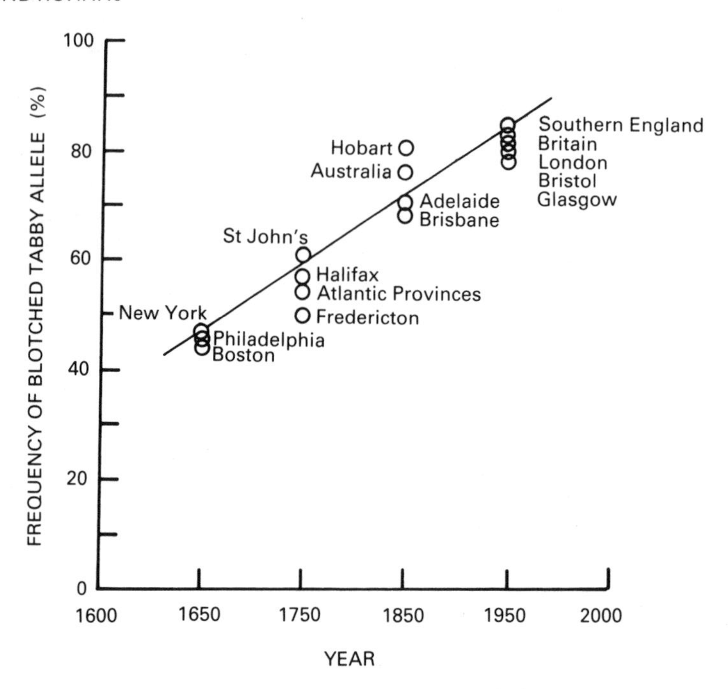

Figure 8.5 The frequency of the blotched tabby allele (t^b) in the cat populations of Britain and former British colonies. The current frequency of the blotched tabby allele in former colonies reflects the frequency in the British population at the time of human colonisation (Todd, 1977) (A.K.)

in the numbers of leopard skins imported into the USA from 10,000 to 2,000 per year (Myers, 1973). As the big cat population diminished, so younger and smaller cats were trapped (Figure 8.6) (Myers, 1973), so that even more pelts were needed to make a single garment. The widespread availability of poisons also allowed the removal of troublesome big cats from many areas, providing effective control even if the victims were elusive (Myers, 1973). Poisons probably did more than anything to bring about the eradication of big cat populations during the second half of the twentieth century. By 1975 the Convention on International Trade in Endangered Species had been formulated, but it was not until 1977 that it was ratified, and many countries have only ratified CITES even more recently. Many cats were included on Appendix 1 of CITES which effectively banned any commercial trade (Table 8.2) and the rest were placed on Appendix 2 in 1980 so that trade in all cats is effectively controlled. This has recently been revised again to put all populations of the ocelot and margay on Appendix 1.

With the loss of trade in big cats, furriers switched their attention to the smaller spotted cats, the Canada lynx and the bobcat. The ocelot and margay suffered in particular. In one year 140,000 ocelot skins were imported by the USA. This level of exploitation was bound to take its toll in a short time. The trouble with trade in

Figure 8.6 A shotgun trap set to kill a jaguar (Mondolfi and Hoogesteijn, 1986) (S.A.)

smaller cat skins is that many more skins are needed to make the same-sized garment. For example, up to 25 pelts are needed to make a garment out of wildcat skins, compared with three for a leopard; other examples are given in Table 8.3 (McMahan, 1986).

It was not long before ocelot and margay populations were affected and consequently there was a second revival in the trade in bobcat and Canada lynx furs during the 1970s (McMahan, 1986). However, although trade in these species continued to be strong into the 1980s, there has recently been a trend towards trade in the furs of even smaller cats from South America (e.g. Geoffroy's cat, tiger cat), Asia and Africa (e.g. wildcat, jungle cat, leopard cat). There seems to be a switch in exploitation to ever smaller species which is very reminiscent of the continued exploitation of the great whales.

Table 8.2 Status and legal protection of the Felidae

Scientific and common names	Status[a] Red Data Book (date)	CITES Appendix No. (date listed)	U.S. Endangered Species[b] List (date listed)
Acinonyx jubatus (cheetah)	V (1972); E for *A. j. venaticus* (1972)	I (1975)	E (1970)
Felis aurata (golden cat)	not rated	II (1977)	not listed
Felis badia (bay cat)	R (1978)	II (1977)	E (1976)
Felis bengalensis (leopard cat)	not rated	I for *F. b. bengalensis* (1977) II for others (1977)	E for *F. b. bengalensis* (1976)
Felis bieti (Chinese desert cat)	not rated	II (1977)	not listed
Felis caracal (caracal)	R for *F. c. michaelis* (1978)	I for Asian population only (1979), II for others	not listed
Felis chaus (jungle cat)	not rated	II (1977)	not listed
Felis colocolo (pampas cat)	not rated	II (1977)	not listed
Felis concolor (puma, cougar)	E for *F. c. cougar* and *F. c. coryi* (1981)	I for *F. c. coryi*, *F. c. costaricensis* and *F. c. cougar* (1975); II for others (1977)	E for *F. c. coryi*, *F. c. costaricensis* (1976), and *F. c. cougar* (1973)
Felis geoffroyi (Geoffroy's cat)	not rated	II (1977)	not listed
Felis guigna (kodkod)	not rated	II (1977)	not listed
Felis iriomotensis (Iriomote cat)	E (1978)	II (1977)	E (1979)
Felis jacobita (mountain cat)	R (1981)	I (1975)	E (1976)
Felis manul (steppe cat)	not rated	II (1977)	not listed
Felis margarita (sand cat)	E for *F. m. scheffeli* (1978)	II (1977)	not listed
Pardofelis marmorata (marbled cat)	I (1978)	I (1975)	E (1976)
Felis nigripes (black-footed cat)	not rated	I (1975)	E (1976)
Felis pardalis (ocelot)	V (1978)	I for *F. p. mearnsi* and *F. p. mitis* (1975), II for others (1975) I for all populations (1989)	E (1972, 1982)
Felis planiceps (flat-headed cat)	I (1978)	I (1975)	E (1976)
Felis rubiginosa (rusty-spotted cat)	not rated	I for Indian population only (1979), II for others (1977)	not listed
Felis serval (serval)	not rated	II (1975)	E for *F. s. constantina* (1970)
Felis silvestris (wild cat)	not rated	II (excluding *F. catus*)	not listed

Felis temminckii (Temminck's cat)	I (1978)	I (1975)	E (1976)
Felis tigrina (little tiger cat)	V (1981)	I for *F. t. oncilla* (1975) II for others (1975); all populations I (1989)	E (1972)
Felis viverrina (fishing cat)	not rated	II (1977)	not listed
Felis wiedii (margay)	V (1981)	I for *F. w. nicaraguae* and *F. w. salvinia* (1975) II for others (1975) I for all populations (1989)	E (1970)
Felis yagouaroundi (jaguarundi)	I (1981)	I for *F. y. cacomitli*, *F. y. fossata*, *F. y. panamensis*, and *F. y. tolteca* (1975); II for others (1975)	E for *F. y. cacomitli*, *F. y. fossata*, *F. y. panamensis*, and *F. y. tolteca* (1976)
Lynx canadensis (Canadian lynx)	not rated	II (1977)	not listed
Lynx lynx (European lynx)	not rated	II (1975, 1977)	not listed
Lynx l. pardinus (Spanish lynx)	E (1978)	II (1977) I (1989)	E as *F. lynx pardina* (1970)
Lynx rufus (bobcat)	not rated	I for *L. r. escuinapae* (1975), II for others (1977)	E for *L. r. escuinapae* (1976)
Neofelis nebulosa (clouded leopard)	V (1978)	I (1975)	E (1972)
Panthera leo (lion)	E for *P. l. persica* (1978)	I for *P. l. persica* (1977), II for others (1977)	E for *P. l. persica* (1970)
Panthera onca (jaguar)	V (1981)	I (1975)	E (1972)
Panthera pardus (leopard)	E for *P. p. panthera*, *P. p. orientalis*, *P. p. nimr*, *P. p. tulliana*, and *P. p. jarvisi* (1972); V (1972) for others	I (1975)	T for populations south of sub-Saharan Africa (1981); E for others (1970)
Panthera tigris (tiger)	E (1978)	I (1975) except II for *P. t. altaica* (1975)	E (1970)
Panthera uncia (snow leopard)	E (1978)	I (1975)	E (1972)

Notes

a E = 'Endangered'—Taxa in danger of extinction and whose survival is unlikely if the causal factors continue operating; V = 'Vulnerable'—Taxa believed likely to move into the endangered category in the near future if the causal factors continue operating; R = 'Rare'—Taxa with small populations that are not at present endangered or vulnerable, but are at risk; I = 'Indeterminate'—Taxa *known* to be endangered, vulnerable, or rare but where there is not enough information to say which of the three categories is best.

b U.S. Fish and Wildlife Service 1982a. E = endangered and T = threatened.

Source: McMahan (1986) (modified)

Table 8.3 The number of pelts needed to make garments from different species of small wild cat.

Species	Pelts/garment	Species	Pelts/garment
Ocelot	12.9	Jungle cat	12.0
Tiger cat	24.0	Leopard cat	15.0
Margay	15.0	Serval	13.0
Geoffroy's cat	25.0	Caracal	20.0
Pampas cat	5.3	Canada lynx	7.5
Wildcat	30.0	Bobcat	9.4

Source: McMahan (1986)

However, some cat species do seem to be able to withstand a high level of exploitation for their fur. In China, about 100,000 leopard cat pelts are harvested each year, with no apparent sign of population reduction (Lu and Sheng, 1986b). Indeed in some areas the adaptable leopard cat has increased at the expense of larger competitors such as the clouded leopard and Temminck's golden cat. Trade in bobcat fur seems to be effectively controlled by CITES (Funderbunk, 1986). Just under 100,000 bobcats are trapped (51 per cent in the USA) or shot (49 per cent), of which 75 per cent are exported, mostly to Europe (75 per cent to the Federal Republic of Germany). This level of cropping is regulated by each state at a level of about 10 per cent of the standing population. Documentation of trappers' returns and CITES have apparently resulted in a steady level of harvesting of bobcats each year.

The trapping of Canada lynx is affected by snowshoe hare cycles, resulting in a cyclic yield of furs every ten to eleven years. Severe trapping during the period of a low lynx population density has resulted in the non-recovery of the lynx population in some areas as the hare population has begun to rise. This has been recorded for lynx populations in Newfoundland and Alberta in Canada, and Alaska (Brand and Keith, 1979; Bergerud, 1983; Bailey *et al.*, 1986). Brand and Keith (1979) recommend that trapping should be stopped during the three years of a low lynx population, so that it is at a fairly high level when the hares recover. In this way fur production could be maximised.

Although CITES does seem to limit legal large-scale trade in Appendix 1 species, it does not limit trade in Appendix 2 species. Data for 1979 and 1980 show that although almost all trade in Appendix 1 cats was for translocation between zoos for the purposes of captive breeding, virtually all the trade in Appendix 2 cats was as

Table 8.4 International trade in the pelts of wild cats for 1979 and 1980

Scientific and common name	1979	1980
Felis geoffroyi (Geoffroy's cat)	114,456	145,358
Lynx rufus (bobcat)	125,714	140,330
Felis chaus (jungle cat)	89,760	79,306
Felis tigrina (little tiger cat)	46,420	69,525
Lynx canadensis (Canadian lynx)	42,061	64,493
Felis silvestris (wildcat)	7,461	60,109
Felis pardalis (ocelot)	21,796	44,872
Felis wiedii (margay)	20,430	29,785
Felis colocolo (pampas cat)	9,911	26,810
Felis bengalensis (leopard cat)	13,089	8,548
Felis caracal (caracal)	904	5,895
Lynx lynx (Eurasian lynx)	8,701	4,740
Felis manul (Pallas's cat)	2,279	2,051
Felis serval (serval)	1,708	1,770
Felis concolor (puma)	1,258	423
Felidae	70	182
Panthera leo (lion)	151	182
Felis spp.	230	171
Felis aurata (golden cat)	1	3
Felis yagouaroundi (jaguarundi)	0	3
Felis margarita (sand cat)	0	3
Lynx spp.	0	3
Panthera tigris altaica (Siberian tiger)	2	2
Totals	506,402	684,564

Source: McMahan (1986) (modified)

skins and garments for the fur trade (Table 8.4) (McMahan, 1986). However cruel and repulsive the trade in animal furs is, there is sadly no sign of any diminution in the demands made by humans to wear these furs.

EXPLOITATION—THE SWEET SMELL OF SUCCESS

Several studies during the 1970s looked at the possibility of using carnivore odours to repel deer from forestry plantations. For example, the dung of several carnivores was effective in deterring black-tailed deer from feeding (Müller-Schwarze, 1972). Among the most effective was dung from puma and coyote, which are sympatric with the deer, but surprisingly lion dung was also as effective. Snow leopard and tiger dung were slightly less so.

There has, however, been less success with predator urine. Stoddart (1980) found that jaguar urine deterred some rodents from entering traps, but not others. Similar results were found with grysbok, duikers and rodents in South Africa using leopard and caracal urine (Novellie, Bigalke and Pepler, 1982).

Recently, an apparently successful attempt has been made to extract the active ingredient from lion dung, which deer find so repellent (Abbott *et al.*, 1990). In fact, several active ingredients were identified, but it was found that they worked best if used together rather than on their own. It seems hardly likely that deer in Britain have an innate memory of lion dung dating back to the Pleistocene when lions used to roam Europe. It seems more likely that they are probably identifying an odour which is common to all predators, which is produced as a result of the digestion of a diet rich in fat and meat.

CONSERVATION

Almost every species of wild cat is threatened in some way with extinction. Most of the commoner species are subject to the vicissitudes of the fur trade coupled with an acceleration in habitat loss. This is particularly devastating to rainforest and island species which have nowhere else to go. Only about 100 Iriomote cats remain and the clouded leopard is probably now extinct on Taiwan, a victim of the medicine and culinary trades, as are many Asian mammals (Rabinowitz, 1988). The Indian lion is confined to the Gir Forest Reserve in India where about 250 animals survive.

In the 1970s, Project Tiger became one of the most important and successful conservation projects seen anywhere in the world. The population of Indian tigers had fallen to less than 2,000 individuals from a peak of perhaps 140,000. By setting up an effective series of reserves throughout India and protecting the remaining tigers from persecution, the population had more than doubled by the 1980s. In the Sundarbans of India and Bangladesh, tigers have become so abundant that they are a threat to local fishermen and honey collectors. India, Nepal and Bangladesh have decided to sacrifice land that might otherwise feed hungry mouths and the few people who are victims of the odd man-eater in order to ensure the survival of this big cat. The benefits come from the increase in tourism to see the tiger and other Asian wildlife. However, many of the reserves hold too few tigers for maintaining viable populations. This will, therefore, require the transfer of tigers between reserves to ensure the genetic integrity of tiger populations throughout the Indian subcontinent.

Loss of genetic diversity due to inbreeding can also be a problem in zoos, where captive breeding programmes are becoming increasingly important as a means of maintaining populations in captivity for 300 to 1,000 years, until such a time as it is possible to reintroduce them to the wild. This period of survival in captivity has been called the demographic winter and is generally believed to be the amount of time it will take for the human population to reach its maximum and begin to fall again. Captive breeding programmes or Species Survival Plans have been or are being developed for different subspecies of the tiger, the Asian lion, the clouded leopard, cheetah and snow leopard (Foose, 1987). Of particular concern in the cat world is the maintenance of a viable captive population of Siberian tigers. Only about 450 to 500 individuals of the world's largest cat survive in the wild, which means that a captive population of about 600 individuals is a very important backup in the event of any natural disaster. Unfortunately, this captive population is descended from only about 25 wild animals, which means that a considerable amount of the initial genetic diversity (in this case heterozygosity) has been lost due to inbreeding and genetic drift (Figure 8.7) (Ralls and Ballou, 1986; Foose, 1987). Although more wild Siberian tigers are kept in captivity, only a few individuals have contributed to today's captive gene pool. Not only is the genetic diversity of the captive population less than that of the wild one, but it will take a much larger captive population to retain 90 per cent of

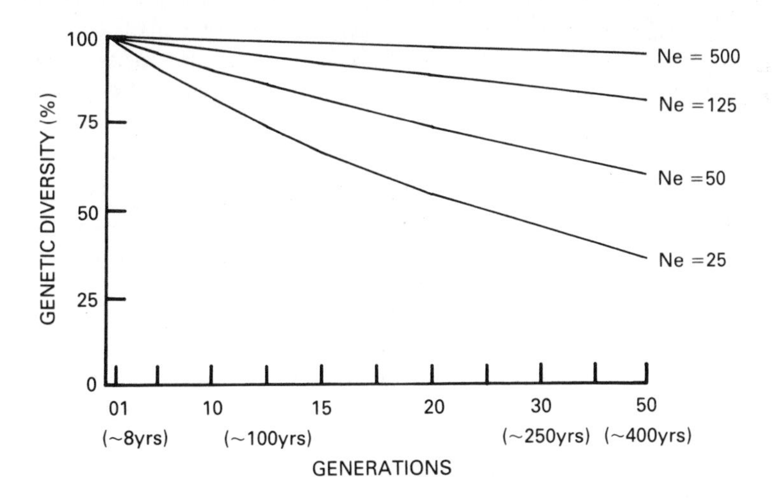

Figure 8.7 Decline in genetic diversity is greater over time for smaller effective population sizes (Ne) (Foose, 1987). The real population size (N) has the same rate of change in genetic diversity or degree of inbreeding as an ideal population, which is called the effective population size (Ne). Ne is usually less than N and is dependent on the sex ratio, social organisation, etc. (A.K.)

the remaining genetic diversity for 300 years or more (Figure 8.7). But does inbreeding matter? Ralls and Ballou (1986) found that inbreeding in mammals results in a loss in fecundity and poor survival of the young. Genetic drift in small populations means that deleterious or lethal genes may become fixed so that the captive population may be doomed to extinction. Species Survival Plans (SSPs) attempt to equalise the genetic contributions of the original founders and maintain an effective population size so that loss in genetic diversity is minimised. This can result in the prevention of some animals from breeding because their ancestors are over-represented in the gene pool, or even the culling of excess individuals to make room for genetically different animals. Fortunately detailed studbooks have been maintained for most endangered species so that it is fairly straightforward to formulate Species Survival Plans if all institutions are prepared to cooperate. In the future, space in zoos could become critical and decisions are even now being made as to which species can be saved in this way.

By maintaining the maximum amount of genetic diversity in the population, it gives any reintroduced population the maximum adaptability to survive in changing conditions in the wild.

THE PROBLEMATICAL CHEETAH

The cheetah has always been difficult to breed in captivity, and when it has been achieved the survival of its young has always been low. In the sixteenth century the Mogul emperor Akbar kept 1,000 cheetahs mainly for hunting, but only one litter of three was ever born (Divyabhanusinh, 1987). Recent attempts at captive breeding did not succeed until 1956, at Philadelphia Zoo. Since then, breeding has become more commonplace, notably at Whipsnade Zoo in Bedfordshire where more than 130 cubs have been born since 1967. However, it is still by no means easy to breed cheetahs in captivity. Part of the reason is that female cheetahs will not come into oestrus if kept permanently with the male. By keeping them separate and reintroducing them, the female is stimulated into oestrus. Some zoos have also found that competition between several males is important in promoting captive breeding. To some extent this parallels the wild situation, where females are solitary and nomadic, and males are territorial and live in coalitions (Caro and Collins, 1987a, b; Frame, 1984).

O'Brien *et al.* (1985) looked at captive cheetah populations to see if there was any reason for this lack of breeding success. The signs were that this was the result of inbreeding. Although litter size was often large, sperm concentration and viability was low (29 per cent viable compared to 71 per cent in the domestic cat). Survival of the cubs was often less than 40 per cent, which is very low for a captive situation. Also cheetah skulls are highly asymmetrical, especially when compared with other cats such as ocelots and margays, providing yet another indicator of inbreeding (Wayne *et al.*, 1986). O'Brien, Modi and O'Brien (1986) found that captive cheetahs were almost identical genetically, so that it was possible to carry out successful skin grafts between any two cheetahs (Figure 8.8). However, skin grafts from domestic cats were rejected. They found that only two gene loci showed very low levels of polymorphism, the rest were monomorphic. This represents less than 2 per cent genetic diversity compared with a minimum of 10 per cent for other wild cats. In fact cheetahs are more inbred than some strains of highly inbred laboratory mice. But the problem was not a result of captive breeding from a small founder population because wild

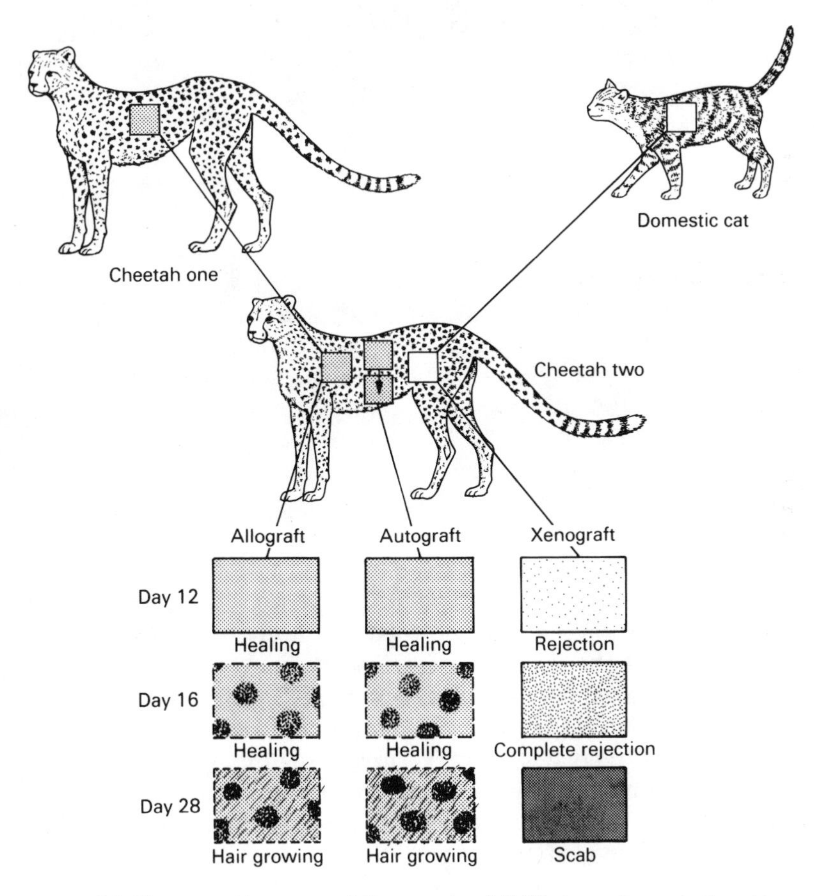

Figure 8.8 The major histocompatibility complex (MHC) determines whether skin grafts are accepted or not. All cheetahs have the same MHC genes. After 28 days, skin grafts from another cheetah (allograft) or the same cheetah (autograft) are both accepted, showing the MHC to be identical. The skin graft from a domestic cat (xenograft) was rejected, showing that the MHC complex was functional in both cheetahs (O'Brien *et al.*, 1986) (S.A.)

cheetahs from South Africa were also virtually genetic clones of each other.

O'Brien *et al.* (1987b) also looked at cheetahs from East Africa and found that the genetic diversity of these animals was about 4 per cent, but they were still very inbred (Figure 8.9). The breeding success of cheetahs at Whipsnade Zoo in recent years has been put down in part to the crossing of South and East African cheetahs and O'Brien *et al.* (1987b) have suggested that this is a good way of promoting the survival of captive populations. The two populations are so inbred that it suggests that the existence of separate subspecies of East and South African cheetahs is very doubtful.

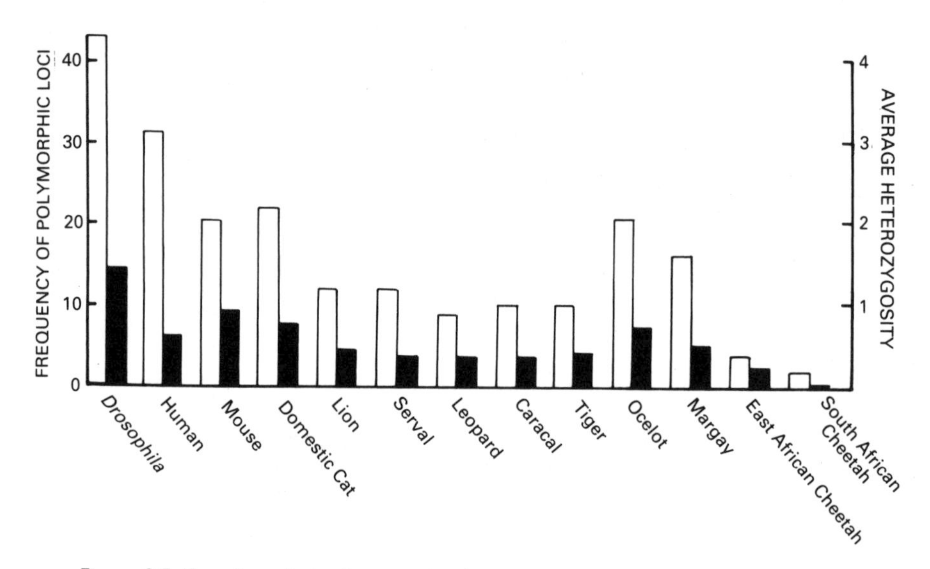

Figure 8.9 Genetic variation in cats, mice, humans and the fruit fly (*Drosophila*), based on allozyme electrophoretic surveys (O'Brien *et al.*, 1987b). Open bars show percentage of polymorphic gene loci, solid bars show the proportion of heterozygosity (A.K.)

The conclusion from the study by Stephen O'Brien and his coworkers was that the once cosmopolitan cheetah had suffered a severe population bottleneck about 10,000 years ago at the end of the last Ice Age. This drastic population reduction and its subsequent slow recovery is very similar to the problems faced by captive mammals where genetic diversity is quickly lost if the founder population is too small. During the nineteenth century, the South African population suffered another population bottleneck, possibly due to persecution by humans or loss of prey due to disease. This has resulted in a further reduction in genetic diversity in South Africa.

This suggests that the only way that the cheetah can survive is by maintaining very high populations which can withstand huge losses in reproductive output. However, the picture may not be as gloomy as it seems. Because the first population bottleneck occurred so long ago, most of the lethal and deleterious genes originally in the population seem to have been lost. Secondly, cheetah populations in Namibia and Zimbabwe are increasing at such a rate that they have almost become a pest and farmers have asked for permission to control their numbers.

In East Africa, Myers (1973) made some very gloomy predictions about the survival of the cheetah there and in the rest of Africa due to human encroachment onto its habitat. However, a recent study in Kenya suggests that cheetahs can live with humans far better than the lion and leopard, and may actually benefit from the exclusion of the other two species (Hamilton, 1986). Cheetahs very rarely scavenge food, so that they are less susceptible to poisoning than the other big cats. In Kenya cheetahs were found in virtually all the areas recorded in the early 1970s and were also found to occur at higher population densities in some areas than previously recorded. It was discovered that cheetahs were not restricted to chasing prey in daylight across the open savanna, but were more adaptable hunters. Far from being a specialised cursorial hunter, Hamilton (1986) found that cheetahs would hunt like other cats by stalking at night and would feed on much smaller prey such as hares and guineafowl. Thus the cheetah is far more suited to the pastoral lifestyle of humans in much of Kenya than its bigger cousins which come into conflict with local humans by preying on livestock.

MAN-EATERS

Stories abound about the incidence of fearsome man-eating big cats, but by far the most frequent perpetrator of these deeds is the tiger. In reality, man-eating tigers are a problem in the Indian subcontinent, especially since the success of Project Tiger. In the past, control of man-eaters was achieved by tracking down the rogue individuals and shooting them. It was always assumed that old tigers with broken teeth or some other permanent injury were responsible for man-eating, but in fact many perfectly healthy animals indulge in this habit. It seems that these tigers lose their fear of humans and young tigers are educated by their mothers to recognise humans as legitimate prey.

However, man-eating is not a problem throughout the tiger's range. In the Caspian region and most of southeastern Asia man-eating is virtually unrecorded (McDougal, 1987). It did, however, become a local problem in the Arakan region of Burma during the Second World War, when tigers fed on the corpses of soldiers who died during the retreat of 1942 (McDougal, 1987). In some areas there has ceased to be a problem because the tigers are no longer extant. This includes, for example, southern China, where up to 60 people from one village were killed by tigers in six weeks during 1922 (McDougal, 1987). Other former problem areas include Singapore and Manchuria.

India has always had problems with man-eating tigers. In the Khandesh District, 500 people and 20,000 cattle were said to have been killed by wild beasts, mostly tigers, during 1822 (McDougal, 1987). The problem was solved mainly by killing all the tigers and destruction of tiger habitat. It seems that man-eating occurs if the tigers do not have an adequate natural food supply and are encroached upon by humans while still retaining a good breeding population (McDougal, 1987). In the Caspian region and other non-problem areas, the tigers appear to have been removed at the same rate as their habitat has been lost, so that they were never brought into conflict with humans.

The highest density of wild tigers occurs in the mangrove forest swamps of the Sundarbans of India and Bangladesh and until recently the local human population suffered a very high incidence of man-eating. A tiger reserve was established in the Sundarbans of about 2,600 km^2, of which a core area of 1,300 km^2 was designated as a National Park. The rich prey base, particularly of wild boar, and legal protection has helped the tiger population to expand from 135 in 1973 to 264 in 1983, an average increase of 10.5 per cent per year (Sanyal, 1987). Conflicts arose between humans and tigers because the people exploited the mangroves for firewood and honey and some people were inevitably eaten by tigers—although less than 50 per year. As a result, a detailed study of the incidence of man-eating was carried out and a management plan implemented for the benefit of people and tigers. It was discovered that only 5 per cent of the tigers were man-eaters and that most people were killed by tigers in the early morning, late afternoon and around midnight (Sanyal, 1987). It was recommended that buffer zones be set up along the forest fringes where the needs of the local people for firewood could be met, so lessening the pressure on the forests. Export of wood to metropolitan areas was banned. An alternative fuel supply was provided so that people did not use the preferred denning sites of tigers and so did not disturb the forest. Management extended also to man-eating tigers. Stray tigers were tranquillised and returned to the forest, and farm-bred wild pigs were released into the buffer zones to provide sufficient prey. The most innovative measure involved setting up clay models of honey collectors/woodcutters/fishermen connected to a 230 volt electricity supply. This has apparently been highly successful in deterring man-eaters from attacking people, and the average rate of kill has dropped from 45 people per year to only 21 per year since 1983, when the new management procedures were implemented. It is clear that the most successful measure has been the use of electric shocks to deter man-eaters. By 1985 honey collectors were reassured of their safety and have now doubled the annual crop of honey harvested from the Sundarbans (Sanyal, 1987).

REINTRODUCTIONS TO THE WILD

In the 1960s and 1970s there were several successful attempts by the late George and Joy Adamson to reintroduce individual lions, cheetahs and leopards to the wild in East Africa. Indeed George Adamson successfully established a pride of lions in the Kora Game Reserve in northern Kenya. Similarly Billy Arjan Singh has had some success with tigers and leopards in India. However, the only cat that has been systematically reintroduced to the wild, with only mixed success, is the Eurasian lynx in western Europe.

The lynx, like the wolf, was exterminated throughout most of Europe because of its imagined threat to humans and its real threat to domestic livestock. Only in Slovakia did the Eurasian lynx survive in Central Europe until the middle of the twentieth century (Gossow and Honsig-Erlenburg, 1986). During the 1960s the lynx spread naturally from Slovakia to the densely forested border regions of Czechoslovakia and Bavaria, where it spread into the Bayerischer Wald National Park. This natural spread of the lynx inspired attempts for its reintroduction to several areas in western Europe, which could only be achieved because of a change in attitude to the way that game populations should be managed. In the past it was believed that predator eradication would result in more game for humans to kill, but now it is known that predators have an insignificant or small effect on game populations, and may even help improve the quality of the game by removing diseased, injured or old individuals.

It was intended to introduce lynx to the Harz Mountains of West Germany, but the individuals intended for this destination were introduced instead into northern Slovenia in Yugoslavia in 1973, where they have since established a successful population. Lynx were also successfully introduced to the Canton of Obwalden in Switzerland. Good populations of lynx now survive there and in the Jura region, and are continuing to spread into the Central Alps (Gossow and Honsig-Erlenburg, 1986; Haller and Breitenmoser, 1986).

However, there have been some failures or less successful attempts, including attempted reintroductions into the Gran Paradiso National Park (Italy), the Swiss National Park, the Bayerischer Wald National Park (Germany) and France. The attempt to supplement the natural return of the lynx to the Bayerisicher Wald National Park failed because of hunting or a high incidence of road casualties, and the natural population has again declined in that part of Europe. In 1977 the last major attempt at lynx reintroduction was started in Austria.

Gossow and Honsig-Erlenburg (1986) have reviewed the problems and successes of introducing and managing the lynx populations of Austria. The main fear was that the lynx would reduce the roe deer population and feed on sheep. However, it was shown from studies of their food requirements that the lynx would have a negligible effect on the overabundant deer population, and that sheep farmers in Switzerland were known to be very keen on the compensation scheme introduced there for sheep killed by lynx. The only fear that could not be allayed for certain was that the hunting of the lynx might affect the behaviour of the deer, thus affecting the ease with which humans could hunt the deer.

One surprise that came from the release of lynx into Austria was that diet varied in different areas. In the Turrach area most prey were roe deer as expected, but in the Flattnitz-Felfernig Range most prey were red deer. The preference for red deer is thought to be because the lynx occupy the less dense forest at higher elevations in the Felfernig Range where red deer occur because of supplemental winter feeding by deer managers. Most of the red deer that are killed are yearlings and hinds, and many of these are in poor condition. During a typical winter ten red deer are killed by lynx on a typical range. In former years up to ten red deer would starve to death during during the winter in the same area, but in recent years none have done so, suggesting that lynx are weeding out these weak individuals. In their first year the introduced lynx population ranged over 7,500 hectares, but now the population occupies 600,000 hectares. The population is, however, not evenly distributed, but is dispersed or moves randomly through the area.

In Switzerland, Haller and Breitenmoser (1986) found that lynx spreading into new areas in the Valais (Central Alps) occupied smaller home ranges (46 km^2 for a female) than an established lynx population in the northern Alps, where males occupied home ranges of 275–450 km^2 and females covered 86–135 km^2. The difference appears to occur because in new areas prey species are easy to capture, since they lack experience in dealing with predators. As they adapt to the presence of the lynx, so the predator needs a larger area to gain sufficient food.

Recently, captive-bred servals were introduced to the Rustenburg Nature Reserve in South Africa (van Aarde and Skinner, 1986). By radiotracking the servals, it was possible to see how they dispersed from the release point and look at their daily activity cycle. As expected, the servals were mainly nocturnal, but showed some activity in late afternoon. They foraged over distances of 90 to 463 metres per hour during the night, particularly in the riverine habitat. The males occupied home ranges of 2.08 and 2.7 km^2 respectively, which they established within a period of 20 to 30 days. Most of their activity was centred on the release site, thereby showing that this must be carefully evaluated in future introductions if the survival of reintroduced servals is to be assured.

Reintroductions will become increasingly important in the future, especially in the genetic management of tiger populations. We must learn as many lessons as possible from these first attempts to ensure a high success rate in the future.

THE FUTURE

The future for wild cats throughout the world is not likely to be a good one at the present rate of habitat loss, particularly in tropical forests. It is to be hoped that the trade in cat fur, particularly in the tropics, will be increasingly controlled and preferably, if attitudes can be changed in the western world, will be stopped as fur coats fall out of fashion. But the main threat to all wild cats for some time to come will still be habitat loss.

We can only hope that the increased awareness of the problems that face our environment and the need to do something now rather than later will also benefit the plight of the most specialised predators the world has ever seen.

MEASUREMENTS AND BODY WEIGHTS OF THE FELIDAE

Table A.1 Body measurements (mm) and weights (kg) of cats from the ocelot lineage. When a mean figure is given, the range is given in parentheses

Species/ Locality	Sex	n	TOL	n	HB	n	TL	n	WT	Ref.
Felis pardalis										
—	—	—	920–1367			—	270–400	—	6.6–10.9	a
Venezuela	M	14	1078 (1000–1220)			14	350 (300–410)	8	10 (7–14.5)	b
Venezuela	F	7	1022 (905–1140)			7	322 (260–365)	5	8.8 (7–10.8)	b
Surinam	F	2	1055–1105			2	305	2	10–11	c
Peru	M							5	10.4 (9–12)	d
Peru	F							4	8.3 (7–10.6)	d
—	—			—	550–1000	—	300–450	—	11.3–15.8	e
—	M	—	1143 (950–1367)			—	330 (280–480)	—	13.2 (11.3–15.8)	f
—	F	—	1041 (920–1209)			—	305 (270–371)			f
Argentina	—	—	889 (800–1000)			—	406 (350–450)			f
F. wiedii										
—	M			9	533	9	353			g
	F			5	501	5	342			g
—	—			—	463–790	—	331–510			e
—	M	—	862–1300			—	331–510			f
	F	—	805–1029			—	342–440			f

Table A.1 Cont.

Species/ Locality	Sex	n	TOL	n	HB	n	TL	n	WT	Ref.
F. tigrina										
Costa Rica	M	2	505–721			2	255–290			h
	F	3	734 (697–790)			3	256 (245–276)			h
Venezuela	M	3	813 (805–830)			3	336 (317–360)			b
	F	2	772 (763–780)			2	288 (270–305)	1	2	b
Colombia	M	1	820			1	330			b
—	M			5	528	5	274			g
	F			2	482	2	249			g
—	—			—	400–550	—	250–400			e
	M							1	2.75	e
	F							1	1.75	e
—	—	—	705–970			—	255–420			f
—	—			—	533 (500–550)	—	330 (250–420)			i
F. geoffroyi										
—	—			—	450–700	—	260–350			e
—	—			—	584 (550–650)	—	350			f
—	M	—	790–940			—	270–280			j
—	F	—	736–910			—	250–275			j
F. guigna										
ssp. *guigna*				—	406 (390–450)	—	203 (195–230)			i
ssp. *tigrillo*				—	493	—	229			i
F. colocolo										
Argentina	—			—	635 (600–700)	—	300			i
Chile	—			—	584 (567–642)	—	305 (295–322)			i
F. jacobita										
Peru	M	1	990			1	413	1	4	k
—	—			—	600	—	350			i

Notes
n—sample size
TOL— total length
HB—head and body length
TL—tail length
WT—weight

References:
(a) Tewes and Schmidly, 1987
(b) Mondolfi, 1986
(c) Husson, 1978
(d) Emmons, 1988
(e) Nowak and Paradiso, 1983
(f) Hall, 1981
(g) Kitchener, unpublished
(h) Gardner, 1971
(i) Guggisberg, 1975
(j) Ximinez, 1975
(k) Kuhn, 1973

Table A.2 Body measurements (mm) and weights (kg) of cats from the domestic cat lineage. When a mean figure is given, the range is given in parentheses

Species/ Locality	Sex	n	TOL	n	HB	n	TL	n	WT	Ref.
Felis silvestris										
Scotland	M			13	591–622	13	305–343	13	4.5–7	a
	F			2	546–584	2	305–330	2	4–4.3	a
	M			22	572–635	22	286–370			b
	F			14	495–571	14	257–320			b
	M			102	589 (365–653)	102	315 (210–342)	102	5.1 (3–6.9)	c
	F			5	571 (545–584)	5	311 (293–331)	5	4.3 (4–4.5)	c
	M			3	579–635	3	305–360			d
	F			6	502–571	6	279–340			d
	M			26	564 (515–650)	26	307 (235–356)	26	4.7 (3.5–7.1)	e
	F			16	543 (507–595)	16	293 (240–360)	16	3.9 (2.5–5.6)	e
	M			9	574 (547–632)	9	293 (267–315)	9	4.4 (3.8–5)	f
	F			7	563 (521–580)	7	289 (258–300)	7	4 (3.5–4.6)	f
France	M			114	440–650	114	215–345	114	5(–7.7)	g
	F			72	430–570	72	215–340	72	3.5	g
Germany	M			15	585–670	15	280–345	15	3.8–6.2	h
	F			14	405–640	14	230–310	14	1.6–4.9	h
Rumania	M			24	470–981	24	240–385	24	2.3–7.7	i
	F			11	430–700	11	220–340	11	2.1–7.7	i
Czecho-	M			98	510–780	98	240–380	98	1–8	j
slovakia	F			51	470–690	51	230–360	98	1.8–7.1	j
Caucasus	M			—	686 (630–750)	—	305 (300–340)	—	6	k
	F			—	610 (580–630)	—	305 (270–330)	—	4.5	k
F. s. lybica										
—	—			—	550 (470–660)	—	340 (200–380)			l
	M							—	5(3.7–6.5)	l
	F							—	4.3(3–5.5)	l
West Africa	—			6	513 (434–660)	6	307 (292–336)			m
F. s. griselda										
Botswana	M	32	920 (850–1005)			32	344 (320–375)	32	5.1 (3.8–6.4)	n
	F	27	886 (820–947)			27	336 (310–370)	26	4.2 (3.2–5.5)	n
—	—			11	518–625	11	290–376			d

Table A.2 Cont.

Species/ Locality	Sex	n	TOL	n	HB	n	TL	n	WT	Ref.
F. s. cafra										
Cape Province	M			21	601 (545–665)	21	305 (275–360)	10	4.9 (4–6.2)	n
	F			15	550 (460–621)	16	295 (250–355)	10	3.7 (2.4–5)	n
	—			14	478–599	14	295–381			d
F. s. ugandae										
—	—			8	480–590	8	340–420			d
F. s. rubida										
—	—			5	422–533	5	310–320			d
F. s. ornata										
Pakistan	—			4	497 (470–542)	4	249 (219–290)	1	4	p
—	M			8	497 (439–597)	8	276 (254–310)			d
	F			3	488 (444–564)	3	270 (254–295)			d
India	—			—	584 (500–650)	—	279 (250–300)	—	3.2(3–4)	q
Turkmenistan	—			—	584 (550–700)	—	279 (250–330)			q
F. s. tristrami										
	M	2	860–888			2	315–390			r
	F	2	753–770			2	282–290			r
F. s. gordoni										
Oman	M	3	743,813,846			2	291,316			r
	F	2	669–787			1	263			r
F. s. sarda										
Corsica	M			1	650	1	250	1	2.8	s
	F			1	640	1	240	1	2.0	s
F. catus										
Marion I.	M							—	4.5	t
	F							—	2.5	t
Macquarie I.	M			31	522+/–20	31	269+/–16	31	4.5+/–0.5	u
	F			13	478+/–21	13	252+/–12	13	3.3+/–0.8	u
—	—			—	460	—	300			v
F. bieti										
—	—			—	762 (685–840)	—	305 (290–350)			q
—	—			8	775 (691–853)	8	376 (325–559)			d

Table A.2 Cont.

Species/ Locality	Sex	n	TOL	n	HB	n	TL	n	WT	Ref.
F. chaus										
—	—			—	500–750	—	250–290	—	4–16	v
—	—			8	774 (690–853)	8	376 (325–559)			d
India	—			—	600	—	229 (228–254)	—	7.3(5–9)	q
Central Asia	—			—	737 (730–750)	—	279 (250–290)			q
Pakistan	M			1	803	1	250	1	9	p
	F			1	618	1	256			p
Iraq (*furax*)	M	1	1049			1	272			r
	F	2	880–917			2	256–270			r
Russia	M			—	730–750	—	250–290			w
	F			—	630	—	250			w
ssp. *affinis*										
—	M			7	674 (635–721)	7	294 (213–335)	5	6.8 (5.4–8.6)	d
—	F			7	618 (579–650)	7	263 (229–284)	5	5.4 (4.8–6.4)	d
ssp. *kelaarti*										
—	M			5	626 (620–640)	5	281 (264–310)	4	5.4 (5–5.9)	d
—	F			4	608 (569–635)	4	248 (229–262)	1	3.1	d
ssp. *kutas*										
—	M			6	624 (569–711)	6	260 (244–279)	6	4.9 (3.9–5.9)	d
	F			4	572 (554–594)	4	249 (234–259)	4	3.1 (2.6–3.6)	d
F. margarita										
—	—			—	508 (450–572)	—	305 (280–348)			d
Pakistan	M			1	570	1	280			p
Sahara	—			—	390–420	—	245–270			m
Arabia	M	2	702–730			2	250–260			r
Russia	M			1	530	1	260			w
	—	1	750	1	572	2	290–348			w
—	M			2	518 (457–579)	2	318 (279–356)			d
	F			2	432 (396–467)	2	272 (249–295)			d
F. nigripes										
—	M			—	457 (425–500)	—	178 (150–200)	—	1.5–2.75	q
—	F			—	356 (337–368)	—	152 (157–170)			q

Table A.2 Cont.

Species/ Locality	Sex	n	TOL	n	HB	n	TL	n	WT	Ref.
F. nigripes										
Botswana	M	5	579 (540–631)			5	177 (164–198)	5	1.6 (1.5–1.7)	n
	F	3	513 (495–530)			3	153 (126–170)	3	1.1 (1–1.4)	n
Southern Africa	—			3	354 (337–368)	3	165 (157–170)			x
F. manul										
—	—			—	559 (500–650)	—	254 (210–310)	—	3.1 (2.5–3.5)	q
Ladakh	—			1	519					p
Baluchistan	—			1	636	1	294	—	2.7–3.4	p
—	—			3	639 (635–648)	3	300 (292–305)			d

Notes
n—sample size
TOL— total length
HB—head and body length
TL—tail length
WT—weight
References:
(a) Kirk, 1935
(b) Tetley, 1941
(c) Kirk and Wagstaffe, 1943
(d) Pocock, 1951
(e) Kolb, 1977
(f) Corbett, 1979
(g) Schauenberg, 1981
(h) Braunschweig, 1963; Piechoki, 1973
(i) Vasiliu and Almasan, 1969
(j) Sladek *et al.*, 1971
(k) Novikov, 1962
(l) Kingdon, 1977
(m) Rosevear, 1974
(n) Smithers, 1983
(p) Roberts, 1977
(q) Guggisberg, 1975
(r) Harrison, 1968
(s) Arrighi and Salotti, 1988
(t) van Aarde, 1980
(u) Jones, 1977
(v) Nowak and Paradiso, 1983
(w) Ognev, 1935
(x) Pocock, 1907

Table A.3 Body measurements (cm) and weights (kg) of cats from the pantherine lineage. When a mean figure is given, the range is given in parentheses

Species/ Locality	Sex	n	TOL	n	HB	n	TL	n	WT	Ref.
Felis bengalensis										
Malaya	—			2	52–56	2	11–12	2	2.6–3.1	a
Pakistan	F			1	54	1	27			b
—								—	2.4–5	b
	—			—	44–107	—	23–44	—	3–7	c
Thailand	—			—	44–55	—	23–29	—	3–5	d
India	—			—	64 (61–66)	—	28 (28–30)			e
Borneo	—			2	40–44	2	17–22			aa
ssp. *euptilura*										
Russia	—	2	109–140			1	44			f
	—			2	75–90	2	35–38			f
	—			—	81(75–90)	—	36(35–37)			e

Table A.3 Cont.

Species/ Locality	Sex	n	TOL	n	HB	n	TL	n	WT	Ref.
F. rubiginosa										
	—			—	35–48	—	15–25			c
	—			—	43(41–46)	—	24			e
F. planiceps										
Malaya	—			2	46–48	2	13	2	1.6–2.1	a
	—			—	41–50	—	13–15			c
Thailand	—			—	46–48	—	13	—	1.5–2.2	d
	—			—	56(53–61)	—	18(15–20)			e
Borneo	—			6	44–50	6	13–17			aa
F. viverrina										
Pakistan	M			1	72	1	29	1	11.3	b
India	M	1	83			1	23	1	3.6	b
	F	1	102			1	26	1	6.8	b
	—			—	75–86	—	26–33	—	7.7–14	c
	—			—	81(75–86)	—	30(25–33)	—	11.4 (7.7–14)	e
Thailand	—			—	72–78	—	25–29	—	7–11	d
F. iriomotensis										
Iriomote	M			—	51–56	—	25–30	—	4.1	g
	F							—	3.2	g
	M			—	58+/–2			—	4.2+/–0.5	h
	F			—	53+/–2			—	3.2+/–0.3	h
F. aurata										
	M			4	66	4	35			i
	F			3	86	3	30			i
	M			9	74(62–86)	9	31(25–37)	4	9.1 (5.3–12)	j
	F			3	71(66–74)	3	30(28–30)	1	6.2	j
	—			—	62–102	—	16–46	—	5.3–16	k
	—			—	74(65–85)	—	30(24–40)	—	15	e
F. temminckii										
	—			—	73–105	—	43–56	—	12–15	c
	—			—	79	—	48			e
China				—	89 (73–105)	—	51 (48–56)			e
Thailand	—			—	76–82	—	43–49	—	12–15	d
F. badia										
Borneo	—			—	53(50–60)	—	38(35–40)			e
Borneo	–			2	63–69	2	37–43			aa

Table A.3 Cont.

Species/ Locality	Sex	n	TOL	n	HB	n	TL	n	WT	Ref.
F. serval										
Zimbabwe	M	23	111 (96–120)			23	31(28–38)	20	11.1 (8.6–13.5)	l
	F	23	110 (97–123)			23	29(25–33)	23	9.7 (8.6–11.8)	l
	–			—	67–100	—	24–35			k
	M							—	13(10–18)	k
	F							—	11 (8.7–12.5)	k
	—			—	81(70–95)	—	41(36–45)	—	16(13–18)	e
F. caracal										
	—			—	76(60–92)	—	23–31			k
	M							—	14.5–19	k
Botswana/	M	6	112 (106–123)			6	30(26–32)	6	13.8 (11.5–17)	l
Zimbabwe	F	3	106 (102–111)			3	29(28–30)	3	11.9 (10.9–11.5)	l
Cape Prov.	M	15	111 (80–126)			15	28(20–29)	15	9.5 (4.1–14.5)	l
	F	23	98 (81–122)			23	25(21–30)	23	7.8 (4.2–14.5)	l
Cape Prov.	M	46	117 (102–127)					46	14.5 (8.6–20)	m m
	F	32	109 (99–119)					32	11 (8.6–14.5)	m
Pakistan	—			—	74	—	23			b
Russia	—			—	65–75	—	25–27			f
Arabia	F	1	86			—	23			n
	—			—	71(66–76)	—	22	—	17(15–18)	e
F. yaguarondi										
	—	—	89–137			—	33—61	—	4.5–9	o
Venezuela	M	5	114 (101–138)			5	41(38–44)	3	2.9 (2.1–3.5)	p
	F	5	95 (92–102)			5	40(39–41)	1	2.6	p
	—			—	66(55–77)	—	46(33–60)	—	7.3 (4.5–9)	e
F. concolor										
Idaho, USA	M							4	72	q
	F							6	46	q
Idaho, USA	M							—	68(59–82)	e
	F							—	45	e
Central								—	25	e
America	M			—	150 (105–196)	—	71(66–78)	—	87 (67–103)	e
	F			—	119 (97–152)	—	66(53–82)	—	49(36–60)	e

Table A.3 Cont.

Species/ Locality	Sex	n	TOL	n	HB	n	TL	n	WT	Ref.
—	M	—	220–230					—	55–65	r
—	F	—	220–210					—	35–45	r
Acinonyx jubatus										
Kenya	M	4	212 (201–224)					4	61(58–65)	s
	F	2	(191–236)					1	63	s
Namibia	M	7	206 (191–221)			7	72(65–76)	7	53.9 (39–59)	l
	F	6	190 (184–196)			6	67(63–69)	6	43(36–48)	l
			—			—	112–150	—	35–72	c
ssp. *venaticus*										
Russia	—			—	104					f
Lynx l. lynx										
Sweden	M							13	17.9 (13–24)	t
	F							10	16.8(–22)	t
	—			—	80–130	—	11–24	—	8–38	c
Northern Europe	—	—	91–154			—	11–24			u
Russia	—			—	82–105	—	20–31			u
Russia	F			3	85–91	1	35	2	10–15	f
Iraq	—	1	106			1	18			n
L. l. isabellinus										
Pakistan	—			—	86	—	20	—	25	e
	M			1	103	1	22			b
L. l. pardinus										
	—			—	96 (85–110)	—	13(12–13)			e
L. canadensis										
				—	82–95	—	10–12	—	6.8–18.1	v
	—			—	80–100	—	5–14	—	5.1–17.2	c
North America	—	—	82–95			—	9–12	—	4–17.3	u
Newfound-land	M	—	89(74–107)					—	11(6–13)	w
	F	—	84(76–96)					—	8(5–12)	w
Alberta	—	—	78–105			—	8–12			u
Alaska	—	—	67–85			—	7–13			u
	M	—	87+/–4					—	10+/–1	x
	F	—	82+/–3					—	9+/–1	x
Cape Breton I.	M	—	80+/–4					—	9+/–2	y
	F	—	76+/–3					—	8+/–1	y

Table A.3 Cont.

Species/ Locality	Sex	n	TOL	n	HB	n	TL	n	WT	Ref.
L. canadensis										
Ontario	M	—	85+/–3							x
	F	—	81+/–3							x
Manitoba	M	—	84+/–3							x
	F	—	80+/–2							x
L. rufus										
				—	79–125	—	13–20			v
				—	651	—	11–19	—	4.1–15.3	c
—	M	—	82–95			—	13–16	—	12.7(5–31)	z
	F	—	73–85			—	12–14	—	6.8(4–15)	z

Notes
n—sample size
TOL— total length
HB—head and body length
TL—tail length
WT—weight
References:
(a) Muul and Lim, 1970
(b) Roberts, 1977
(c) Nowak and Paradiso, 1983
(d) Lekagul and McNeely, 1977
(e) Guggisberg, 1975
(f) Ognev, 1935
(g) Yasuma, 1988
(h) Izawa and Ono, 1989
(i) Rosevear, 1974
(j) van Mensch and van Bree, 1969
(k) Kingdon, 1977
(l) Smithers, 1983
(m) Pringle and Pringle, 1979
(n) Harrison, 1968
(o) Tewes and Schmidly, 1987
(p) Mondolfi, 1986
(q) Hornocker, 1970
(r) Currier, 1983
(s) Meinertzhagen, 1938
(t) Haglund, 1966
(u) Tumlison, 1987
(v) Hall, 1981
(w) Saunders, 1963a
(x) Quinn and Parker, 1987
(y) Parker *et al.*, 1983
(z) Rolley, 1987
(aa) Payne *et al.*, 1985

Table A.4 Body measurements (cm) and weights (kg) of cats from the *Panthera* group. When a mean figure is given, the range is given in parentheses

Species/ Locality	Sex	n	TOL	n	HB	n	TL	n	WT	Ref.
Pardofelis marmorata										
Thailand	—			—	45–53	—	48–55	—	2–5	a
India	—	—	90			—	45			b
—	—			—	53(46–61)	—	38(35–41)			c
Borneo	—			2	46–49	2	48–50	2	2.4–2.5	r
Neofelis nebulosa										
—	—			—	62–107	—	55–91	—	16–23	d
India	—	—	195					—	18–20	b
Thailand	—			—	75–95	—	55–80	—	16–23	a
Panthera uncia										
—	—			—	100–130	—	80–100	—	25–75	e
—	—			3	100–110	3	84			f
	F							1	36	f

Table A.4 Cont.

Species/ Locality	Sex	n	TOL	n	HB	n	TL	n	WT	Ref.
Panthera uncia										
India	—			—	100–110			—	–40	b
Russia	—			—	130	—	90			g
Kashmir	M			1	112	1	91			c
Kashmir	F			1	99	1	83			c
P. pardus										
—	—			—	91–191	—	58–110			d
	M							—	37–90	d
	F							—	28–60	d
Zimbabwe	M	13	211 (201–236)					13	60(52–71)	h
	F	7	185 (178–188)					7	32(28–35)	h
Cape Prov.	M		(92–125)	21	111 (92–125)	20	68(51–80)	27	31(20–45)	h
	F			8	103 (95–105)	8	68(64–74)	9	21(17–26)	h
ssp. *panthera*										
	M			2	132–135	2	76–86			i
Ethiopia	M			3	150 (146–152)	3	95 (99–102)			i
	F			2	113–114	2	84–85			i
India	M	—	215–245					—	52–68	b
	F	—	215–230					—	39–50	b
Thailand	—			—	108–130	—	80–100	—	45–65	a
ssp. *jarvisi*										
Sinai	—	1	211			1	74			j
ssp. *tulliana*										
Jordan	F	1	206			1	75			j
Russia	—			—	138–183	—	109–116			g
ssp. *saxicolor*										
Iraq	M	2	226–259			2	91–94			j
ssp.*nimr*										
	F	1	178			1	74			j
	—	4	168–201			4	66–81			j
ssp. *orientalis*										
	M			4	107–145	4	85–95	2	32.2–32.8	g
—	—			—	122 (107–136)	—	86(82–90)	—	37(32–40)	c
ssp. *japonensis*										
	—			—	117–151	—	75–82	—	–130	g
P. onca										
—	—			—	112–185	—	45–75	—	36–158	d
Brazil	M	6	205 (194–212)			6	61(55–66)	6	95(79–119)	k

Table A.4 Cont.

Species/ Locality	Sex	n	TOL	n	HB	n	TL	n	WT	Ref.
P. onca										
	F	3	194 (186–203)			3	60(59–60)	3	78(73–85)	k
Venezuela	M			8	148 (129–170)	7	63(60–68)	9	95(67–121)	l
	F			3	129 (122–133)	3	59(54–63)	3	56(43–65)	l
Belize	M	6	190			6	56	6	57	m
P. tigris										
ssp. *altaica*										
—	M	—	270–330					—	180–306	n
	F	—	240–275					—	100–167	n
ssp. *tigris*										
—	M	—	270–310					—	180–258	n
	F	—	240–265					—	100–160	n
	M	—	267–312					—	191 (160–234)	o
	F	39	239–277					39	131	o
	M							26	219(–248)	o
	F							20	155 (159–268)	o
India	M	—	275–290					—	180–230	b
	F	—	260					—	45	b
ssp. *corbetti*										
—	M	—	255–285					—	150–195	n
	F	—	230–255					—	100–130	n
Thailand	—			—	170–229	—	95–119	—	180–245	a
Malaysia										
(Terengganu)	F	8	221 (178–246)	8	148 (119–170)	8	72 (53–79)	8	72 (24–89)	s
(Kelanatan)	F	5	222 (183–249)	5	145 (122–165)	5	77 (61–86)	5	97 (53–156)	s
ssp. *virgata*										
—	M	—	270–295					—	170–240	n
	F	—	240–260					—	85–135	n
Russia	—			—	161–270	—	90–110			g
ssp. *amovensis*										
—	M	—	230–265					—	130–175	n
	F	—	220–240					—	100–115	n
ssp. *sumatrae*										
Sumatra	M	—	220–255					—	100–140	n
	F	—	215–230					—	75–110	n
ssp. *sondaica*										
Java	M	1	248					—	100–141	n
	F	—	—					—	75–115	n

Table A.4 Cont.

Species/ Locality	Sex	n	TOL	n	HB	n	TL	n	WT	Ref.
P. tigris ssp. *balica*										
Bali	M	—	220–230					—	90–100	n
	F	—	190–210					—	65–80	n
P. leo										
—	M	—	170–250	—	90–105	—	123	—	150–250	d
	F	—	140–175	—	70–100	—	107	—	120–182	d
Kenya	M	14	246–284					14	172 (150–189)	p
	F	5	241–269					5	151 (122–182)	p
Zimbabwe	M							4	167,171, 176, 184	q
	F							2	137,145	q
India	M	—	275–290						180–230	b

Notes
n—sample size
TOL— total length
HB—head and body length
TL—tail length
WT—weight

References:
(a) Lekagul and McNeely, 1977
(b) Prater, 1974
(c) Guggisberg, 1975
(d) Nowak and Paradiso, 1983
(e) Hemmer, 1972
(f) Roberts, 1977
(g) Ognev, 1935
(h) Smithers, 1983
(i) Pocock, 1932
(j) Harrison, 1968
(k) Schaller and Vasconcelos, 1978
(l) Mondolfi and Hoogesteijn, 1986
(m) Rabinowitz and Nottingham, 1986
(n) Mazak, 1981
(o) Schaller, 1967
(p) Meinertzhagen, 1938
(q) Schaller, 1972
(r) Payne *et al.*, 1985
(s) Khan, 1987

BIBLIOGRAPHY

ANONYMOUS (1986) Onza specimen obtained—identity being studied. *The ISC Newsletter* *5*: 1–6

VAN AARDE, R.J., (1979) Distribution and density of feral cats on Marion Island. *S. Afr. J. Antarc. Res.* *9*: 14–19.

VAN AARDE, R.J., (1980) The diet and feeding behaviour of feral cats, *Felis catus* at Marion Island. *S. Afr. J. Wildl. Res.* *10*: 123–128.

VAN AARDE, R.J. and VAN DYK, A. (1986) Inheritance of the king coat colour pattern in cheetahs *Acinonyx jubatus*. *J. Zool. Lond.* *209*: 573–578.

VAN AARDE, R.J. and SKINNER, J.D. (1986). Pattern of space use by relocated sevals *Felis serval*. *Afr. J. Ecol.* *24*: 97–101.

ABBOTT, D.H., BAINES, D.A., FAULKES, C.G., LEWIS, E., NING, P.C.Y.K. and TOMLINSON, A.J. (1990) A natural deer repellent: Chemistry and behaviour. In Macdonald, D.W., Müller-Schwarze, D. and Natynczuk, S. (Eds). *Chemical signals in vertebrates*, *V*: 602–612, Oxford University Press.

ACKERMAN, B.B., LINDZEY, F.G. and HEMKER, T.P. (1986) Predictive energetics model for cougars. In S.D. Miller and D.D. Everett (eds.) *Cats of the World*, National Wildlife Federation, Washington D.C.: 333–352.

ADAMS, D.B. (1979) The cheetah: Native American. *Science* *205*: 1155–1158.

AKERSTEN, W.A. (1985) Canine function in *Smilodon*. *Contrib. Sci., Nat. Hist. Mus. Los Angeles Cty.* 356: 1–22.

ALBONE, E.S. and GRÖNNEBERG, T.O. (1977) Lipids of the anal sac secretions of the red fox, *Vulpes vulpes*, and of the lion, *Pathera leo*. *J. Lipid Res.* *18*: 474–479.

ALBONE, E.S. and SHIRLEY, S.G. (1984) *Mammalian semiochemistry*. John Wiley, Chichester.

ALCALA, A.C. and BROWN, W.C. (1969) Notes on the food habits of three Philippine wild mammals. *Silliman J.* *16*: 91–94.

Alexander, R.McN., Bennett, M.B. and Ker, R.F. (1986) Mechanical properties and function of the paw pads of some mammals. *J. Zool. Lond.* *209*: 405–419.

Ansell, W.F.H. and Dowsett, R.J. (1988) *Mammals of Malawi.* Trendine Press, St. Ives.

Apps, P.J. (1983) Aspects of the ecology of feral cats on Dassen Island, South Africa. *S. Afr. J. Zool.* *18*: 393–399.

Apps, P.J. (1966) Home ranges of feral cats on Dassen Island. *J. Mammal.* *67*: 199–200.

Armitage, P.L. and Clutton-Brock, J. (1981) A radiological and histological investigation into the mummification of cats from Ancient Egypt. *J. Arch. Sci.* *8*: 185–196.

Arrighi, J. and Salotti, M. (1988) Le chat sauvage (*Felis silvestris* Schreber, 1777) en Corse. Confirmation de sa presence et approche taxonomique. *Mammalia* *52*: 123–125.

Artois, M. (1985) Utilisation de l'éspace et du temps chez le renard (*Vulpes vulpes*) et la chat forestier (*Felis silvestris*) en Lorraine. *Gibier Faune Sauvage* *3*: 33–57.

Ashman, D. (1975) Mountain lion investigations. Nevada Fish and Game Department, Pittman-Robertson Project W–148–6.

Ashman, D. (1978) Mountain lion investigations. Nevada Fish and Game Department, Pittman-Robertson Project W–48–9.

Ashman, D. (1979) Mountain lion investigations. Nevada Fish and Game Department, Pittman-Robertson Project W–48–10.

Ashton, D.G. and Jones, D.M. (1980) Veterinary aspects of the management of non-domestic cats. In J. Barzdo (ed.) *Management of wild cats in captivity*. Proceedings of Symposium 4 of the Association of British Wild Animal Keepers. Association of British Wild Animal Keepers, Whipsnade: 12–22.

Aymerich, M. (1982) Etude comparative des régimes alimentaires du lynx pardelle (*Lynx pardina* Temminck, 1824) et du chat sauvage (*Felis silvestris* Schreber, 1777) au centre de la peninsule Iberique. *Mammalia* *46*: 515–521.

Bailey, T.N. (1974) Social organisation in a bobcat population. *J. Wildl. Mgmt.* *38*: 435–446.

Bailey, T.N., Bangs, E.E., Portner, M.F., Malloy, J.C. and McAvinchey, R.J. (1986) An apparent overexploited lynx population on the Kenai Peninsula, Alaska. *J. Wildl. Mgmt.* *50*: 279–289.

Baldwin, J.A. (1975) Notes and speculations on the domestication of the cat in Egypt. *Anthropos* *70*: 428–448.

Baldwin, J.A. (1980) The domestic cat, *Felis catus* L., in the Pacific Islands. *Carniv. Genet. Newsl.* *4*: 57–66.

Barash, D.P. (1971) Cooperative hunting in the lynx. *J. Mammal.* *52*: 480.

Barrett, P. and Bateson, P. (1976) The development of play in cats. *Behaviour 55*: 106–120.

Baskin, J.A. (1981) *Barbourofelis* (Nimravidae) and *Nimravides* (Felidae) with a description of two new species from the late Miocene of Florida. *J. Mammal. 62*: 122–139.

Bavin, R. (1976) Mountain lion research. New Mexico Game and Fish Department, Pittman-Robertson project W–93–18, Performance report.

Bavin, R. (1978) Mountain lion research. New Mexico Game and Fish Department, Pittman-Robertson project W-124-R-1.

Bayly, C.P. (1976) Observations on the food of the feral cat (*Felis catus*) in an arid environment. *Sth. Aust. Nat. 51*: 22–24.

Beasom, S.L. and Moore, R.A. (1977) Bobcat food habit response to a change in prey abundance. *SW Nat. 21*: 451–457.

de Beaumont, G. (1964) Remarques sur la classification des Felidae. *Ecol. Geol. Helvet. 51*: 837–845.

Bekoff, M. (1989) Behavioural development of terrestrial carnivores. In J.L. Gittelman (ed.) *Carnivore behaviour, ecology and evolution.* Chapman and Hall, London: 89–124.

Bennett, E.T. (1833) Characters of a new species of cat (*Felis, Linn.*) from the continent of India, presented by J.M. Heath esq. *Proc. Zool. Soc. Lond. 1833*: 68–69.

Berg, W.E. (1979) Ecology of bobcats in northern Minnesota. In *Bobcat research conference: Current research on biology and management of Lynx rufus.* National Wildlife Federation Tech. Ser. 6. National Wildlife federation, Washington: 55–61.

Bergerud, A.T. (1971) The population dynamics of Newfoundland caribou. *Wildl. Mgmt. 25*: 1–55.

Bergerud, A.T. (1983) Prey switching in a simple ecosystem. *Sci. Am. 241*: 116–125.

Berrie, P.M. (1973) Ecology and status of the lynx in interior Alaska. In R.L. Eaton (ed.) *The world's cats, Vol. 1.* World Wildlife Safari, Oregon: 4–41.

Berta, A., (1983) A new species of small cat (Felidae) from the late Pliocene-early Pleistocene (Uquian) of Argentina. *J. Mammal. 64*: 720–725.

Bertram, B.C.R. (1975a) Social factors in lion reproduction. *J. Zool. Lond. 177*: 463–482.

Bertram, B.C.R. (1975b) The social system of lions. *Sci. Am. 232*: 54–65.

Bertram, B.C.R. (1976) Kin selection in lions and in evolution. In P.P.G. Bateson and R.A. Hinde (eds.), *Growing points in ethology.* Cambridge University Press, Cambridge: 281–301.

Bertram, B. (1978) *Pride of lions.* Dent, London.

Bertram, B.C.R. (1979) Serengeti predators and their social systems. In A.R.E. Sinclair and M. Norton-Griffiths (eds.) *Serengeti Dynamics of an ecosystem.* Chicago University Press, Chicago: 223–248.

Bertram, B.C.R. (1983) Cooperation and competition in lions. *Nature 302*: 356.

Bewick, T. (1807) *A general history of the quadrupeds.* Bewick and Hodgson, Newcastle upon Tyne.

Birkeland, K.H. and Myrberget, S. (1980) The diet of the lynx *Lynx lynx* in Norway. *Fauna Norv. Ser. A. 1*: 24–28.

Bisbal, F.J. (1986) Food habits of some neotropical carnivores in Venezuela (Mammalia, Carnivora). *Mammalia 50*: 329–339.

de Boer, J.N. (1977) The age of olfactory cues functioning in the chemocommunication among male cats. *Behav. Process. 2*: 209–225.

Bothma, J. du P. and Le Riche, E.A.N. (1986) Prey preference and hunting efficiency of the Kalahari Desert leopard. In S.D. Miller and D.D. Everett (eds.) *Cats of the World.* National Wildlife Federation, Washington D.C.: 389–414.

Brahmachary, R.L. and Dutta, J. (1987) Chemical communication in the tiger and the leopard. In R.L. Tilson and U.S. Seal (eds.) *Tigers of the world.* Noyes, New Jersey: 296–303.

Brand, C.J. and Keith, L.B. (1979) Lynx demography during a snowshoe hare decline in Alberta. *J. Wildl. Mgmt 43*: 827–849.

Brand, C.J., Keith, L.B. and Fischer, C.A. (1976) Lynx responses to changing snowshoe hare densities in Central Alberta. *J. Wildl. Mgmt 40*: 416–428.

Braunschweig, A. (1963) Untersuchungen an Wildkatzen und diesen änhlichen Hauskatzen. *Z. Jagdwiss 9*: 109–112.

Breeden, S. (1989) The happy fisher. *BBC Wildl. Mag. 7*: 238–241.

le Brun, A., Cliuzan, S., Davis, S.J.M., Hansen, J. and Renault-Miskovsky, J. (1987) Le neolithique preceramique de Chypre. *L'Anthropologie 91*: 283–316.

Burton, J.A. and Pearson, B. (1987) *Rare mammals of the world.* Collins, London.

Burton, M. (1978) Cats In: *Encyclopedia of the Animal World Vol. 1.* 346–351.

Bygott, J.FD., Bertram, B.C.R. and Hanby, J.P. (1979) Male lions in large coalitions gain reproductive advantages. *Nature 282*: 839–841.

Cade, C.E. (1968) A note on breeding the caracal lynx at Nairobi Zoo. *Int. Zoo Yearb. 8*: 45.

Caraco, T. and Wolf, L.L. (1975) Ecological determinants of group sizes of foraging lions. *Am. Nat. 109*: 343–352.

Caro, T.M. (1979) *The development of predation in cats.* PhD. Thesis, University of St. Andrews (not seen).

Caro, T.M. (1981) Predatory behaviour and social play in kittens. *Behav. 76*: 1–24.

Caro, T.M. (1987) Indirect costs of play: cheetah cubs reduce maternal hunting success. *Anim. Behav. 35*: 295–297.

Caro, T.M. and Collins, D.A. (1987a) Ecological characteristics of territories of male cheetah (*Acinonyx jubatus*). *J. Zool. Lond.* *211*: 89–105.

Caro, T.M. and Collns, D.A. (1987b) Male cheetah social organization and territoriality. *Ethology* *74*: 52–64.

Cherfas, J. (1987) How to thrill your cat this Christmas. *New Sci* *116*: 42–45.

Childs, J.E. (1986) Size-dependent predation on rats (*Rattus norvegicus*) by house cats (*Felis catus*) in an urban setting. *J. Mammal.* *67*: 196–199.

Churcher, P.B. and Lawton, J.H. (1987) Predation by domestic cats in an English village. *J. Zool. Lond.* *212*: 439–455.

Clark, J.M. (1976) Variation in coat colour gene frequencies and selection in the cats of Scotland. *Genetica* *46*: 401–412.

Clutton-Brock, J. (1981) *Domesticated animals from early times.* Heinemann/British Museum (Natural History), London.

Collier, G.E. and O'Brien, S.J. (1985) A molecular phylogeny of the Felidae immunological distance. *Evolution* *39*: 473–487.

Coman, B.J. and Brunner, H. (1972) Food habits of the feral house cat in Victoria. *J. Wildl. Mgmt* *36*: 39–53.

Conde, B. Nguyen-Thi-Thu-Cuc, Vaillant, F. and Schauenberg, P. (1972) Le régime alimentaire du chat forestier (*F. silvestris* Schr.) en France. *Mammalia* *36*: 112–119.

Cook, L.M. and Yalden, D.M. (1980) A note on the diet of feral cats on Deserta Grande. *Bocagiana* *52*: 1–4.

Corbet, G.B. (1978) *The mammals of the Palaearctic region.* British Museum (Natural History), London.

Corbet, G.B. and Hill, J.E. (1986) *A world list of mammalian species.* British Museum (Natural History), London.

Corbett, L.K. (1979) *Feeding ecology and social organisation of wildcats* (Felis silvestris) *and domestic cats* (Felis catus) *in Scotland.* Unpublished Ph.D. thesis, University of Aberdeen.

Crowe, D.M. (1975) Aspects of ageing, growth and reproduction of bobcats in Wyoming. *J. Mammal.* *56*: 177–198.

Currier, M.J.P. (1976) *Characteristics of the mountain lion population near Cannon City, Colorado.* M.S. Thesis, Colorado State University, Fort Collins (not seen).

Currier, M.J.P. (1983) *Felis concolor. Mammalian Species* *200*: 1–7.

Dards, J.L. (1978) Home ranges of feral cats in Portsmouth Dockyard. *Carniv. Genet. Newsl.* *3*: 242–255.

Dards, J.L. (1981) Habitat utilisation by feral cats in Portsmouth Dockyard. In *The ecology and control of feral cats.* UFAW, Potters Bar: 30–46.

Darie, E.M. and Boersma, P.D. (1984) Why lionesses copulate with more than one male. *Am. Nat* *123*: 594–611.

Davis, S.J.M. (1987) *The archaeology of animals.* Batsford, London.

Deag, J.M., Lawrence, C.E. and Manning, A. (1987) The consequences of differences in litter size for the nursing cat and her kittens. *J. Zool. Lond. 213*: 153–179.

DEAG, J.M., MANNING, A. and LAWRENCE, C.E. (1988) Factors influencing the mother-kitten relationship. In D.C. Turner and P. Bateson (eds.) *The domestic cat: The biology of its behaviour*. Cambridge University Press, Cambridge: 23–40.

DELIBES, M. (1980) Feeding ecology of the Spanish lynx in the Coto Donaña. *Acta Theriol. 25*: 309–324.

DIAMOND, J.M. (1988) Why cats have nine lives? *Nature 332*: 586–587.

DIDIER, R. (1949) Etude systematique de l'os penien des mammiferes. *Mammalia 13*: 17–37.

DIVYABHANUSINH (1987) Record of two unique observations of the Indian cheetah in *Tuzuk-I-Jahangari. J. Bombay Nat. Hist. Soc. 84*: 269–274.

DODD, G.H. and SQUIRRELL, D.J. (1980) Structure and mechanism in the mammalian olfactory system. *Symp. Zool. Soc. Lond. 45*: 35–56.

DURANT, S.M., CARO, T.M., COLLINS, D.A., ALAWI, R.M. and FITZGIBBON, C.D. (1988) Migration patterns of Thomson's gazelles and cheetahs on the Serengeti Plains. *Afr. J. Ecol. 26*: 257–268.

EATON, R.L. (1968) Group interactions, spacing and territoriality in cheetahs. *Z. Tierpsychol. 27*: 481–491.

EATON, R.L. (1974) *The cheetah*. Van Nostrand Reinhold, New York.

EATON, R.L. (1976a) Why some felids copulate so much. In R.L. Eaton (ed.) *The world's cats, Vol. 3*. Carnivore Research Institute, Washington D.C.: 74–94.

EATON, R.L. (1976b) The evolution of sociality in the Felidae. In R.L. Eaton (ed.) *The World's Cats, Vol. 3*. Carnivore Research Institute, Washington D.C.: 95–142.

EISENBERG, J.F. (1973) Mammalian social systems: Are primate social systems unique? *Symp. IVth Int. Congr. Primat. 1*: 232–249, Kager, Basel.

EISENBERG, J.F. (1990) *Mammals of the northern Neotropics, vol. 1*. Chicago University Press, London.

EISENBERG, J.F. and LEYHAUSEN, P. (1972) The phylogenesis of predatory behaviour in mammals. *Z. Tierpsychol. 30*: 59–93.

EISENBERG, J.F. and LOCKHART, M. (1972) An ecological reconnaissance of Wilpattu National Park, Ceylon. *Smithson. Contrib. Zool. 101*: 1–118.

ELLIOTT, J.P., MCTAGGART COWAN, I. and HOLLING, C.S. (1977) Prey capture by the African lion. *Can. J. Zool 55*: 1811–1828.

ELOFF, F.C. (1973) Ecology and behaviour of the Kalahari lion. In R.L. Eaton (ed.) *The World's Cats, Vol 1*. Carnivore Research Institute, Washington D.C.: 90–126.

ELTON, C. and NICHOLSON, M. (1942) The ten-year cycle in numbers of the lynx in Canada. *J. Anim. Ecol. 11*: 215–244.

EMERSON, S.B. and RADINSKY, L. (1980) Functional analysis of sabertooth cranial morphology. *Paleobiol.* *6*: 295–312.

Emmons, L.H. (1987) Comparative feeding ecology of felids in a neotropical rainforest. *Behv. Ecol. Sociobiol.* *20*: 271–283.

Emmons, L.H. (1988) A field study of ocelots (*Felis pardalis*) in Peru. *Rev. Ecol. (Terre Vie)* *43*: 133–157.

Erlinge, S., Goransson, G., Hogstedt, G. Jansson, G., Liberg, O., Loman, J., Nilsson, I.N., von Schantz, T. and Sylven, M. (1984) Can vertebrate predators regulate their prey? *Am. Nat.* *123*: 125–133.

Errington, P.L. (1936) Notes on the food habits of southern Wisconsin house cats. *J. Mammal.* *17*: 64–65.

Estes, R.D. (1972) The role of the vomeronasal organ in mammalian reproduction. *Mammalia* *36*: 315–341.

Ewer, R.F. (1973) *The carnivores.* Cornell University Press, New York.

Fagen, R. (1981) *Animal play behaviour.* Oxford University Press, Oxford.

Fendley, T.T. and Buie, D.E. (1986) Seasonal home range and movement patterns of the bobcat on the Savannah River Plant. In S.D. Miller and D.D. Everett (eds.) *Cats of the World.* National Wildlife Federation, Washington D.C.: 237–259.

Fitzgerald, B.M. (1978) Feeding ecology of feral house cats in New Zealand forest. *Carniv. Genet. Newsl.* *4*: 67–71.

Fitzgerald, B.M. (1988) Diets of domestic cats and their impact on prey populations. In D.C. Turner and P. Bateson (eds.) *The domestic cat: The biology of its behaviour.* Cambridge University Press, Cambridge: 123–150.

Fitzgerald, B.M. and Karl, B.J. (1979) Foods of feral house cats (*Felis catus* L.) in forest of the Orongorongo Valley, Wellington. *NZ J. Zool.* *6*: 107–126.

Flynn, J.J. and Galiano, H. (1982) Phylogeny of early Tertiary Carnivora, with a description of a new species of *Protictis* from the middle Eocene of northwestern Wyoming. *Am. Mus. Novit.* *2725*: 1–64.

Foose, T.J. (1987) Species survival plans and overall management strategies. In R.L. Tilson and U.S. Seal (eds.) *Tigers of the world.* Noyes, New Jersey: 304–316.

Fox, J.L. and Chundawat, R.S. (1988) Observations of snow leopard stalking, killing and feeding behaviour. *Mammalia* *52*: 137–140.

Frame, G.W. (1984) Cheetah. In D. Macdonald (ed.) *The encyclopaedia of mammals, Vol 1.* Unwin, London: 40–43.

Fritts, S.H. and Sealander, J.A. (1978a) Diets of bobcats in Arkansas with special reference to age and sex differences. *J. Wildl. Mgmt* *42*: 533–539.

Fritts, S.H. and Sealander, J.A. (1978b) Reproductive biology and population characteristics of bobcats (*Lynx rufus*) in Arkansas. *J. Mammal.* *59*: 347–353.

FULLER, T.K., BIKNEVICIUS, A.R. and KAT, P.W. (1988) Home range of an African wild cat, *Felis silvestris* (Schreber) near Elmenteita, Kenya. *Z. Säugetierk.* *53*: 380–381.

FUNDERBUNK, S. (1986) International trade in US and Canadian bobcats, 1977–1981. In S.D. Miller and D.D. Everett (eds.) *Cats of the World.* National Wildlife Federation, Washington D.C.: 489–501.

GARDNER, A.L. (1971) Notes on the little spotted cat, *Felis tigrina oncilla* Thomas, in Costa Rica. *J. Mammal.* *52*: 464–465.

GEERTSEMA, A.A. (1976) Impressions and observations on serval behaviour in Tanzania, East Africa. *Mammalia* *40*: 13–19.

GEERTSEMA, A.A. (1985) Aspects of the ecology of the serval *Leptailurus serval* in the Ngorongoro Crater, Tanzania. *Neth. J. Zool.* *35*: 527–610.

GEIST, V. (1987) Bergmann's Rule is invalid. *Can J. Zool.* *65*: 1035–1038.

GITTELMAN, J.L. and HARVEY, P.H. (1982) Carnivore home-range size, metabolic needs and ecology. *Behav. Ecol. Sociobiol.* *10*: 57–63.

GONYEA, W.J. (1976a) Adaptive differences in the body proportions of large felids. *Acta Anat.* *96*: 81–96.

GONYEA, W.J. (1976b) Behavioural implications of saber-toothed felid morphology. *Paleobiol.* *2*: 332–342.

GONYEA, W. and ASHWORTH, R. (1975) The form and function of retractile claws in the Felidae and other representative carnivorans. *J. Morph.* *145*: 229–238.

GOSLING, L.M. (1982) A reassessment of the function of scent marking in territories. *Z. Tierpsychol.* *60*: 89–118.

GOSSOW, H. and HONSIG-ERLENBURG, P. (1986) Management problems with reintroduced lynx in Austria. In S.D. Miller and D.D. Everett (eds.) *Cats of the World.* National Wildlife Federation, Washington D.C.: 77–83.

GOSZCZYSKI, J. (1986) Locomotor activity of terrestrial predators and its consequences. *Acta Theriologica* *31*: 79–95.

GOTCH, A.F. (1975) *Mammals—Their Latin names explained.* Blandford, Poole.

GREAVES, W.S. (1983) A functional analysis of carnassial biting. *Biol. J. Linn. Soc.* *20*: 353–363.

GREENWELL, R. (1987) Is this the beast the Spaniards saw in Montezuma's Zoo? *BBC Wildl. Mag.* *5*: 354–359.

GRIFFITH, M.A. and FENDLEY, T.T. (1986a) Influence of density of movement behaviour and home range size of adult bobcats on the Savannah River Plant. In S.D. Miller and D.D. Everett (eds.) *Cats of the World.* National Wildlife Federation, Washington D.C.: 261–275.

GRIFFITH, M.A. and FENDLEY, T.T. (1986b) Pre and post dispersal movement behaviour of subadult bobcats on the Savannah River Plant. In S.D. Miller and D.D. Everett (eds.) *Cats of the World.*

National Wildlife Federation, Washington D.C.: 277–289.

GROBLER, J.H. (1981) Feeding behaviour of the caracal *Felis caracal* Schreber 1776 in the Mountain Zebra National Park. *S. Afr. J. Zool. 16*: 259–262.

GUGGISBERG, C.A.W. (1975) *Wild cats of the world.* David and Charles, Newton Abbott.

HAGLUND, B. (1966) Winter habits of the lynx (*Lynx lynx* L.) and wolverine (*Gulo gulo* L.) as revealed by tracking in the snow. *Viltrevy 4*: 81–310.

HALL, E.R. (1981) *Mammals of North America,* 2nd edn. John Wiley, New York.

HALL, H.T. and NEWSOM, J.D. (1978) Summer home ranges and movements of bobcats in bottomland hardwoods of southern Louisiana. *Proc. Ann. Conf. Southeast Assoc. Fish and Wildl. Agencies 30*: 427–436 (not seen).

HALLER, H. and BREITENMOSER, U. (1986) Zur Raumorganisation der in den Schweizer Alpen wiederangesiedelten Population des Luchses (*Lynx lynx*), *Z. Säugetierk.* 51: 289–311.

HALTENORTH, T. (1953) *Die Wildkatzen der alten Welt: eine Ubersicht uber die Untergattung* Felis. Leipzig.

HAMILTON, P.H. (1986) Status of the cheetah in Kenya with reference to sub-Saharan Africa. In S.D. Miller and D.D. Everett (eds.) *Cats of the World.* National Wildlife Federation, Washington D.C.: 65–76.

HAPPOLD, D.C.D. (1987) *The mammals of Nigeria.* Oxford University Press, Oxford.

HARRISON, D.L. (1968) *The mammals of Arabia vol. II.* Benn, London.

HELLER, S.P. and FENDLEY, T.T. (1986) Bobcat habitat on the Savannah River Plant. In S.D. Miller and D.D. Everett (eds.) *Cats of the World.* National Wildlife Federation, Washington D.C.: 415–423.

HEMKER, T.P., LINDZEY, F.G. and ACKERMAN, B.B. (1984) Population characteristics and movement patterns of cougars in southern Utah. *J. Wildl. Mgmt. 48*: 1275–1284.

HEMKER, T.P., LINDZEY, F.G., ACKERMAN, B.B. and BUTTON, A.J. (1986) Survival of cougar cubs in a non-hunted population. In S.D. Miller and D.D. Everett (eds.) *Cats of the World.* National Wildlife Federation, Washington D.C.: 327–332.

HEMMER, H. (1972) *Uncia uncia. Mammalian species 20*: 1–5.

HEMMER, H. (1976a) Fossil history of living Felidae. In R.L. Eaton (ed.) *The World's Cats, Vol. 3.* Carnivore Research Institute, Washington D.C.: 1–14.

HEMMER, H. (1976b) Gestation period and postnatal development in felids. In R.L. Eaton (ed.) The World's Cats, Vol. 3. Carnivore Research Institute, Washington D.C.: 143–165.

HEMMER, H. (1978a) Nachweis der Sandkatze (*Felis margarita harrisoni,* Hemmer, Grubb and Groves, 1976) in Jordanien. *Z. Saugetierk. 43*: 62–64.

HEMMER, H. (1978b) Were the leopard cat and the sand cat among the ancestry of domestic cat races? *Carnivore 1*: 106–108.

HEMMER, H. (1987) The phylogeny of the tiger (*Panthera tigris*). In R.L. Tilson and U.S. Seal (eds.) *Tigers of the world.* Noyes, New Jersey: 28–35.

HEWSON, R (1983) The food of wild cats (*Felis silvestris*) and red foxes (*Vulpes vulpes*) in west and north-east Scotland. *J. Zool. Lond. 200*: 283–289.

HILDEBRAND, M. (1959) Motions of the running cheetah and horse. *J. Mammal. 40*: 481–495.

HILDEBRAND, M. (1961) Further studies on locomotion of the cheetah. *J. Mammal. 42*: 84–91.

HOPKINS, R.A., KUTILEK, M.J. and SHREVE, G.L. (1986) Density and home range characteristics of mountain lions in the Diablo Range of California. In S.D. Miller and D.D. Everett (eds.) *Cats of the World.* National Wildlife Federation, Washington D.C.: 223–235.

HOPPE-DOMINIK, B. (1984) Etude du spectre des proies de la panthere, *Panthera pardus,* dans le Parc National de Tai en Cote d'Ivoire. *Mammalia 48*: 477–487.

HOPWOOD, A.T. (1947) Contributions to the study of some African mammals III. Adaptations in the bones of the forelimb in the lion, leopard and cheetah. *J. Linn. Soc. Zool. 41*: 259–271.

HORNOCKER, M.G. (1969) Winter territoriality in mountain lions. *J. Wildl. Mgmt. 33*: 457–464.

HORNOCKER, M.G. (1970) An analysis of mountain lion predation upon mule deer and elk in the Idaho Primitive Area. *Wildl. Monogr. 21*: 1–39.

HORNOCKER, M. and BAILEY, T. (1986) Natural regulation in three species of felids. In S.D. Miller and D.D. Everett (eds.) *Cats of the World.* National Wildlife Federation, Washington D.C.: 211–220.

HOUSTON, A., CLARK, C., MCNAMARA, J. and MANGEL, M. (1988) Dynamic models in behavioural and evolutionary ecology. *Nature 332*: 29–34.

HOUSTON, D.C. (1988) Digestive efficiency and hunting behaviour in cats, dogs and vultures. *J. Zool. Lond. 216*: 603–605.

HUBBS, E.L. (1951) Food habits of feral house cats in the Sacramento Valley. *Calif. Fish and Game 37*: 177–189.

HUGHES, A. (1976) A supplement to the cat schematic eye. *Vision Res. 16*: 149–154.

HUGHES, A. (1977) The topography of vision in mammals of contrasting lifestyle: Comparative optics and retinal organisation. In F. Crescitelli (ed.) *The visual system in vertebrates,* Handbook of sensory physiology, vol. 7, part 5. Springer, Berlin: 613–756.

HUGHES, A. (1985) New perspectives in retinal organisation. *Prog. Ret. Res. 4*: 243–314.

HUNT, R.M. (1974) The auditory bulla in Carnivora: An anatomical basis for reappraisal of carnivore evolution. *J. Morphol. 143*:

21–76.
Husson, A.M. (1978) *The mammals of Suriname.* Brill, Leiden.

Imaizumi, Y. (1967) A new genus and species of cat from Iriomote, Ryukyu Islands. *J. Mammal. Soc. Japan 3*: 75–105.
Inoue, T. (1972) The food habit of the Tsushima leopard cat, *Felis bengalensis* ssp., analysed from their scats. *J. Mammal. Soc. Jap. 5*: 155–169.
Ishunin, G.I. (1965) On the biology of *Felis chaus chaus* in south Uzbekistan. *Zool. Zh. 44*: 630–632.
Iverson, J.B. (1978) The impact of feral cats and dogs on populations of the West Indian rock iguana (*Cyclura carinata*). *Biol. Conserv. 13–14*: 63–73.
Izawa, M. and Ono, Y. (1986) Mother-offspring relationship in the feral cat population. *J. Mammal Soc. Jap. 11*: 27–34.
Izawa, M. and Ono, Y. (1989) Social system of the Iriomote cat (*Felis iriomotensis*). Abstract, 5th International Theriological Congres, Vol. 2, p. 608. Rome, 27–29 August, 1989.

Jackson, P. (1989) New cat discovered. *Cat news 10.*
Jehl, J.R. and Parkes, K.C. (1983) 'Replacements' of landbird species on Socorro Island, Mexico. *The Auk 100*: 551–559.
Jenkins, D., Watson, A. and Miller, G.R. (1964) Predation and red grouse populations. *J. Appl. Ecol. 1*: 183–195.
Jenkinson, R.D.S. (1983) The recent history of the northern lynx (*Lynx lynx* Linne) in the British Isles. *Quat. Newsl. 41*: 1–7.
Jones, E. (1977) Ecology of the feral cat, *Felis catus* (L.), (Carnivora: Felidae) on Macquarie Island. *Aust. Wildl. Res. 4*: 249–262.
Jones, E. and Coman, B.J. (1981) Ecology of the feral cat, *Felis catus* (L.) in South-Eastern Australia. *Aust. Wildl. Res. 8*: 537–547.
Jones, J.H. and Smith, N.S. (1979) Bobcat density and prey selection in Central Arizona. *J. Wildl. Mgmt. 43*: 666–672.

Khan, A.A. and Beg, M.A. (1986) Food of some mammalian predators in the cultivated areas of Punjab. *Pakistan J. Zool. 18*: 71–79.
Khan, M.K. bin M. (1987) Tigers in Malaysia: Prospects for the future. In R.L. Tilson and U.S. Seal (eds.) *Tigers of the world.* Noyes, New Jersey: 75–84.
Kiley-Worthington, M. (1976) The tail movements of ungulates, canids and felids with particular reference to their causation and function as displays. *Behaviour 56*: 69–115.
Kiltie, R.A. (1984) Size ratios among sympatric neotropical cats. *Oecologia 61*: 411–416.
Kingdon, J. (1977) *East African Mammals, Vol. 3A.* Academic Press, London.
Kirk, J.C (1935) Wild and domestic cat compared. *Scott. Nat.* (1935): 161–169.

KIRK, J.C. and WAGSTAFFE, R. (1943) A contribution to the study of the Scottish wildcat (*Felis silvestris grampia* Miller), *N.W. Nat. 18*: 271–275.

KIRKPATRICK, R.D. and RAUZON, M.J. (1986) Foods of feral cats *Felis catus* on Jarvis and Howland Islands, Central Pacific Ocean. *Biotrop. 18*: 72–75.

KITCHINGS, J.T. and STORY, J.D. (1978) Preliminary studies of bobcat activity patterns. *Proc. S.E. Assoc. Fish and Wildl. Agencies 32*: 53–59.

KLIMOVA, V.N. and CHERNYS, A.G. (1980) Adaptive peculiarities of adventitious acoustic cavities in the family Felidae (Carnivora). *Zool. Zh. 59*: 762–770.

KOFORD, C.B. (1977) *Status and welfare of the puma* (Felis concolor) *in California, 1973–1976.* Final report to the Defenders of Wildlife and the National Audubon Society (not seen).

KOFORD, C.B. (1978) The welfare of the puma in California, 1976. *Carnivore 1*: 92–96.

KOLB, H.H. (1977) Wild cat. In G.B. Corbet and H.H. Southern (eds.) *The Handbook of British Mammals.* Blackwell Scientific, Oxford: 375–382.

KRATOCHVIL, J. (1982) The karyotype and system of the family Felidae (Carnivora, Mammalia). *Fol. Zool. 31*: 289–304.

KRATOCHVIL, J. and KRATOCHVIL, Z. (1976) The origin of the domesticated forms of the genus *Felis* (Mammalia). *Zool. Listy 25*: 193–208.

KRUUK, H. (1972) Surplus killing by carnivores. *J. Zool. Lond. 166*: 233–244.

KRUUK, H. (1986) Interactions between Felidae and their prey species: A review. In S.D. Miller and D.D. Everett (eds.) *Cats of the World.* National Wildlife Federation, Washington D.C.: 353–374.

KRUUK, H. and TURNER, M. (1967) Comparative notes on predation by lion, leopard, cheetah and wild dog in the serengeti area, East Africa. *Mammalia 31*: 1–27.

KUHN, H.-J. (1973) Zur Kenntniss der Andenkatze, *Felis (Oreailurus) jacobita* Cornalia, 1865. *Säugetierk. Mitt. 21*: 359–364.

KURTEN, B. (1968) *Pleistocene mammals of Europe.* Weidenfeld and Nicolson, London.

KURTEN, B. (1973a) Pleistocene jaguars in North America. *Comment. Biologic. 62*: 1–23.

KURTEN, B. (1973b) Geographic variation in size in the puma. *Comment. Biologic. 63*: 1–8.

KURTEN, B. (1976) Fossil puma (Mammalia: Felidae) in North America. *Neth. J. Zool. 26*: 502–534.

KURTEN, B. and ANDERSON, E. (1980) *Pleistocene mammals of North America.* Columbia University Press, New York.

KUTILEK, M.J., HOPKINS, R.A. and SMITH, T.E. (1980) *Second annual progress report on the ecology of mountain lion* (Felis concolor) *in the Diablo Range of California.* San Jose State University, Department of Biological Science, San Jose, California.

LANCIA, R.A., WOODWARD, D.K. and MILLER, S.D. (1986) Summer movement patterns and habitat use by bobcats on Croatan National Forest, North Carolina. In S.D. Miller and D.D. Everett (eds.) *Cats of the World*. National Wildlife Federation, Washington D.C.: 425–436.

LAWHEAD, D.N. (1978) *Home range, density and habitat preference of the bobcat on the Three Bar Wildlife Area, Arizona*. M.S. Thesis, University of Arizona (not seen).

LAY, D.M., ANDERSON, J.A.W. and HASSINGER, J.D. (1970) New records of small mammals from West Pakistan. *Mammalia 34*: 98–106.

LEKAGUL, B. and NCNEELY, J. (1977) *Mammals of Thailand*. Bangkok.

LEMBECK, M. (1986) Long term behaviour and population dynamics of an unharvested bobcat population in San Diego County. In S.D. Miller and D.D. Everett (eds.) *Cats of the World*. National Wildlife Federation, Washington D.C.: 305–310.

LEOPOLD, B.D. and KRAUSMAN, P.R. (1986) Diets of 3 predators in Big Bend National Park, Texas. *J. Wildl. Mgmt. 50*: 290–295.

LEVER, C. (1985) *Naturalised mammals of the world*. Longman, London.

LEYHAUSEN, P. (1979) *Cat behaviour*. Garland STPM Press, New York.

LEYHAUSEN, P. (1988) The tame and the wild—another Just So story? In D.C. Turner and P. Bateson (eds.) *The domestic cat: The biology of its behaviour*. Cambridge University Press, Cambridge: 57–66.

LEYHAUSEN, P. and WOLFF, R. (1959) Das Revier einer Hauskatze. *Z. Tierpsychol. 16*: 666–670.

LIBERG, O. (1980) Spacing patterns in a population of rural free-roaming domestic cats. *Oikos 35*: 336–349.

LIBERG, O. (1984a) Home range and territoriality in free-ranging house cats. *Acta Zool. Fenn. 171*: 283–285.

LIBERG, O. (1984b) Food habits and prey impact by feral and house-based domestic cats in a rural area of southern Sweden. *J. Mammal. 65*: 424–432.

LIBERG, O. and SANDELL, M. (1988) Spatial organisation and reproductive tactics in the domestic cat and other felids. In D.C. Turner and P. Bateson (eds.) *The domestic cat: The biology of its behaviour*. Cambridge University Press, Cambridge: 83–98.

LINDEMANN, W (1953) Einiges über die Wildkatzes der Ostkarpathen (*Felis s. silvestris* Schreber, 1777). *Säugetierk. Mitt. 1*: 73–74.

LITVAITIS, J.A., CLARK, A.G. and HUNT, J.H. (1986) Prey selection and fat deposits of bobcats (*Felis rufus*) during autumn and winter in Maine. *J. Mammal. 67*: 389–392.

LITVAITIS, J.A., MAJOR, J.T. and SHERBURNE, J.A. (1987) Influence of season and human-induced mortality on spatial organisation of bobcats (*Felis rufus*) in Maine. *J. Mammal. 68*: 100–106.

LITVINOV, V.P. (1981) Food habits of the jungle cat in the bird winter quarters of eastern Transcaucasia. *Byull. Mosk. Obshch. Ispyt. Prir. (Otd. Biol.) 86*: 19–23.

LOCKET, N.A. (1977) Adaptations to the deep-sea environment. In F.

Crescitelli (ed.) *The visual system in vertebrates.* Springer, Berlin: 67–192.

LOVERIDGE, G.G. (1986) Body weight changes and energy intake of cats during gestation and lactation. *Anim. Technol. 37*: 7–15.

LU HOUJI and SHENG HELIN (1986b) The status and population fluctuation of the leopard cat in China. In S.D. Miller and D.D. Everett (eds.) *Cats of the World.* National Wildlife Federation, Washington D.C.: 59–62.

MCCORD, C.M. (1974) Selection of winter habitat by bobcats (*Lynx rufus*) on the Quabbin Reservation, Massachusetts. *J. Mammal. 55*: 428–437.

MACDONALD, D.W. (1985) The carnivores: Order Carnivora. In R.E. Brown and D.W. Macdonald (eds.) *Social odours in mammals.* Clarendon Press, Oxford: 619–722.

MACDONALD, D.W. and APPS, P.J. (1978) The social behaviour of a group of semi-independent farm cats *Felis catus*: A progress report. *Carniv. Genet. Newsl. 3*: 256–268.

MCDOUGAL, C. (1987) The man-eating tiger in geographical and historical perspective. In R.L. Tilson and U.S. Seal (eds.) *Tigers of the world.* Noyes, New Jersey: 435–448.

MCDOUGAL, C. and SMITH, J.L.D. (1986) Scent marking in tigers. In S.D. Miller and D.D. Everett (eds.) *Cats of the World.* National Wildlife Federation, Washington D.C.: 221.

MCMAHAN, L.R. (1986) The international cat trade. In S.D. Miller and D.D. Everett (eds.) *Cats of the World.* National Wildlife Federation, Washington D.C.: 461–487.

MAEHR, D.S. and BRADY, J.R. (1986) Food habits of bobcats in Florida, *J. Mammal. 67*: 133–138.

MAJOR, J.T., SHERBURNE, J.A., LITVAITIS, J.A. and HARRISON, D.J. (1986) Resource use and interspecific relationships between bobcats and other large mammalian predators in Maine. In S.D. Miller and D.D. Everett (eds.) *Cats of the World.* National Wildlife Federation, Washington D.C.: 291.

MARTIN, L.D., GILBERT, B.M. and ADAMS, D.B. (1977) A cheetah-like cat in the North American Pleistocene. *Science 195*: 981–982.

MARTIN, L.D. (1980) Functional morphology and the evolution of cats. *Trans. Nebrask. Acad. Sci. 8*: 141–154.

MARTIN, L.D. (1989) Fossil history of the terrestrial Carnivora. In J.L. Gittelman (ed.) *Carnivore behaviour, ecology and evolution.* Chapman and Hall, London: 536–568.

MARTIN, P. (1982) Weaning and behavioural development of the cat. Ph.D. Thesis, University of Cambridge (not seen).

MARTIN, P. and BATESON, P. (1988) Behavioural development in the cat. In D.C. Turner and P. Bateson (eds.) *The domestic cat: The biology of its behaviour.* Cambridge University Press, Cambridge: 9–22.

MATJUSHKIN, E.N., ZHIVOTCHENKO, V.I. and SMIRNOV, E.N. (1977) *The*

Amur tiger in the USSR. Unpublished report of the IUCN, Morges, Switzerland. (not seen; cited in Sunquist, 1981).

MAZAK, V. (1981) *Panthera tigris. Mammalian Species 152*: 1–8.

MECH, L.D. (1980) Age, sex, reproduction and spatial organisation of lynxes colonising North East Minnesota. *J. Mammal. 61*: 261–267.

MEINERTZHAGEN, R. (1938) Some weights and measurements of large mammals. *Proc. Zool. Soc. Lond. 108*: 433–439.

VAN MENSCH, P.J.A. and VAN BREE, P.J.H. (1969) On the African golden cat *Profelis aurata* (Temminck, 1827). *Biologica Gabonica 5*: 235–269.

MILLER, G.J. (1980) Some new evidence in support of the stabbing hypothesis for *Smilodon californicus* Bovard. *Carnivore 3*: 8–26.

MITCHELL, B., SHENTON, J. and UYS, J. (1965) Predation on large mammals in the Kafue National Park, Zambia. *Zool. Africana 1*: 297–318.

MONDOLFI, E. (1986) Notes on the biology and status of the small wild cats in Venezuela. In S.D. Miller and D.D. Everett (eds.) *Cats of the World*. National Wildlife Federation, Washington D.C.: 125–146.

MONDOLFI, E. and HOOGESTEIJN, R. (1986) Notes on the biology of the jaguar in Venezuela. In S.D. Miller and D.D. Everett (eds.) *Cats of the World*. National Wildlife Federation, Washington D.C.: 85–123.

MUCKENHIRN, N.A. and EISENBERG, J.F. (1973) Home ranges and predation of the Ceylon leopard. In R.L. Eaton (ed.) *The world's cats, Vol. 1*. World Wildlife Safari, Oregon: 142–173.

MÜLLER-SCHWARZE, D. (1972) Responses of young black-tailed deer to predator odors. *J. Mammal. 53*: 393–394.

MURRAY, J.D. (1988) How the leopard gets its spots. *Sci. Am. 258*: 62–69.

MUUL, I. and LIM, B.-L. (1970) Ecological and morphological observations of *Felis planiceps*. *J. Mammal. 51*: 806–808.

MYERS, N. (1973) The spotted cats and the fur trade. In R.L. Eaton (ed.) *The world's cats, vol. 1*. World Wildlife Safari, Oregon: 276–326.

NASILOV, S.B. (1972) Feeding of the wild cat in Azerbaijan. *Ekologiya 2*: 101–102.

NATOLI, E. and DE VITO, E. (1988) The mating system of feral cats living in a group. In D.C. Turner and P. Bateson (eds.) *The domestic cat: The biology of its behaviour*. Cambridge University Press, Cambridge: 99–110.

NELLIS, C.H. and KEITH, L.B. (1968) Hunting activities and success of lynxes in Alberta. *J. Wildl. Mgmt. 32*: 718–722.

NELLIS, C.H., WETMORE, S.P. and KEITH, L.B. (1972) Lynx-prey interactions in central Alberta. *J. Wildl. Mgmt. 36*: 320–329.

NOVAK, M., BAKER, J.A., OBBVARD, M.E. and MALLOCH, B. (1987a) *Wild furbearer management and conservation in North America*. Ministry of Natural Resources, Ontario.

NOVAK, M., OBBARD, M.E., JONES, J.G., NEWMAN, R., BOOTH, A.,

Satterthwaite, A.J. and Linscombe, G. (1987b) *Furbearer harvests in North America 1600–1984*. Ministry of Natural Resources, Ontario.

Novellie, P., Bigalke, R.C. and Pepler, D. (1982) Can predator urine be used as a buck or rodent repellent. *Sth. Afr. For. J. 123*: 51–55.

Novikov, G.A. (1962) *Carnivorous mammals of the fauna of the USSR*. Israel Program for Scientific Translations, Jerusalem.

Nowak, R.M. and Paradiso, J.L. (1983) *Walker's mammals of the world*, 4th edn. John Hopkins University Press, Baltimore.

O'Brien, S.J., Roelke, M.E., Marker, L., Newman, A., Winkler, C.A., Meltzer, D., Colly, L., Evermann, J.F., Bush, M. and Wildt, D.E. (1985) Genetic basis for species vulnerability in the cheetah. *Science 227*: 1428–1434.

O'Brien, S.J., Wildt, D.E. and Bush, M. (1986) The cheetah in genetic peril. *Sci. Am. 254(5)*: 68–76.

O'Brien, S.J., Collier, G.E., Benveniste, R.E., Nash, W.G., Newman, A.K., Simonson, J.M., Eichelberger, M.A., Seal, U.S., Janssen, D., Bush, M. and Wildt, D.E. (1987a) Setting the molecular clock in the Felidae: The great cats, *Panthera*. In R.L. Tilson and U.S. Seal (eds.) *Tigers of the world*. Noyes, New Jersey: 10–27.

O'Brien, S.J., Wildt, D.E., Bush, M., Caro, T.M., Fitzgibbon, C., Aggunday, I. and Leakey, R.E. (1987b). East African cheetahs: evidence for two population bottlenecks? *Proc. Nat. Acad. Sci. USA, Biol. Sci. 84*: 508–511.

O'Connor, R.M. (1986) Reproduction and age distribution of female lynx in Alaska, 1961–1971—preliminary results. In S.D. Miller and D.D. Everett (eds.) *Cats of the World*. National Wildlife Federation, Washington D.C.: 311–325.

Oftedal, O.T. and Gittelman, J.L. (1989) Patterns of energy output during reproduction in carnivores. In J.L. Gittelman (ed.) *Carnivore behaviour, ecology and evolution*. Chapman and Hall, London: 355–378.

Ognev, S.I. (1935) *Mammals of the USSR and adjacent countries, Vol. 3: Carnivora*. Moscow, English translation, Jerusalem, 1962.

van Orsdol, K.G. (1984) Foraging behaviour and hunting success of lions in Queen Elizabeth National Park, Uganda. *Afr. J. Ecol. 22*: 79–99.

van Orsdol, K.G. (1986) Feeding behaviour and food intake of lions in Ruwenzori National Park, Uganda. In S.D. Miller and D.D. Everett *Cats of the World*. National Wildlife Federation, Washington D.C.: 377–388.

Packer, C. (1986) The ecology of sociality in felids. In D.I. Rubenstein and R.W. Wrangham (eds.) *Ecological aspects of social evolution*. Princeton University Press, New Jersey: 429–451.

Packer, C. and Pusey, A (1982) Cooperation and competition within coalitions of male lions: Kin selection or game theory. *Nature 296*: 740–742.

PACKER, C. and PUSEY, A. (1983a) Adaptations of female lions to infanticide by incoming males. *Am. Nat.* *121*: 716–728.

PACKER, C. and PUSEY, A. (1983b) Cooperation and competition in lions. *Nature* *302*: 356.

PACKER, C. and PUSEY, A.E. (1984) Infanticide in carnivores. In G. Haufsater and S.B. Hrdy (eds.) *Infanticide*. Aldine, New York: 31–42.

PACKER, C. and RUTTAN, L. (1988) The evolution of cooperative hunting. *Am. Nat.* *132*: 159–198.

PALEN, G.F. and GODDARD, G.V. (1966) Catnip and oestrus behaviour in the cat. *Anim. Behav.* *14*: 372–377.

PANAMAN, R. (1981) Behaviour and ecology of free-ranging female farm cats. *Z. Tierpsychol.* *56*: 59–73.

PARKER, G.R., MAXWELL, J.W. and MORTON, L.D. (1983) The ecology of the lynx (*Lynx canadensis*) on Cape Breton Island. *Can. J. Zool.* *61*: 770–786.

PAYNE, J., FRANCIS, C.M. and PHILLIPPS, K. (1985) *A field guide to the mammals of Borneo*. Sabah Society/WWF Malayasia, Kota Kinabalu.

PEARSON, O.P. (1964) Carnivore-mouse predation: An example of its intensity and bioenergetics. *J. Mammal.* *45*: 177–188.

PEARSON, O.P. (1966) The prey of carnivores during one cycle of mouse abundance. *J. Anim. Ecol.* *35*: 217–233.

PIECHOKI, R. (1973) Schutz und Hege der Wildkatze (*Felis silvestris*). In H. Stubbe (ed.) *Buch der Hege.* *1*. VEb, Deutsch, Landwirtschaftsuerlag, Berlin: 342–372.

DE PIENAAR, U. (1969) Predator-prey relations amongst the larger mammals of the Kruger National Park. *Koedoe.* *12*: 108–176.

POCOCK, R.I. (1907) Notes upon some African species of the genus *Felis*, based upon specimens recently exhibited in the Society's Gardens. *Proc. Zool. Soc. Lond.* *1907*: 656–677.

POCOCK, R.I. (1916a) On some external characters of *Cryptoprocta. Ann. Mag. Nat. Hist.* *17*: 413–425.

POCOCK, R.I. (1916b) On some of the cranial and external characters of the hunting leopard or cheetah (*Acinonyx jubatus*). *Ann. Mag. Nat. Hist.* 18: 419–429.

POCOCK, R.I. (1917) On the external characters of the Felidae, *Ann. Mag. Nat. Hist.* *19*: 113–136.

POCOCK, R.I. (1927) Description of a new species of cheetah (*Acinonyx*). *Proc. Zool. Soc. Lond.* *1927*: 245–252.

POCOCK, R.I. (1932) The leopards of Africa. *Proc. Zool. Soc. Lond.* *1932*: 543–594.

POCOCK, R.I. (1951) *Catalogue of the genus* Felis. British Museum (Natural History), London.

POOLE, T. (1985) *Social behaviour in mammals.* Blackie, London.

PRATER, S.H. (1974) *The book of Indian animals*, 4th edn. Bombay Natural History Society, Bombay.

PRINGLE, J.A. and PRINGLE, V.L. (1979) Observations on the lynx *Felis caracal* in the Bedford district. *Sth. Afr. J. Zool.* 14: 1–4.

Provost, E.E., Nelson, C.A. and Marshall, A.D. (1973) Population dynamics and behaviour in the bobcat. In R.L. Eaton (ed.) *The World's Cats, Vol. 3*. Carnivore Research Institute, Washington D.C.: 42–67.

Pulliainen, E. (1981) Winter diet of *Felis lynx* L. in SE Finland as compared with the nutrition of other northern lynxes. *Z. Saugetierk. 46*: 249–259.

Pusey, A.E. and Packer, C. (1987) The evolution of sex-biased dispersal in lions, *Behaviour 101*: 275–310.

Quinn, N.W.S. and Gardner, J.F. (1984) Relationships of age and sex to lynx pelt characteristics. *J. Wildl. Mgmt. 48*: 753–756.

Quinn, N.W.S. and Parker, G. (1987) Lynx. In M. Novak, J.A. Baker, M.E. Obbard and B. Malloch (eds.) *Wild furbearer management and conservation in North America*. Ministry of Natural Resources, Ontario: 682–695.

Rabinowitz, A. (1988) The clouded leopard in Taiwan. *Oryx 22*: 46–47.

Rabinowitz, A., Andau, P. and Chai, P.P.K. (1987) The clouded leopard in Borneo. *Oryx 21*: 107–111.

Rabinowitz, A.R. and Nottingham, B.G. (1986) Ecology and behaviour of the jaguar (*Panthera onca*) in Belize, Central America. *J. Zool. Lond. A 210*: 145–159.

Radinsky, L. (1975) Evolution of the felid brain. *Brain Behav. Evol. 11*: 214–254.

Radinsky, L. (1977) Brains of early carnivores. *Paleobiol. 3*: 333–349.

Radinsky, L. (1978) Evolution of brain size in carnivores and ungulates. *Am. Nat. 112*: 815–831.

Radinsky, L.B. (1981) Evolution of skull shape in carnvores, 1. Representative modern carnivores. *Biol. J. Linn. Soc. 15*: 369–388.

Radinsky, L.B. (1982) Evolution of skull shape in carnivores. 3. The origin and early radiation of the modern carnivore families, *Paleobiol. 8*: 177–195.

Ragni, B. (1978) Observations on the ecology and behaviour of the wild cat (*Felis silvestris* Schreber, 1777) in Italy. *Carniv. Genet. Newsl. 3*: 270–274.

Ragni, B. and Randi, E. (1986) Multivariate analysis of craniometric characters in European wild cat, domestic cat and African wild cat (genus *Felis*). *Z. Säugetierk. 51*: 243–251.

Ralls, K. and Ballou, J. (1986) Captive breeding programs with a small number of founders. *Trends Ecol. Evol. 1*: 19–22.

Rieger, I. (1979) Scent rubbing in carnivores. *Carnivore 2*: 17–25.

Roberts, T. (1977) *Mammals of Pakistan*. Benn, London.

Robinson, I.H. and Delibes, M. (1988) The distribution of faeces by the Spanish lynx (*Felis pardina*). *J. Zool. Lond. 216*: 577–582.

Robinson, R. (1969a) The white tigers of Rewa. *Carniv. Genet. Newsl. 8*: 192–193.

Robinson, R. (1969b) The breeding of spotted and black leopards. *J. Bombay Nat. Hist. Soc. 66*: 423–429.

Robinson, R. (1970) Homologous mutants in mammalian coat colour variation. *Symp. Zool. Soc. Lond. 26*: 251–269.

Robinson, R. (1976a) Homologous genetic variation in the Felidae. *Genetica 46*: 1–31.

Robinson, R. (1976b) Cytogenetics of the Felidae. In R.L. Eaton *The World's Cats, Vol. 3*. Carnivore Research Institute, Washington D.C.: 15–28.

Robinson, R. (1984) Cat. In I.L. Mason (ed.) *Evolution of domestic animals*. Longman, London: 217–224.

Rolley, R.E. (1987) Bobcat. In M. Novak, J.A. Baker, M.E. Obbard and B. Malloch (eds.) *Wild furbearer management and conservation in North America*. Ministry of Natural Resources, Ontario: 670–681.

Rosenzweig, M.L. (1966) Community structure in sympatric Carnivora. *J. Mammal. 47*: 602–612.

Rosevear, D.R. (1974) *The carnivores of West Africa*. British Museum (Natural History), London.

Rowe-Rowe, D.T. (1978) The small carnivores of Natal. *The Lammergeyer 25*: 1–45.

Rudnai, J. (1973) Reproductive biology of lions (*Panthera leo massaica* Neumann) in Nairobi National Park. *E. Afr. Wildl. J. 11*: 241–253.

Rudnai, J. (1974) The pattern of lion predation in Nairobi Park. *E. Afr. Wild. J. 12*: 213–225.

Rudnai, J. (1979) Ecology of lions in Nairobi National Park and the adjoining Kitengala Conservation Unit in Kenya. *Afr. J. Ecol. 17*: 85–95.

Ryder, M. (1973) *Hair*. Arnold, London.

Sales, G. and Pye, D. (1974) *Ultrasonic communication by animals*. Chapman and Hall, London.

Sanyal, P. (1987) Managing the man-eaters in the Sundarbans Tiger Reserve of India—A case study. In R.L. Tilson and U.S. Seal (eds.) *Tigers of the world*. Noyes, New Jersey: 427–434.

Sapozhenkov, Y.F. (1961) On the ecology of *Felis lybica* Forst, in eastern Kara-Kumy. *Zool. Zh. 40*: 1585–1586.

Sapozhenkov, Y.F. (1962) The ecology of the caracal (*Felis caracal* Mull.) in the Karakum. *Zool. Zh.* 41: 1110–1112.

Saunders, J.K. (1963a) Food habits of the lynx in Newfoundland. *J. Wildl. Mgmt 27*: 384–390.

Saunders, J.K. (1963b) Movement and activities of the lynx in Newfoundland. *J. Wildl. Mgmt 27*: 390–400.

Savage, R.J.G. (1977) Evolution in carnivorous mammals. *Palaeontology 20*: 237–271.

Savage, R.J.G. and Long, M.L. (1987) *The evolution of mammals*. British

Museum (Natural History), London.

SCHALLER, G.B. (1967) *The deer and the tiger.* Chicago University Press, Chicago.

SCHALLER, G.B. (1972) *The Serengeti lion.* Chicago University Press, Chicago.

SCHALLER, G.B. (1977) *Mountain monarchs.* Chicago University Press, Chicago.

SCHALLER, G.B. AND CRAWSHAW, P.G. (1980) Movement patterns of jaguar. *Biotropica 12*: 161–168.

SCHALLER, G.B. and VASCONCELOS, J.M.C. (1978) Jaguar predation on capybara. *Z. Saugetierk. 43*: 296–301.

SCHAUENBERG, P. (1981) Elements d'ecologie du chat forestier d'Europe *Felis silvestris* Schreber, 1777. *Rev. Ecol. (Terre et Vie) 35*: 3–36.

SCHNEIRLA, T.C., ROSENBLATT, J.S. and TOBACH, E. (1966) Maternal behaviour in the cat. In H.L. Rheingold (ed.) *Maternal behaviour in mammals.* John Wiley, London: 122.

SCOTT, P.P. and LLOYD-JACOBS, M.A. (1959) Reduction in the anoestrus period of laboratory cats by increased illumination. *Nature 184*: 2022.

SEIDEL, B. and WISSER, J. (1987) Clinical diseases of captive tigers—European literature. In R.L. Tilson and U.S. Seal (eds.) *Tigers of the world.* Noyes, New Jersey: 205–230.

SEIDENSTICKER, J. (1976) On the ecological separation between tigers and leopards. *Biotropica 8*: 225–234.

SEIDENSTICKER, J.C., HORNOCKER, M.G., WILES, W.V. and MESSICK, J.P. (1973) Mountain lion social organisation in the Idaho Primitive Area. *Wildl. Monogr. 35*: 1–60.

SHARMA, I.K (1979) Habitats, feeding, breeding and reaction to man of the desert cat *Felis libyca* (Gray) in the Indian desert. *J. Bombay Nat. Hist. Soc. 76*: 498–499.

SHAW, H.G. (1973) Ecology of the mountain lion in Arizona. In *Wildlife research in Arizona.* Arizona Game and Fish Department, Phoenix: 77–107.

SHUKER, K.P.N. (1989) *Mystery cats of the world.* Hale, London.

SITTON, L. and WALLEN, S. (1976) *California mountain lion study.* California Department of Fish and Game, Sacramento. (Not seen)

SITTON, L., WALLEN, S., WEAVER, R.A. and MACGREGOR, W.G. (1976) *California mountain lion study.* California Department of Fish and Game, Sacramento. (Not seen)

SKINNER, J.D. (1979) Feeding behaviour in caracal *Felis caracal. J. Zool. Lond. 189*: 523–557.

SLADEK, J. (1962) Vorlaufige Augaben uber die ernahrung der Wildkatze in der Slovakei auf Grund der Magenhuntersuchungen. *Symp. Therio. Brno.*, pp. 286–289. (Not seen)

SLADEK, J., MOSANSKY, A. and PALASTHY, J. (1971) Variability of external quantitative characteristics of the wildcat. *Biol. Bratsilava 26*: 881–825. (Not seen)

SMITH, J.L.D., MCDOUGAL, C. and MIQUELLE, D. (1988) Scent marking in free-ranging tigers, *Panthera tigris*. *Anim. Behav.* *36*: 1–10.

SMITHERS, R.H.N. (1971) *The mammals of Botswana*. National Museums of Rhodesia, Salisbury.

SMITHERS, R.H.N. (1983) *Mammals of the southern African subregion*. Mammal Research Institute, Pretoria.

SMITHERS, R.H.N. and WILSON, V.J. (1979) *Checklist and atlas of the mammals of Zimbabwe Rhodesia*. National Museums of Zimbabwe Rhodesia, Salisbury.

STEIN, B.E., MAGALHAES-CASTRO, B. and Kruger, L. (1976) Relationship between visual and tactile representations in cat superior colliculus. *J. Neurophsiol.* *39*: 401–419.

STODDART, D.M. (1980) Some responses of a free-living community of rodents to the odors of predators. In D. Müller-Schwarze and R.M. Silverstein (eds.) *Chemical signals: Vertebrates and aquatic invertebrates*, Plenum, New York: 1–10.

STUART, C.T. (1977) Analysis of *Felis lybica* and *Genetta genetta* scats from the central Namib Desert, South West Africa. *Zool. Afr.* *12*: 239–241.

STUART, C.T. (1988) The incidence of surplus killing by *Panthera pardus* and *Felis caracal* in Cape Province, South Africa. *Mammalia* *50*: 556–558.

SUNQUIST, M.E. (1981) The social organization of tigers (*Panthera tigris*) in Royal Chitawan National Park, Nepal. *Smith Contrib. Zool.* *336*: 1–98.

SWART, J., PERRIN, M.R., HEARNE, J.W. and FOURIE, L.J. (1986) Mathematical model of the interaction between rock hyrax and caracal lynx, based on demographic data from populations in the Mountain Zebra National Park, South Africa. *S. Afr. J. Sci.* *82*: 289–294.

TABOR, R. (1983) *The wild life of the domestic cat*. Arrow, London.

TAYLOR, C.R. and ROWNTREE, V.J. (1973) Temperature regulation and heat balance in running cheetahs: a strategy for sprinters? *Am. J. Physiol.* *224*: 848–851.

TEHSIN, R.H. (1979) Do leopards use their whiskers as wind detector? *J. Bombay Nat. Hist. Soc.* *77*: 128–129.

TETLEY, H. (1941) On the Scottish wild cat. *Proc. Zool. Soc. Lond. Ser. B* *1941*: 13–23.

TEWES, M.E. and EVERETT, D.D. (1986) Status and distribution of the endangered ocelot and jaguarundi in Texas. In S.D. Miller and D.D. Everett (eds.) *Cats of the World*. National Wildlife Federation, Washington D.C.: 147–156.

TEWES, M.E. and SCHMIDLY, D.J. (1987) The neotropical felids: Jaguar, ocelot, margay and jaguarundi. In M. Novak, J. Baker, M.E. Obbard and B. Malloch (eds.) *Wild furbearer management and conservation in North America*. Ministry of Natural Resources, Ontario: 697–711.

THAPAR, V. (1986) *Tiger: Portrait of a predator*. Collins, London.

THENIUS, E. (1967) Zur Phylogenie der Feliden (Carnivora, Mamm.). *Zeit. Zool. Syst.* *5*: 129–143.

THENIUS, E. (1969) *Phylogenie der Mammalia*. de Gruytere, Berlin.

TODD, N.B. (1962) Inheritance of the catnip response in domestic cats. *J. Hered.* *53*: 54–56.

TODD, N.B. (1977) Cats and commerce. *Sci. Am.* *237*: 100–107.

TODD, N.B. (1978) An ecological, behavioural genetic model for the domestication of the cat. *Carnivore* *1*: 52–60.

TUMLISON, R. (1987) *Felis lynx*. *Mamm. Spec.* *269*: 1–8.

TURNER, D.C. and BATESON, P. (eds.) (1988) *The domestic cat: The biology of its behaviour*. Cambridge University Press, Cambridge.

VAKKUR, G.J. and BISHOP, P.O. (1963) The schematic eye in the cat. *Vision Res.* *3*: 357–381.

VAN VALKENBURGH, B. and RUFF, C.B. (1987) Canine tooth strength and killing behaviour in large carnivores. *J. Zool. Lond.* *212*: 1–19.

VASILIU, G.D. and ALMASAN, H. (1969) Contributii la cunoasterea taxonometriei unor mamifere (Carnivora) din Romania. *Studii Communic. Muz. Stintele Nat. Bacau*, pp. 283–291 (not seen; cited in Corbett, 1979).

VERBENE, G. and DE BOER, J.N. (1976) Chemocommunication among domestic cats mediated by the olfactory and vomeronasal senses, I. Chemocommunication. *Z. Tierpsychol.* *42*: 86–109.

WALLS, G.L. (1942) *The vertebrate eye*. Harper, New York.

WAYNE, R.K., MODI, W.S. and O'BRIEN, S.J. (1986) Morphological variability and asymmetry in the cheetah (*Acinonyx jubatus*), a genetically uniform species. *Evolution* *40*: 78–85.

WAYNE, R.K., BENVENISTE, R.E., JANCZEWSKI, N. and O'BRIEN, S.J. (1989) Molecular and biochemical evolution of the Carnivora. In J.L. Gittelman (ed.) *Carnivore behaviour, ecology and evolution*. Chapman and Hall, London: 465–494.

WEBSTER, D.B. (1962) A function of the enlarged middle-ear cavities of the kangaroo rat. *Dipodomys Physiol. Zool.* *35*: 248–255.

WEIGEL, I. (1961) Das Fellmuster der wildlebenden Katzenarten und der Hauskatze in vergleichender und stammesgeschichtlicher Hinsicht. *Säugetierk. Mitt.* *9*: 1–120.

WEIGEL, I. (1975) Small felids and clouded leopards. In B. Gzrimek (ed.) *Gzrimek's animal life encyclopedia, vol. 12*, Van Nostrand Reinhold, New York: 281–332.

WEMMER, C. and SCOW, K. (1977) Communication in the Felidae with emphasis on scent marking and contact patterns. In T.A. Sebeok (ed.) *How animals communicate*. Indiana University Press, Bloomington: 749–766.

WEN, G.Y., STURMAN, J.A. and SHEK, J.W. (1985) A comparative study of the tapetum, retina and skull of the ferret, dog and cat. *Lab. Anim. Sci.* *35*: 200–209.

WERDELIN, L. (1981) The evolution of lynxes. *Ann. Zool. Fenn. 18*: 37–71.

WERDELIN, L. (1983) Morphological patterns in the skulls of cats. *Biol. J. Linn. Soc. 19*: 375–391.

WERDELIN, L. (1985) Small Pleistocene felines of North America. *J. Vert. Paleont. 5*: 194–210.

WILSON, P. (1984) Puma predation on guanacos in Torres del Paine National Park, Chile. *Mammalia 48*: 515–522.

WOZENCRAFT, W.C. (1989) Appendix: Classification of the Recent Carnivora. In J.L. Gittelman (ed.) *Carnivore behaviour, ecology and evolution*. Chapman and Hall, London: 569–594.

WROGEMANN, N. (1975) *Cheetah under the sun*. McGraw-Hill, Johannesberg.

XIMINEZ, A. (1975) *Felis geoffroyi. Mamm. Spec. 54*: 1–4.

XIMINEZ, A. (1988) Notas sobre felidos neotropicales, IX. *Felis (Leopardus) pardalis mitis* F. Cuvier, 1820 en el Uruguay (Mammalia: Carnivora: Felidae). *Comm. Zool. Mus. Hist. Nat. Montevideo 12*: 1–7.

YAMAYA, S. and YASUMA, S. (1986) Coleoptera from droppings of the Iriomote cat, *Prionailurus iriomotensis* (Imaizumi). In *Papers on entomology presented to Prof. Takehiko Nakane in commemoration of his retirement*, Japan Society of Coleopterology, Tokyo: 181–193.

YANEZ, J.L., CARDENAS, J.C., GEZELLE, P. and JAKSIC, F.M. (1986) Food habits of the southernmost mountain lions (*Felis concolor*) in South America: Natural versus livestocked ranges, *J. Mammal. 67*: 604–606.

YASUMA, S. (1981) Feeding behaviour of Iriomote cat (*Prionailurus iriomotensis* Imaizumi, 1967). *Bull. Tokyo Univ. Forests. 70*: 81–140.

YASUMA, S. (1988) Iriomote cat: King of the night. *Anim. King. Nov/Dec 1988*: 12–21.

YORK, W. (1973) A study of serval melanism in the Aberdares and some general behavioural information. In R.L. Eaton (ed.) *The World's Cats, Vol. 1*. Carnivore Research Institute, Washington D.C.: 191–197.

ZEUNER, F.E. (1963) *A history of domesticated animals*. Hutchinson, London.

ZEZULAK, D.S. and SCHWAB, R.G. (1979) A comparison of density, home range and utilization of bobcat populations on Lava Beds and Joshua Tree National Monument, California. In L. Blum and P. Escherich (eds.) *Proc. Bobcat Res. Conf. NWF Sci. and Tech. Ser. 6*: 74–79.

ZHELTUKHIN, A.S. (1986) Biocoenotic relationships of the European lynx (*Lynx lynx*) in the southern taiga of the Upper Volga. *Zool. Zh. 65*: 259–271.

INDEX